Introducing the Oscillations Based Paradigm

Darius Plikynas

Introducing the Oscillations Based Paradigm

The Simulation of Agents and Social Systems

 Springer

Darius Plikynas
Research and Development Center
Kazimieras Simonavičius University
Vilnius
Lithuania

and

Institute of Mathematics and Informatics
Vilnius University
Vilnius
Lithuania

ISBN 978-3-319-81801-6   ISBN 978-3-319-39040-6   (eBook)
DOI 10.1007/978-3-319-39040-6

Foreword by Prof. Andrea Omicini

What happens when one tries to link together the research on neuroscience, artificial intelligence, multi-agent systems, and social networks? After decades when multi-, inter-, and trans-disciplinary research were mostly the result of scattered, isolated activity, or, the rare consequence of some specific, application-oriented research project, research on complex systems demonstrated how all the issues and problems arising from the diversity of the conceptual landscapes, the methods, and the peculiar languages could (and should) actually be resolved. Two decades ago, the emergence of complex networks made it apparent how a single common viewpoint over heterogeneous fields such as biological systems, computer science, economics, and social sciences, to name a few, could lead to spectacular results.

Nowadays, social systems represent a novel frontier for inter-disciplinary studies, as witnessed, for instance, by the surprisingly successful application of statistical mechanics to economics. Generally speaking, their multi-faceted articulation first primarily motivates regrouping of models and methods across all of those disciplines involving complex systems. Also, the huge access to data, along with the lack of reliable models allowing for actual predictability in other words, an abundance of evidence devoid of clear explanations motivates combining and integrating a spectrum of approaches, models, and methods from the hard and soft sciences. Then, the many diverse interpretations to which social systems are amenable, encourage the applications of different methods from a heterogeneous collection of fields: for instance, socio-technical systems, like social networks, are at the same time software systems as well as social systems in their own right, but they could also be modeled as economical systems, statistical mechanics systems, organizational type, and so on.

Well-placed along this train of ideas, the book Introducing the Oscillations-Based Paradigm: the Simulation of Agents and Social Systems, by Prof. Darius Plikynas, provides us with quite an original viewpoint towards social system simulation, by exploiting a mixture of research methods and approaches to generate the desired mechanisms. The aim of the book is to create an oscillation-based paradigm for simulation of agents and multi-agent systems that shares some of the features that are typical of realizable social systems. The objective of the book is extremely ambitious, and invites the reader to be willing to cross disciplinary boundaries, as well as to establish connections that may sometimes appear to be somewhat counterintuitive.

The first point of interest of this book is that it reviews a wide range of theories across various disciplines, while looking for some common field-theoretical elements. In this way, it outlines the fundamental principles of coordination and self-organization as they emerge at the quantum, cellular, and social levels. The findings bring to the forefront a number of principles deemed to be universal, and which are further adopted to formulate the main premises and postulates for the proposed OSIMAS (oscillation-based multi-agent systems) simulation paradigm.

When I first met the author, during his visit in Cesena in November 2014, the first thing that struck me was the originality of his way of thinking. In this book, the reader will find that sort of originality, at least from two different perspectives. On the one hand, the book can be properly read as a (much welcome) call for a survey of cross-disciplinary work, providing the research-oriented reader with some tools for simulation for a variety of topics. On the other hand, this work connects several disciplinary domains that can be very difficult to interconnect, in general. In doing so, the author has produced an impressive intellectual accomplishment that seems quite remarkable within the landscape of contemporary social science.

The book draws specifically on artificial intelligence, cognitive science, neuroscience, quantum physics, finance and economics, and mathematics. Despite the traditional difficulty in dealing with such highly formalized fields, the author does a commendable job in making most of the technical content of the book accessible to scientists, in a way that can be more or less independent of their disciplinary background.

Notwithstanding the appealing empirical power of the currently applied methods, modern social research is still fragmented with a lack of a cohesive modeling media for understanding and simulating the highly complex nature of social behavior. This book addresses that problem the lack of a multidisciplinary connecting paradigm by bringing together a mosaic of research in the fields of neuroscience, artificial intelligence, multi-agent systems, and social network domains. The need for a common multidisciplinary research framework arises essentially because these fields share a common objective of investigation and simulation, that is, the study of individual and collective behavior. The tremendous advance in computational power over recent years has greatly widened the scope of simulation options available to social scientists: since it is quite apparent that many of them have not realized this yet, this book could also be seen as a way forward to demonstrate the significant potential that lies right at their fingertips.

First of all, the book includes a number of introductory chapters aimed at establishing the groundwork for the OSIMAS approach; then it moves to applications, drawing upon different approaches and a diversity of disciplines, which are exemplified in the later chapters. The appendices appear particularly valuable in that they describe a number of key theoretical paradigms that are embedded into the scope of topics. It is worth noting that these appendices read very well, even though they cover quite a broad range of topics that are at the same time philosophically dense and somewhat off the mainstream. Originality, indeed, has its price: setting the theoretical backbone of the book in the appendices and not at the beginning will not appear familiar to most readers; thus some may be a little unsettled. The most common approach, of course, is to first establish the theoretical/conceptual framework, and then construct within that framework. But originality can pay dividends as well: after reading the book, I found the current arrangement of topics to be working quite well since much of the conceptual content is stationed outside of the main text, so as to leave the necessary space to cover technicalities, thus adding to the fluency of reading.

From the viewpoint of methodological approaches, this book has a lot to offer, not only to social scientists but also to a much broader audience. The current fragmented state of our science is more and more challenged by multimodal complexity and inter-relationships of concepts: as such, it implies a vital need for the deeper and the simultaneously wider understanding of how common fundamental organizational principles may establish inter-disciplinary connections. Such a formidable challenge has been undertaken in this multidisciplinary work, which at the same time incorporates the qualities of a novel, conceptually integrative, and experimentally based simulation research framework.

Prof. Andrea Omicini
Alma Mater Studiorum—Università di Bologna
Bologna/Cesena, Italy

Foreword by Prof. John T. Cacioppo

Oscillations-based paradigm starts from a review of a wide range of theories across various disciplines in search of some common field-like fundamental principles of coordination and self-organization existing on the quantum, cellular, and social levels. Findings that are described as suggesting universal principles are employed to formulate main premises and postulates for the proposed OSIMAS (oscillation-based multi-agent system) simulation paradigm. The OSIMAS draws on three conceptual models—oscillating agent model (OAM), pervasive information field (PIF) and wave-like interaction mechanism (WIM)—as well as work on the oscillating nature of brain activity and of the nonlocal field-like self-organization properties of modern information societies. Arguments based on conceptual analyses, experimental evidence, and simulations are used to support the premises underlying the OSIMAS paradigm.

This book is not for the reader who seeks a well-developed and empirically supported overview of the science of social behavior. This book is controversial, provocative, speculative, and sometimes maddening, but it is also stimulating. For instance, OSIMAS is based on the notion that the physical basis of consciousness is a bioelectromagnetic field that exhibits wave-mechanical dynamics. It is certainly the case that any electrical current produces a magnetic field. The neural activity underlying human thought produces a bioelectromagnetic field, and the miniscule currents produced by neuronal activity produce miniscule magnetic fields that exert forces and torques on each other. We do not live in isolation but rather we live in the context of and interact with other people. Professor Plikynas argues that fields produced by individual brains can be modeled as a system of bioelectromagnetic fields. I found more controversial the contention that the miniscule bioelectromagnetic fields produce societal-level bioelectromagnetic field that form some sort of coherent field of common mental activity, and I suspect many readers may share this skepticism.

However, this book is not designed to communicate the established; its purpose is to stimulate novel conceptual and methodological approaches to the scientific study of the physical basis of social cognition and behavior. There are risks when new scientific fields are explored or at least new connections between academic fields are pursued. Hints for applications are provided, which need to be tested.

The goal of the simulation framework outlined in this book is to provide a common platform for addressing prospective applications in the field of social neuroscience, distributed cognition, multi-agent systems, computational intelligence, team neurodynamics, neuroeconomics, etc. The lack of an integrative research framework is one of the core problems in social theory. Agent-based modeling systems for investigating social behavior exist, but none to my knowledge that attempt to span as many disciplines as OSIMAS.

The perspective adopted by the author is absolutely original. Although the results it seeks to describe and organize are drawn from the extant scientific literature, the approach is really new and provocative. I thought the book was so unique in its viewpoint that I found it well worth reading even though I repeatedly found myself disagreeing with premises and details. There is an important role for controversial books like this one in science. Whether or not you agree with the specific processes described in this book, the originality of some of the ideas about material basis for behavior at the societal level may well leave you ruminating about new questions and possibilities. And, to me, that is the strength of this book.

Prof. John T. Cacioppo, Ph.D.
Tiffany and Margaret Blake Distinguished Service
Professor of Psychology and of Psychiatry and Behavioral Neuroscience
and
Director
Center for Cognitive and Social Neuroscience
The University of Chicago
Chicago, IL, USA

Preface

This book presents a conceptually original, oscillations-based paradigm aimed at the modeling of agents and their systems as coherent, stylized, neurodynamic processes. Putting into perspective the various pathways of research as they are interrelated, the book provides a philosophical underpinning, experimental background and some modeling tools. In this effort, some general fundamental design principles of the field-theoretical view of the oscillating agent are outlined, as well as of coherent economic or social systems in general.

The conceptual insights presented here are employed for the formulation of the main postulates and principles of the proposed Oscillation-Based Multi-Agent System (OSIMAS) simulation paradigm.[1] The OSIMAS provides a foundation for the creation of an oscillations-based social simulation research framework. It explores various stylized oscillating agent models (OAM), which can be further expanded to the multi-agent system (MAS) simulation model. For that reason, a canonical pervasive information field (PIF) and wave-like interaction models (WIM) are also employed.

Consequently, this study not only discusses the major conceptual assumptions of the proposed (OSIMAS) paradigm, but also presents an electroencephalography (EEG)-based experimental validation framework and some prior empirical results to validate or disprove the OSIMAS assumptions. Based on the conceptual, experimental and simulation findings, a modeling framework is constructed aimed at the prospective development of the proposed pioneering paradigm.[2] This book also systemizes some other studies and applications, which were most relevant to the work presented here.

However, this is a challenging multidisciplinary research area, which is only starting to evolve following recent technological advances. The major problem concerning these fundamental and applied research areas is related to the lack of a

---

[1]The OSIMAS project's website is available at http://osimas.ksu.lt.

[2]The OSIMAS simulation platform and interactive models are available at http://vlab.vva.lt; for login, enter the username Guest and password guest555.

conceptual unifying theory, which could provide a bridge for a coherent research framework. The author hopes that this multidisciplinary book, via the unique conceptual, experimental and modeling framework, helps to bridge that gap.

In fact, in recent years there has been a noticeable surge of interest in the multidisciplinary areas, bridging the gaps between diverse fragmented research in terms of disciplines, methods, and scale. In this regard, multidisciplinary research is always packed with some element of risk, as it is very difficult to not only attempt to master a few research areas, but also to combine them so that one can propose certain experimental validation in an effort to resolve the long-standing fundamental issues confronting us in these areas of research, not only individually but also collectively. It is a delicate balancing act, especially when one is attempting to simultaneously address what appear to be related, but conflicting fundamental issues in such seemingly diverse disciplines as quantum mechanics, neuroscience, artificial intelligence, cognitive and behavioral sciences, etc.

Accordingly, the mainstay of this book is to systematically review and synthesize related fundamental, experimental and simulation approaches stemming from various multidisciplinary areas of research. Thorough analyses are performed using universal principles of field-based self-organization on various scales, starting from the micro level, i.e., quantum physics, where we can find a great number of theoretical and experimental studies suggesting that field-theoretical properties at the micro level occur not at the quantum level with subatomic or atomic particles, but also in the world of large molecules, the brain's neural networks, and presumably on the social scale too.

In fact, the proposition that the physical manifestation of consciousness is a bioelectromagnetic field (BEM) that exhibits wave-mechanical dynamics has profound implications for our understanding of the phenomenon of consciousness. Moreover, even our societies (the macro-world) can no longer be viewed as entities totally separate from the quantum effects taking place in the heads of a society's members. At least theoretically, we can acknowledge and arrive at the viewpoint that societies can also be understood as collective consciousness processes emerging from the coordinated behavior of the conscious and subconscious bioelectromagnetic brain fields of individual members of society. In this way, we can hypothetically admit that emergent social behavior could be at least partly associated with some coherent, collective, bioelectromagnetic brain field, characterized by some synchronous collective mind state processes and, therefore, can inherit some degree of field-like behavior.

By reflecting a subject's visceral connection to the world and to other people, research into collective brain activations may reveal the states of a subject that are evident in overt behavior, but are connected to collective mental states and are thus helpful for a more holistic understanding of a subject's conduct. We have to be aware that studying the mind-field basis of social cognition and interaction in terms of two- or multi-person neuroscience may shift the focus of traditional research into social communication from basic sensory functions in individual subjects toward the study of interconnected mind-fields.

From another perspective, social networks, digital services, and the network economy at large, are highly heterogeneous with many links and complex inter-relations. Uncoupled and indirect interactions among agents require the ability to affect and perceive information in a field-broadcasted context. In this sense, there is a need to look for ways to model the information network as a virtual information field, where each network node receives pervasive (broadcasted) information-field values. Such an approach is targeted to enforce indirect and uncoupled (contextual) interactions among agents in order to represent contextual broadcasted information in a form locally accessible and immediately usable by network agents. In this regard, there is a clear need to develop a collective mind-field paradigm capable of simulating some complex social cognitive and behavioral phenomena, such as collective states' temporal circadian rhythms, spatial clustering, herd effects, convergence effects, social networking, media and political campaigns impact on collective mind states, propaganda and information war effects on perception of threats, etc.

The implications of collective behavior modeling in the simulated agent-based social mediums play a paramount role in achieving a better understanding of the behavior of modern digitally interconnected economic systems. It provides a unique possibility to define some novel ways to manage the core individual and collective parameters, which impede or enable the observation of local and nonlocal emergent complex social phenomena. In a wider perspective, such modeling provides means to simulate and investigate various sensitivity, fragility, and contagion processes in different social mediums.

In this way, via agent states-related diffusion properties, we could investigate complex economic phenomena like the spread of stock market crashes, currency crises, speculative oscillations (bubbles and crashes), social unrest, recessionary effects, sovereign defaults, etc. All these effects are closely associated with social fragility, which is affected by and follows seasonal, production, political, business, financial, and other cycles. However, there is a long way to go and the author believes this manuscript can contribute in that way via some of the original conceptual insights, experimental frameworks and social simulation models provided.

The multidisciplinary OSIMAS paradigm, not only via some theoretical insights but also via an empirical and simulation framework, contributes to a new way for modeling the complex dynamics of individual and collective mind states, where emergent complex social behavior is understood as a consequence of coherent individual mind-field effects.

We hope that approaches based on the proposed research framework will help to reveal new frontiers of multidisciplinary research. However, we also admit that extended additional conceptual, experimental and simulation research needs to be done to examine, in detail, the issues and criteria that will help improve the OSIMAS paradigm and associated models. This could yield new knowledge and surprising perspectives, allowing a better understanding of social agents, their social organization principles, and corresponding simulation models of individual and collective behavioral patterns. In a way, this book with its presented pioneering paradigm is just a beginning, which reveals a perspective of a vast new multidisciplinary research area.

Hence, because of its highly multidisciplinary nature, this book might be of interest to scholars in many disciplines, e.g., the artificial intelligence (AI) community, multi-agent systems (MAS) research, neuroscience and cognitive science, psychology, social and economic sciences, applied quantum physics, biological systems research, etc. Students from various disciplines may also find a lot of inspiring ideas here about some unconventional research topics.

This book has been appreciatively initiated by the "Creation of oscillations-based social simulation paradigm" (OSIMAS) research project, funded by the European Social Fund following the Global Grant measure program (project No. VP1-3.1-SMM-07-K-01-137). The author is grateful to all who have participated in the project and preparation of the book as well. Especially during the course of the related studies, the author is thankful for the collaboration with Profs. habil. Dr. Sarunas Raudys, James F. Glazebrook, and Dragan Milovanovic, who have actively participated in the research process, reviewed the monograph, and prepared their substantial contribution in some chapters and appendixes. Besides this, the author is deeply grateful to the OSIMAS research team members Andrej Cholscevnikov, Saulius Masteika, Darius Kezys, Karolis Tamosiunas, Mindaugas Baranauskas, Jonas Matuzas, Aistis Raudys, and Mantas Grigalavicius. I am thankful for the fruitful discussions from David W. Orme-Johnson, Arvydas Tamulis, Fred Travis, Dean Radin, Giuseppe Vizzari, Andrea Omicini, Eric Dietrich, Davide Secchi, Algirdas Laukaitis, Saulius Maskeliunas, and other researchers. Thanks are due also to anonymous reviewers who helped to improve the final stages of the book prepraration. I am also thankful for the proofreading efforts and suggested edits by the publisher.

In general, the content of this book is organized so that the reader would first become familiar with the philosophical and conceptual ideas stemming from the great diversity of related research approaches. Afterwards it introduces the OSIMAS research conceptual framework, which proceeds with the experimental chapters that pertain to the neuroscience related EEG (electroencephalography) research of human brain oscillations in the form of brainwaves. Further chapters provide various OAM and MAS modeling setups, which culminate with some simulation results. At the end of the book, Appendixes A, B, C, and D review OSIMAS research approaches in the broader light of other related theories and models.

In fact, the structure of this book has been partly formed by a summation and adaptation of the earlier publications. Therefore, some chapters have their own line of thought and can be read separately from the rest of the book. However, all the chapters are closely related to the OSIMAS paradigm and follow its development in a chronological way.

Brief comments regarding the content are provided below, and the first synopsis chapter explains the multidisciplinary scope and major aims of the book in detail. It briefly introduces the related cross-disciplinary research approaches.

The second chapter mainly focuses on (i) identifying the general properties of information-based economic systems, (ii) ways to model the emergent and self-organizing features of social networks, and (iii) discussion on how to simulate complex economic systems using the field-based approach and multi-agent platforms. This chapter provides some ideas and examples on how, not only the communication mechanism, but also the social agents can be simulated as oscillating processes.

The third chapter starts from a review of a wide range of theories across various disciplines in search of some common field-like fundamental principles of coordination and self-organization existing on the quantum, cellular, and social levels. These findings outline some universal principles, which are further employed to formulate the main premises and postulates of the proposed OSIMAS simulation paradigm.

The fourth chapter not only comments on the major conceptual assumptions of the proposed OSIMAS paradigm, but also presents an EEG-based experimental validation framework and some empirical results to validate the major OSIMAS assumptions from the viewpoint of some baseline individual neurodynamics. In other words, here is presented the experimental neuroscience based research framework and some experimental evidences of the oscillatory nature of social agents as approximations of real humans. In this regard, we noticed from the neuroscience domain that basic mind states, which directly influence human behavior, can be characterized by the specific brainwave oscillations. For the experimental validation (or disproof) of the biologically inspired OSIMAS paradigm we have designed a framework of baseline individual and group-wide EEG-based experiments. Initial baseline individual tests of spectral cross-correlations of EEG-recorded brainwave patterns for some mental states have been provided in this chapter. Preliminary experimental results do not refute the main OSIMAS postulates.

The fifth chapter presents a simulation of human brain EEG signal dynamics using a refined coupled oscillator energy exchange model (COEEM). We have created the coupled oscillations-based COEEM model in order to establish a relationship between the experimentally measured EEG signal oscillations and the conceptually described oscillating agents in the OSIMAS paradigm. Investigation and further refinements of this biologically inspired experimental model helped to define the features of our constructed oscillating agent model (OAM) in the OSIMAS paradigm. Hence, the oscillation-based modeling of human brain EEG signals oscillations, using a refined Kuramoto model, not only directly demonstrates validity of the OSIMAS premises but also helps to specify the OAM, but also significantly contributes to EEG signals prognostication research, which is the major topic of this particular direction of neuroscience research.

The sixth chapter describes a conceptually novel modeling approach, based on quantum theory, to the basic human mind states as systems of coherent oscillation. The aim is to bridge the gap between fundamental theory, experimental observation, and the simulation of agents' mind states. The proposed quantum mechanics-based approach reveals possibilities of employing wave functions and quantum operators for a stylized description of basic mind states and the transitions between them in the OAM. The basic mind states are defined using experimentally observed EEG spectra, i.e., brainwaves (delta, theta, alpha, beta, and gamma), which reveal the oscillatory nature of the agents' mind states. Such an approach provides an opportunity to model the dynamics of the basic mind states by employing stylized oscillation-based representations of the characteristic EEG power spectral density (PSD) distributions of the brainwaves observed in experiments.

In the seventh chapter, a description and some explanatory sources pertaining to the OAM are provided, which are an intrinsic part of the oscillations-based OSIMAS paradigm. This chapter presents the OAM simulation results of the basic mind states' (BMS) dynamics. The OAM models a social agent in terms of a set of basic mind-brain states and the transitions between them. The presented OAM model is based on the experimental EEG (electroencephalography) spectral power redistribution processes observed during the different mental states, stylized entropy/negentropy processes and internal energy dynamics. In this way, the OAM paves the way for a putative link between the conceptual oscillations-based OSIMAS framework and the empirically observed coherent bioelectromagnetic oscillations of ensembles of neurons in the brain (EEG spectra).

The eighth chapter deals with the multi-agent system (MAS) simulation setup and analysis of the obtained simulation results. The OAM-based MAS is essentially based on the OAM described in the seventh chapter. In general, the OAM-based MAS construction is aimed at simulating circadian group-wide, i.e., collective mind-brain states and the correspondingly emerging coherent and synergy phenomena. This chapter is aimed towards the simulation setup for a group of agents, which coordinate (synchronize) to some degree in their internal BMS processes not by using pair-to-pair interaction, but a field-like (mean field) coordination that pertains to (i) the common external circadian rhythms, and (ii) some basic collective synergy effects, which conserve the systems' total energy and decrease system's entropy. The OAM-based MAS simulation is oriented to reveal the basic conditions and parameters for the emergence of some coherent group-wide behaviors.

The ninth and tenth chapters were mostly contributed by prof. habil. Dr. Sarunas Raudys. In the ninth chapter, improved cellular mechanism based models of excitable media employed to mimic economic and financial investing units (agents) in rapidly changing environments are presented. We have used an AI-based approach to investigate excitation information propagation in artificial societies. We used a cellular mechanism approach, in which it is assumed that social media is composed of tens of thousands of community agents, where useful (innovative) information can be transmitted to the closest neighboring agents. The model's originality lies in its exploitation of the artificial neuron-based agent schema with a nonlinear activation function to determine the reaction delay, the refractory (agent recovery) period and the algorithms that define the mutual cooperation among the several excitable groups that comprise the agent population. This novel media model allows a methodical analysis of the propagation of temporal and spatial oscillations of several competing innovation/information signals in social media and networks. Depending on the model parameters, we can obtain temporally or spatially oscillating chaotic, spiral, semi-regular, or regular agents' behaviors. The sum of the outputs of the groups of integrated agents fluctuates over time, similar to the observed oscillations in the financial time series.

In the tenth chapter, the spectral representation of the output time series allows classification of the agent groups and determination of the excitable model parameters. The latter fact inspired the creation of the spectra-based clustering of the financial time series. This chapter uses large-scale social media excitation wave-propagation model for the simulation of automated investment strategies. Some OAM, PIF, and WIM principles were integrated in the large-scale AI-based MAS modeling of agent states. The real-life applications use simple wave-like communication mechanism between agents. The spectral representation of the multidimensional time series allows specific features to be generated that are necessary for recognizing the type of chaos-generating model.

At the end of the book the reviewing Appendixes A, B, C, and D were kindly contributed by Prof. James F. Glazebrook. He has focused his attention on certain theories and concepts which are considered to be germane to the tasking mechanisms of the OSIMAS and OAM. These are common to a number of mainstream models of cognition, neurodynamics, and sociobiological/cultural systems. The main topics covered in these four appendixes are: (1) The Maturana–Varela theory of autopoiesis, leading to Varela's foundations for a theory of neurophenomenology, (2) Global Workspace Theory—a principal forerunner in modern theories of cognition, (3) Distributed and Embodied Cognition, (4) Several other approaches interrelated with (1–3) that may also be significant for the OSIMAS and OAM models.

Contents

Abbreviations

ABS	Agent-based system
ACE	Agent-based computational economics
ACT-R	Adaptive character of thought-rational
ART	Adaptive resonance theory
AW	Active wakefulness (BMS)
BDI	Belief–desire–intention
BMS	Basic mind states
CA	Cellular automaton
CEMI	Conscious electromagnetic field theory
CMS	Collective mind states
DCM	Dynamic causal modeling
DS	Deep sleep (BMS)
EEG	Electroencephalography
EM	Electromagnetic field
ESD	Energy spectral density
GESD	Global energy spectral density
GWT	Global workspace theory
HPA	Hypothalamic–pituitary–adrenal
LESD	Local energy spectral density
LIDA	Learning intelligent distribution agent
LISA	Learning and inference with schemes and analogies
MAS	Multi-agent system
MLP	Multilayer perceptron
MMASS	Multilayered multi-agent situated system
MRI	Magnetic resonance imaging
OAM	Oscillating agent model
Orch-OR	Orchestrated objective reduction
OSIMAS	Oscillation-based multi-agent system
PDP	Parallel distributed programming
PIF	Pervasive information field
PPN	Prefrontal parietal network

PSD	Power spectral density
QH	Quantum holographic
RE	Resting (BMS)
RECS	Radical embodied cognitive science
SLP	Single-layer perceptron
SMC	Sensorimotor coupling
SWN	Small world network
TH	Thinking (BMS)
UCM	Unconscious cognitive modules
VCF	Virtual computational field
WIM	Wave-like interaction mechanism
ZPF	Zero point field

Chapter 1
Synopsis

This book addresses a conceptual problem—the lack of a multidisciplinary, connecting paradigm, which could link fragmented research in the fields of neuroscience, artificial intelligence (AI), multi-agent systems (MAS), and the social network domains. The need for a common multidisciplinary research framework essentially arises because these fields share a common object of investigation and simulation, i.e., individual and collective human behavior. Although the fields of research mentioned above all approach this from different perspectives, their common object of investigation unites them.

Investigation into the implicit oscillatory nature of agents and social mediums in general can reveal some new ways of understanding the periodic and nonperiodic fluctuations taking place in real life human individual and group-wide behavior. A closer look at the applied social networks research also reveals some related approaches, which deal, in one way or another, with simulations of the field-like information spreading in social networks. For instance, common behaviors spread through dynamic social networks (Zhang and Wu 2011), the spread of behavior in online social networks (Centola 2010), urban traffic control with coordinating fields (Camurri et al. 2007), mining social networks using wave propagation (Wang et al. 2012), network models of the diffusion of innovations (Valente 1996), oscillations-based simulation of complex social systems (Plikynas 2010; Plikynas et al. 2014b), etc.

The current peer-to-peer-based direct communication approaches in multi-agent systems have been unable to incorporate the huge amount of indirect (contextual) information which prevails in social networks. This is due to the associated complexity and intangibility of the informal information, and the lack of a foundational theory that could create a conceptual framework for the incorporation of implicit information in a more natural way. Thus, there is a need to expand the prevailing agent-based conceptual frameworks in such a way that nonlocal (contextual) interaction and the exchange of information could be incorporated.

However, the complexity of individual and collective human behavior permits myriad theoretical, simulational and experimental techniques of investigation. This

© Springer International Publishing Switzerland 2016
D. Plikynas, *Introducing the Oscillations Based Paradigm*,
DOI 10.1007/978-3-319-39040-6_1

naturally deepens, but also fragments our knowledge. Consequently, any serious attempt to grasp the overall picture requires at least some form of deductive approach, which starts with a broader insight about the fundamental governing principles of human nature and the nature of our society. Therefore, this book first strives to shed new light on research of a common denominator regarding the universal governing principles that drive individual and social behavior as well. Afterwards, it offers some experimental and simulational approaches to validate the proposed conceptual insights.

Below, in a few paragraphs, we briefly review closely related research domains in order to emphasize their common cross-disciplinary origin stemming from the universal principles of self-organization. At the end of the book provided Appendixes A, B, C and D review OSIMAS research approaches in the broader light of other related theories and models like autopoietic systems, global workspace theory, embodied and distributed cognition, adaptive resonance, etc.

In some way, proposed research paradigm is also close to the Dynamic Causal Modelling (DCM) framework, which has been developed recently by the neuroimaging community to explain, using biophysical models, the noninvasive brain imaging data caused by neural processes (David 2007). In DCM, the parameters of biophysical models are estimated from measured data and the evidence for each model is evaluated. This enables one to test different functional hypotheses (i.e., models) for a given data set. The goal is to show that these models can be adapted to get closer to the self-organized and dissipative dynamics of living systems, as covered by formal theories used in system's biology such as autopoiesis. In this sense, from the theoretical standpoint, the OAM approach gets close to the autopoietic systems theoretical framework.

Experimentally driven DCM and conceptually driven autopoietic systems theory can be connected. Based on the DCM framework we construct the OAM, which employs structural and dynamical effects using, e.g., the neuroscience data for modeling agents states' dynamics. From the other side, we can use autopoietic theory to describe the dynamics of mind states' transformation processes, which through their interactions continuously regenerate and realize the network of processes (relations) that produced them (Maturana and Varela 1980).

First, let us briefly discuss new opportunities for the multidisciplinary integration that has emerged following technical neuroscience advancements in the research area of brain activity mapping. This has enabled a qualitatively new level of cognitive, behavioral and computational neuroscience (Bunge and Kahn 2009; Haan and Gunnar 2011). With the increase of computing power, neuroscience methods have crossed the borders of individual brain research. Hardware and software, used for mapping and analysing electromagnetic brain activations, has enabled measurements of brain states across groups of people in real time (Lindenberger et al. 2009; Newandee and Reisman 1996; Nummenmaa et al. 2012; Stevens et al. 2012). This research frontier has made room for emerging multidisciplinary research areas like field-theoretic modeling of consciousness (Libet 2006; McFadden 2002; Pessa and Vitiello 2004; Pribram 1999; Thaheld 2005; Travis and Arenander 2006; Travis and Orme-Johnson 1989; Vitiello 2001), social neuroscience, neuroeconomics, and group neurodynamics (Cacioppo and Decety 2011; Loewenstein et al. 2008), see Table 1.1.

On the other hand, some perspicacious, biologically inspired simulation approaches have emerged in the areas of computational (artificial) intelligence, agent-based and multi-agent systems research (Nagpal and Mamei 2004; Raudys 2004a, b). In turn, these advances have laid the foundations for simulation methods oriented toward intelligent, ubiquitous, pervasive, amorphous, organic computing (Poslad 2009; Servat and Drogoul 2002) and field-based coordination research (Bandini et al. 2006; Camurri et al. 2007; De Paoli and Vizzari 2003; Mamei and Zambonelli 2006), see Table 1.1.

In sum, research trends in neuroscience, AI/MAS and social networks are leading to increasingly complex research approaches in some fascinating ways related with oscillations-based and field-theoretic representations of individual and collective mental and behavioral phenomena.[1] In this regard, the major insights of this research are derived from the novel Oscillation-Based Multi-Agent System (OSIMAS) social simulation paradigm, which links emerging research domains via coherent neurodynamic oscillation-based representations of the individual human mind[2] and society (as a coherent collective mind)[3] states as well.

The major conceptual implications of the paradigm presented in the book are essentially oriented toward the nonmechanistic and in some approaches even close to field-theoretical ways of modeling and simulating individual and collective mind states as well. Whereas, the major prospective practical applications of the OSIMAS paradigm are targeted at the simulations of an agent's behavior and some real social phenomena.[4] The main methodical tools suitable for the appropriate simulations pertain to the agent-based and multi-agent systems (in short, ABS and MAS respectively) approach. Hence, below in a few paragraphs, we introduce how the conceptual ideas presented here can find their way into ABS and MAS applications.

Current ABS and MAS applications are famous for their construction based on the so-called 'bottom to top' principle, which in essence, is explicitly agent-centric with pair-to-pair-based communication protocols between agents. Such models are used for the simulation of local interactions, which are similar to cellular (automata)

---

[1]However strange it may sound to some, humans are most probably not strictly separate, mechanistic points that accidently collide or interact with each other from time to time, but rather, are part of an interrelated web of subtle connections within a framework of collective consciousness.

[2]Some of the electroencephalographic (EEG) experimental evidence provided in the following chapters lead to the idea of interpreting basic human behavioral patterns in terms of mind states, which can be characterized by unique electromagnetic power spectral density distributions across the EEG spectrum.

[3]Such an approach paves the way for the field-theoretic (oscillations based) simulation of complex social networks (Osipov et al. 2007; Pikovsky et al. 2003).

[4]For instance, the contextual (implicit) information spread throughout social media (like propaganda, political campaigns, information wars, etc.), network models of the diffusion of innovations, models of self-excitatory wave propagation in social media, circadian and homeostatic rhytms of coherent mind states, etc.

Table 1 Emerging field-theoretic research approaches in various domains and scales of self-organization

Research domains	Emerging cross-disciplinary field-theoretic research approaches	
	Individual level	Social level
Neuroscience	Consciousness as a coherent electromagnetic field (e.g. represented by the Δ, θ, α, β, γ brain waves)	• Social neuroscience • Neuroeconomics • Group neurodynamics
AI/MAS	Artificial intelligent agent	• Field-based coordination • Ubiquitous intelligence • Pervasive/amorphous computing
Social networks	Social networking agent	Social mediums as excitatory systems, distributed cognition, etc.

mechanism models, except that more sophisticated communication protocols and some intelligent decision algorithms are usually employed (Wooldridge 2009).

Nevertheless, in order to achieve substantial progress in the simulation of complex social phenomena, such models are usually enriched with some 'top to bottom' construction principles like the belief–desire–intention (BDI) approach, agent selection criteria, higher emotional states, altruism, a credit/fines system, added noise in the inputs and outputs, confines for learning speed and acceleration, etc. (Raudys 2004a, b). These principles are applied from some meta-level—in other words, a nonlocal organizational level—which cannot be deduced from the simple agent properties. In this way, the pure 'bottom to top' self-organization principle is lost. This obviously shows that explicit information encoded solely in the properties of individual agents is not enough for the simulation of complex social agent-based systems, as social agents are not so individual after all. They are open systems, intrinsically interwoven in the social networks and therefore, influenced by the external, i.e., not only local, but also regional and global environments at large.

In one way or another, this nonlocal (or 'implicit') self-organizational level is usually introduced artificially, following observed social behavioral patterns. For instance, in communication theory and practice, it is well known that tacit (informal, officially unrecorded) information like the emotional 'atmosphere', working environment, moods, mimicry, gestures, media stories, weather conditions, etc., prevail in social organizations, which all profoundly influence human decisions and social wellbeing in general on an unconscious level. This unconscious (implicit) level of self-organization works in the form of contextual (nonlocal) information shared by all.

Historically, a number of well-known scholars have argued for the existence of a common unconscious. For instance, William James, the founder of psychology as an academic discipline in America, argued that none of the empirical findings of science about the human brain contradict the idea that the brain reflects or transmits a transcendental, infinite continuity of consciousness underlying the phenomenal world instead of producing consciousness de novo, as is commonly assumed in science (Perry 1996). Admittedly, William James was one of the leading thinkers of the late nineteenth century and is believed by many to be one of the most influential philosophers that the United States has ever produced, while others have labeled him, the "Father of American Psychology." Along with Charles Sanders Peirce and John Dewey, he is considered to be one of the major figures associated with the philosophical school known as pragmatism, and is also cited as one of the founders of functional psychology.[5]

Therefore, I see the above-cited William's argument purely from pragmatic grounds. Science is not able to confirm or refute this argument yet. Luckily, science is not a religion. We are open to different viewpoints, argumentation, proving efforts, etc.

In turn, Emile Durkheim, one of the founders of modern sociology, described "collective consciousness" as the mind of society, created when "the consciousness of the individuals, instead of remaining isolated, becomes grouped and combined."

Carl Jung conceived of a collective unconscious as a "reservoir of the experiences of our species": the repository of humanity's collective experience common to everyone. In Jung's view, the collective unconscious embodies archetypal patterns. Jung argued that the common patterns found in art, literature, and cultural artifacts from different parts of the world are evidence for the archetypes of the collective unconscious. He believed that the ideal archetypal patterns inherent in the collective unconscious manifest themselves in cultural traditions, social roles, dreams and intuition to guide the development of the individual mind. Jung's archetypes of the collective unconscious can be thought of as laws of nature in terms of structures of consciousness. In Jung's words, "My thesis then, is as follows: in addition to our immediate consciousness, which is of a thoroughly personal nature and which we believe to be the only empirical psyche (even if we tack on the personal unconscious as an appendix), there exists a second psychic system of a collective, universal, and impersonal nature which is identical in all individuals." According to Carl Jung, this collective unconscious does not develop individually, but is inherited. It consists of pre-existent forms, the archetypes, which can only become conscious secondarily and which give definite form to certain psychic contents (Laszlo 1995).

In sum, they and many other prominent scholars argue that a nonlocal and commonly shared unconscious level of mind exists, which cannot be sensed by the known five senses. The closest fundamental explanations may come from quantum nonlocality, which refers to the quantum mechanical predictions of many-systems measurement correlations for the entangled quantum states (Oppenheim and

---

[5]The citation of William James is not an excursion into the supernatural, unless we admit that all the things we still cannot comprehend about the nature of consciousness belong to the paranormal or supernatural realm. To the author's point of view, it is quite the opposite—the insights of the well known American philosopher William James and many others open up a paradigm shifting viewpoint suggesting a new frontier for the rigorous studies of consciousness as universal entity, similar to the ways we study the laws of matter, energy and space-time.

Wehner 2010).[6] The issue has also been raised that we may have to differentiate between two different types of nonlocality: one related to quantum mechanics and the other to what is termed 'biological nonlocality' (Thaheld 2005).

To date, very few experiments that attempt to explore the possibility of a quantum physics–biology interrelationship have been conducted. The first experiment utilized pairs of human subjects in Faraday cages, where just one of the pair was subjected to photo stimulation, and possible electroencephalographic (EEG) correlations between human brains were investigated (Grinberg-Zylberbaum and Ramos 1987).[7] Later experiments, building upon this pioneering research, continue to corroborate the findings with increasing experimental and statistical sophistication (Radin 2004; Standish et al. 2004).

Experiments which revealed evidence of correlated functional magnetic resonance imaging (fMRI) signals between human brains have also been conducted (Wackermann et al. 2003). These correlations occurred while one subject was being photostimulated and the other subject underwent an fMRI scan. Research was also ongoing at the University of Milan (Pizzi et al. 2004; Thaheld 2005) utilizing pairs of 2 cm diameter basins containing human neurons on printed circuit boards inside Faraday cages placed 20 cm apart. Laser stimulation of just one of the basins revealed consistent waveform autocorrelations between the stimulated and unstimulated basins.

All of these controversial experiments, when taken seriously together, seem to be pointing to a black hole in our understanding of some deep fundamental nature of reality. One needs courage to step away from their comfortable zone and employ out of the box thinking in search of possible solutions. In fact, this can lead us to unusual ways of thinking. For instance, we may imply that some sort of statistically significant, biological entanglement and nonlocality effects may take place between human brains (see the references above and also some additional studies) (Orme-Johnson and Oates 2009; Travis and Orme-Johnson 1989). Naturally, mainstream science usually ignores such inconvenient experimental evidence as not existing or doubtful, as if more accurate measurement might eliminate this misunderstanding. This is usually the case when dealing with controversial findings on the edge of the known due to (i) evident contradictions with the main postulates of established theories, and (ii) problems in replicating the same human mind states in an objective manner.

---

[6]We have to be carefull as there are several different interpretations of quantum mechanics, such as that of the Copenhagen school, modal interpretation, relational interpretation, quantum decoherence, many worlds, information-based interpretations, and quantum Bayesian interpretation, just to name a few.

[7]Faraday cages were used to screen out nearly all relevant electromagnetic, acoustic, and other influences, but not gravity. Researchers noted that 25 % of the time, what they called transferred potential (TP) or simultaneous brainwave events (which others prefer to call correlations or autocorrelations of electrical activities) appeared in the brain's electrical activity or the EEG of an unstimulated subject.

Hence, these and many other controversial findings are usually cast aside as nonsensical speculations without serious consideration. However, we should not throw the baby out with the bathwater, as some so-called "last remaining unresolved controversial issues" are most probably, hints pointing us to some new paradigms, theories and worldviews that show our previous understanding to be a simplified version of the deeper perceptions of reality, unless we want to decide that the current state of science is the last word about reality and there is nothing else beyond it to discover. To the author's point of view, it might be quite the opposite —some controversial experimental findings in the area of consciousness studies are identifying inconsistencies in the current, established understanding of them, which is due to a simplified, reductionist view of reality. A paradigm shift to a new understanding is required, which can deal with the previously mentioned controversial findings in one or another way.

For instance, on behalf of open scientific exploration, we may at least theoretically acknowledge that the idea that contextual implicit information could be distributed in some sort of field (e.g., EM fields), and that fields—although expressing some global information—are locally (unconsciously) perceived by agents. Consequently, it can also lead us to a totally novel understanding of agent mind states as field-like entities, which at least theoretically, could absorb and emit some sort of contextual, e.g., bioelectromagnetic, fields.

Of course, there are also some strong arguments against such quite ambitious assumptions, such as the short, limited range of the EM fields, too noisy of an environment, or the too complex neurodynamics of the brain, etc. All these and many other arguments might be employed to conclude that the assumptions above are baseless. However, this should not preclude us from pursuing further conceptual, experimental and simulation-based exploration of possible alternative ways of understanding. Therefore, considering the current state of research affairs, this pioneering, multidisciplinary research book cannot ambitiously state that mission has been accomplished, proposing final solutions. Nevertheless, it proposes some original, conceptual, experimental simulation methods for further exploration on the edge of known.

The current peer-to-peer-based ABS and MAS direct communication approaches have been unable to incorporate the huge amount of indirect (contextual) information. This is due to the associated complexity and intangibility of the informal information, and the lack of a foundational theory that could create a conceptual framework for the incorporation of implicit information in a more natural way. Thus, there is a need to expand the prevailing ABS/MAS conceptual frameworks in such a way that nonlocal (contextual) interaction and the exchange of information could be incorporated.

It seems plausible that we could introduce local MAS_N and nonlocal MAS_N, layers of self-organization in the prospective ABS/MAS simulation platforms

$$\text{MAS} = (1 - \eta)\text{MAS}_L + \eta\text{MAS}_N, \tag{1.1}$$

where $0 \leq \eta \leq 1$ denotes the degree of nonlocality:

$$\eta \Rightarrow 0, \quad \text{then} \ \text{MAS} = \text{MAS}_L,$$
$$\eta \Rightarrow 1, \quad \text{then} \ \text{MAS} = \text{MAS}_N.$$

In this way, we expand the concept of the ABS/MAS by adding nonlocal levels of self-organization. Starting with $\eta \rightarrow 0$, self-organization could be observed (i) at the local single-agent level, (ii) on the intermediate scale $0 < \eta < 1$ it could be observed in coherent groups and organizations of agents, and (iii) on the global (social continuum) scale ($\eta \rightarrow 1$), it could be observed in large, coherent, societies of agents.

It naturally follows from some real-life observations, e.g., agents interact locally (interchanging information with neighbors), but are also affected by the nonlocal states of the whole system (e.g. traditions, cultures, fashions, national mentalities, political situations, economical/financial situations, etc.). Here the term 'nonlocality,' which we borrowed from quantum physics, could have many social interpretations, but we prefer to understand it as Jung's archetypes of the collective unconscious, which can be thought of as laws of nature in terms of structures of consciousness.[8]

If we admit that these laws of nature can be applied in terms of common structures of consciousness, then our individual mind-field patterns would be cohesively adjusted. This cohesion can be interwoven so deeply in our own reasoning nature, that we could not directly perceive it.[9] In this sense, at least hypothetically we could admit existence of the common mind-field as an intangible medium where the individual cognitive processes are activated and mutually coordinated. We assume, though, that the effects of this very subtle and intangible common mind-field could be experimentally observed and explored in terms of neuroscientifically measured synchronicities and coherence patterns between individual brain wave activations. Social neuroscience has just started exploration of this potentially very perspective area of research.

---

[8]Precedents for the idea that the laws of nature are structures of consciousness can be found throughout Western tradition. Plato, for example, held that the highest and most fundamental kind of reality consists of transcendental forms, which are not known to us through the senses.

[9]Similarly, like we cannot see, smell, or hear air (as a medium), through which other things and people can be seen, heard or smelled. Nerveless, air as a medium exists, despite the fact that our senses do not perceive it directly.

Chapter 2
Towards Wave-like Approach

This chapter discusses some insights about fundamental properties of information rich social systems. It mainly focuses on (i) claiming general properties of information-based social systems, (ii) ways to model emergent and self-organizing features of social networks, (iii) discussion how to simulate complex social systems using field-based approach and multi-agent platforms. Additionally, chapter provides some ideas how to construct field-based communication network of intelligent agent's using currently available computational intelligence methods. A vision of new simulation paradigm offers some useful concepts to transform multidimensional factor space (representing a multiplicity of phenomenal forms and interactions) into the most universal spectral coding system. The chapter gives some ideas and examples how not only the communication mechanism but also the social agents can be simulated as oscillating processes.

2.1 Overview

Notwithstanding the appealing empirical power of the applied methods, modern social research is still fragmented with a lack of cohesive modeling media for understanding and simulating the highly complex nature of social behavior. Nevertheless, our understanding of the traditional economic systems is transforming. For instance, economists and financial specialists have started to doubt about common financial capital theories after several unpredicted financial crises took place during the last decade, e.g., Asia crises, Russia crises, LTCM fund crackup, ENRON financial group crash and the recent global financial meltdown, which has started from mortgage crises in 2007—just to mention a few. The world's largest institutional investors, for example commercial banks, insurance companies, retirement funds, and so on are focused on the development of new concepts.

Interdisciplinary research groups in Santa Fe Institute, MIT, Cambridge, Princeton, Stanford, Yale, Harvard, etc., are increasingly open for the use of

© Springer International Publishing Switzerland 2016
D. Plikynas, *Introducing the Oscillations Based Paradigm*,
DOI 10.1007/978-3-319-39040-6_2

nonlinear, dynamic, and heuristically oriented techniques for the purposes of delivering new concepts, which can deal with simulation of nonlinear dynamics, unexpected price shifts, economic crises, transitional effects, etc. (McCauley 2004). However, constructed models mostly fail to describe such an evolving complexity as they (i) do not have the same order of complexity (or degrees of freedom) and (ii) do not evolve together with the object of investigation.

Today, we appreciate that critical social phenomena, pattern formation, bifurcations, and deterministic chaos occur in both inanimate and animate nature and are implicated in fundamental ways with evolutionary population dynamics and self-organization in biological as well as social development. One of the most widely applicable lessons from other disciplines is that when systems consist of competing elementary forces the tensions that arise create structural complexity. The emergence of complexity, an apparently common phenomenon, and the parallels between the architecture of physical, biological, and social systems only hint at the beginning of a synthetic theory that describes how evolution generates structure. In this book we imply that the employment of the proposed approaches should reveal (i) conditions for appearance of periodic and nonperiodic social cycles, (ii) conditions and features of population and information clustering, (iii) structural invariants of the social evolution.

Much research is concerned with the appropriateness of traditional research approaches and methodologies in the social research domain. There is a growing suspicion that the main hindrances are hidden in modern social processes, which are badly modeled using traditional top–down approaches. With the advent of the new information society, we are increasingly dealing with some modern social phenomena like the network economy, globalization, e-commerce, nonequilibrium dynamics, complex social emergence, predominance of intangible (informational) goods, generative micro–macro relations, etc.

However, there is fast growing empirical literature on multi-agent systems (MAS), which are used to model social phenomena from the bottom-up, like agent-based hybrid intelligent systems (Zhang and Zhang 2004), agent-based computational modeling (Billari 2006; Darley and Outkin 2007), agent-based simulation from modeling to real-world applications, generative social science (Epstein 2006), agent-based computational economics (Tesfatsion and Judd 2006) etc.

Notwithstanding some advances in the use of multi-agent systems, promises were not fully realized for a number of reasons, as it is a truism in the field that the key problems are hiding not only in the agents themselves, but also in the way they communicate, interact and interpret behavioral information available in social markets. This adds a whole new layer of complexity as we see it in the most complex real social networks today (e.g., financial markets) (Hoffmann et al. 2007).

These are highly heterogeneous information networks with many links and complex interrelations. Uncoupled and indirect interactions among agents require the capability of affecting and perceiving broadcasted information context. In this sense, we propose to model the information network as a pervasive information field (PIF). Such an approach gives appropriate means to enforce indirect and uncoupled (contextual) interactions among agents. It is expressive enough to represent contextual broadcasted information in a form locally accessible and immediately usable by network agents.

One of the closest examples in this area is amorphous computing (Nagpal and Mamei 2004). Another interesting proposal in that direction is the multilayered multi-agent situated system (MMASS), which defines a formal and computational framework relying on a layered environmental abstraction (De Paoli and Vizzari 2003). MMASS was related to the simulation of artificial societies and social phenomena for which the physical layers of the environment were also virtual spatial abstractions. In the last decade, a number of other field-based approaches have been introduced, like gradient routing (GRAD), directed diffusion, "Co-Fields" at TOTA programming model, CONRO, etc. (Mamei and Zambonelli 2006).

In fact, almost all proposed systems are either employed for various technological or robotic applications and very few of them like MMASS, agent-based computational demography (ABCD) or agent-based computational economics (ACE Trading World application: the simulation of economic phenomena as complex dynamic systems using large numbers of economic agents involved in distributed local interactions (Tesfatsion and Judd 2006)) are suitable for programmable simulations of social phenomena.

The study of coordinated models goes beyond computer science in that evolutionary computation, behavioral sciences, social sciences, business management, artificial intelligence and logistics also somewhat strictly deal with how social agents can properly coordinate with each other and emerge as globally coherent behaviors from local interactions. The major question we address in this study is: what is the natural and most efficient way of simulating complex information networking and interaction mechanisms in order to reflect the observed multiplicity of modern broadcasting telecommunication systems used by social agents?

To the author's knowledge, the close match to the proposed idea is explored by Mamei and Zambonelli in their study "Field-based coordination for pervasive multi-agent systems" (Mamei and Zambonelli 2006). They have been trying to achieve coordination for multi-robotic applications by means of the field-based (analog of PIF) approach. This is an example of how engineers are starting to understand that, to construct self-organizing and adaptive systems, it may be more appropriate to focus on the engineering of proper interaction mechanisms for components of the system rather than on the engineering of their overall system architecture.

As a matter of fact, effective communication stands among the top most important issues in MAS (multi-agent systems) implementations (Billari 2006). Contractual-based handshaking between two agents shapes peer-to-peer communication in today's MAS systems. Peer-to-peer communication occurs when (1) the connection between sender and receiver is bonded in time and space, (2) two agents exchange ontologically classified semantic information by using a common communication protocol or a matchmaking agent. This approach works well for simple networks. This is not so effective in the complex social networks where the agents' direct and coupled "peer-to-peer" communication model is pushed aside by other noncoupled, indirect, or contextual communication mechanisms.

One obvious example can be seen in modern telecommunication networks, where peer-to-peer connection protocols are no longer prevalent. This happens mainly because they are not efficient enough for multitasking, parallel processing, congested traffic control, conflict resolution, etc., (Panko and Panko 2010).

Not accidentally, there is a striking structural similarity between modern telecommunications and social networks. In fact, the main information traffic in social networks flows through telecommunications networks, which act as a backbone of the modern network-based information economy.

In fact, the information era has shaped efficient protocols for complex information traffic in the telecommunication networks, where (i) each agent can instantly send and receive information simultaneously through multiple communications channels, (ii) information flows are locally managed by the agent's preferences as if having the ability to "tune" to different broadcasting channels, (iii) agents became processing, storing, and retransmitting nodes in the social networks (Nagpal and Mamei 2004). Information is spreading through a multitude of multimedia networks with the speed of light. After all, it does seem like we are immersed in emanating fields of virtual information. The intriguing conceptual question we want to raise for discussion below is whether there is an acceptable way of simulating complex social network phenomena using universal spectral frequency representation.

Hence, this chapter is organized as follows. After formulating major claims and assumptions about fundamental properties of self-organized systems in social domain, we give an example of some fundamental physical models, which can help to model such systems. Next, we elaborate about PIF approach, which gives some clues and basic design ideas for understanding how field-like communication mechanism can be implemented.[1] Afterwards, we discuss wavelike interaction mechanism between heterogeneous agents. Finally, we summarize the conceptual findings and offer future directions for related research.

2.2 Evolving Economic Systems

Following the analogy of telecommunication systems, we assume that the same principles of reductionism and universality should be applied to simulating platforms in the social domain, too. In other words, communication by agents should operate not in the (peer-to-peer) vector-based multidimensional semantic space, but rather directly in the form of multimodal energy (spectra) emanated and absorbed over the social network. The flow of energy (and associated with it, information) in the form of stylized fields, however, requires a somewhat different understanding of the agent's role and their interaction mechanism.

---

[1]On the other hand, in this chapter presented approach is not the only one in this book. In the next chapters, there are presented some other simulation ways too.

In essense, information plays a major role in the economic systems. Information and entropy are the most general estimators of complexity. Admittedly, information does not immediately depend on the physical units and quantities of the chosen reference system, but on the relative frequency of the event occurring in the observed system. We more and more tend to look at information as the third basic quantity beside matter and energy because of central importance of information.[2] In fact, energy is also a quantity similar to information as it has ability to be transported and stored in several physical quantities.

In economics, a new and interesting phenomenon arises if we acknowledge observation that capital is becoming an intangible good. In fact, capital (we assume it as accumulated potential energy) gradually takes form of information in the modern information societies. It means that developed economies have started to capitalize on pure intangible goods or services (Ormerod 1997). In this regard, we can imply that emerging complexity of social networks is gaining ground not in the Newtonian mechanics (statistically described for economic agents as some sort of physical interacting particles), but rather in the form of self-organized information with virtual fields, potential energy states, wave-like interaction mechanisms, etc. Following the above discussion, we can further imply that the modern social order pertains to information fields, ranging from organization of individual agents to the information markets.

In order to proceed further, first we have to name in systemic manner some basic OSIMAS assumptions about underlying principles, which may govern complex information societies. Below we will get to the basic universal laws and starting from there to proceed with some new conceptual proposes.

1 Claim *Effective attempts to simulate and model social behavior most likely pertain to artificially constructed self-organized systems.*

Such systems in depth and spread (comparing with original ones) can evolve in the virtual world settings. Successful imitation depends upon model setup, i.e., its emerging and self-organizing properties.

2 Claim *Modern information-based economic systems (e.g. financial capital markets) and social networks are mostly related with self-organized information.*

For instance, electronic currency and stock exchange (SE) markets deal exclusively with information content. Even more, in some SE electronic markets majority of market deals is made by the sophisticated algorithmic traders. This tendency is expanding exponentially to other electronic markets too. Similarly, social networks pertain to the self-organized information too.

3 Claim *In terms of modeling, the closest fundamental match for the complex self-organized information is energy, which can be stored, transformed and distributed.*

Information being dimensionless expresses itself in a plenty of different forms as its carriers (Arndt 2012). The closest fundamental match for information in terms of

---

[2]The same as energy, information is a quantity that depends on the choice of the system of reference. It can be specified only with the respect to the reference system.

properties is energy. It can and should be employed for modeling self-organized information properties.[3]

4 Claim *In contrast but not in contradiction to the fundamental second law of thermodynamics there are living organisms and their societies, which in the process of self-organization are creating organized structures (negentropy) in expense of increasing entropy in the surrounding environment.[4] In the process of self-organization is produced new information, e.g., new structures or behavioral patterns.*

In the open system, it decreases entropy (disorder) inside the system, but increases chaos outside the system. In this sense, the second law of thermodynamics holds.[5] Emergent structures are maintaining disequilibrium between inner order and outside chaos.

5 Claim *Living beings containing self-organized information are usually not stable because they are exposed to the opposing fundamental processes of entropy (loss of order) and self-organization (creation and maintenance of order).*

In general terms there are at play two universal processes: one for self-organization of information and another for disorganization (entropy governed by the second law of thermodynamics). Both processes interact with each other creating a complex flux of changes.

6 Claim *Information society pertains to a new type of economics where wealth is capitalized in a form of intellectual, commercial and technological information. It is stored in various information markets and has capacity of spreading globally through nets of information channels with the speed of light.*

Modern information economy cannot be easily described by classical Adam Smith's approach as statistically aggregated mechanical approaches simply oversimplify the reality. In short, we are moving to the age of information economics. So, we have to look for fundamentally different modeling tools, which inherit the self-organized properties of economic agents and information society at large.

7 Claim *Modern information societies can be understood and modeled as accumulated potential fields of various information, where energy or excited states are*

---

[3]Information is an intensive quantity (energy is extensive), which means that summation of the two pieces of information leads to accumulated information that is less than or equal to the sum of the two pieces of information. The accumulated information is only equal to the sum of the single pieces of information when there is no mutual information between the subsystems (independence of the random variables).

[4]The second law of thermodynamics states that the entropy has to increase in the closed systems, which indicates that the disorder (disorganization) in closed systems has to increase.

[5]Self-organized systems are not closed systems. So, there is no contradiction with the second law of thermodynamics. There are, though, some limits concerning conservation of information analogous to the conservation of energy, as for example in the time–frequency measurements. This relation shows us that any gain of information that we obtain by smaller sample intervals in time leads to a loss of information of larger frequency intervals in the frequency domain. Obtaining information always means losing it, but in the different scale in time or frequency.

propagated not only in a form of corpuscular kinetic interactions between eco-
nomic agents, but increasingly more like information fields transmitted through the
telecommunication systems or mass media channels.

This gives a good reason to model world capital spatial and temporal dynamics through the means of interfering information waves oscillating in different frequencies, amplitudes, and phases via the geographically distributed population.

In summary, for the effective implementation of the wave-like modeling approach we can employ spectral representation as a universal energy-information warehouse. For that reason we first have to transform all tangible objects-resources into their energy equivalents and then interrelate different types of energy as intangible information stored in the form of corresponding sets of spectral bands. The reasoning behind this is based on the principle of reductionism and universality as we are looking for the means to reduce a multiplicity of forms into one representation.

Thus, according to the proposed approach, agents can be interpreted as processes, which exchange multimodal energy depending on the information they have. We assume that agents are rule-based, input–output reactive systems. Changing an agent's behavior means modifying its characteristic parameters in the behavioral rule set.

2.3 Methodological Framework

To begin with, the central critical question is how we could model the postindustrial (self-organized) information society, which can be characterized by (i) value creation using knowledge (the main capital resource), (ii) a net of various information channels, (iii) on-line processing, storage, and transmitting (OLAP) systems in a chain: data $\rightarrow$ information $\rightarrow$ knowledge. There are no easy answers. However, proceeding comments below may shed some new light.

Much research is concerned recently with an answer as to why economic systems behave chaotically. The understanding might rest on the perception that the ability of the system in chaotic motion to explore wide regions of its phase space, and thus utilize information about a large collection of potential states of the system, can have distinctive positive implications for what the economic system can do in achieving its required functions. Biological or social systems that do not take advantage of chaos in their state spaces may not be able to participate in evolution to the extent we understand that as adaptation to changing environments (Abarbanel 2012).

Social networks are complex systems with inherited chaotic behavior. It means we cannot construct effective long-term forecasting models. Instead, we should look for models which have the same chaotic properties (e.g., invariants) as the original systems (Peters 1996). Such models are not exclusively targeted only for better forecasting. They can target something not less important, i.e., simulation of the geometrical structure of the real data (mimicking chaotic invariants' values). Such approach stresses more on the simulation of complexity itself. In practical terms, it can give close enough estimates for, e.g., average number of peaks or minima via given period, long-range dependencies, self-similarity estimates, major periodic and aperiodic cycles, expectations for critical events, etc.

Let's explore one example in a more formalized way. We are going to construct a physical model related not only with modeling of the set of real people in economic system, but also with a media of interaction transmission. We admit a set $\forall E$ of economic agents as information emitting sources.

The analogous (simplest) physical model can be presented using energy carried by a stretchable string (Bronshtein et al. 2013), see Fig. 2.1a. Here we have just one physical dimension x and economic agents are lined up in it composing a string U. The string of agents lies in between the two stylized fields as shown in the Fig. 2.1a). When a perturbation is present, it pertains to potential and kinetic energy. However, perturbations originate not from the ends of the string (as it is usually implied in the physical string models), but from the both fields as it is represented in Fig. 2.1a)

$$dK = \frac{1}{2}\lambda_0 dx \left(\frac{\partial \eta}{\partial t}\right)^2, \quad K_1 = \frac{dK}{dx} = \frac{1}{2}\lambda_0 \left(\frac{\partial \eta}{\partial t}\right)^2, \tag{2.1}$$

where dK stands for kinetic energy of an element dx of the string ($\{dx\} \to \{\{x\}|x \subset U\}$); K_1 stands for kinetic energy density and $\lambda_0 dx$ is the mass of the element dx. Partial derivatives here denote the dynamics of perturbations.

For negligible Young modulus (i.e., for idealized string with perfect flexibility) the work done to stretch the string an amount Δs against a constant tension τ_0 is $\tau_0 \Delta s$, which therefore equals the gain in potential energy stored in the displaced string.[6] Hence the total potential energy V and local potential energy V_i for the agent i are

$$V = \frac{1}{2}\tau_0 \int_0^l \left(\frac{\partial \eta}{\partial x}\right)^2 dx, \quad V_i = \frac{1}{2}\tau_0 \left(\frac{\partial \eta}{\partial x}\right)^2, \tag{2.2}$$

where l stands for the lenght of the string.

The string model can be helpful realizing how oscillations can represent dynamics and propagation of vibrational states, without simultaneous mass transport. Systems in which waves arise may be envisaged as being composed of infinitely many mutually coupled oscillators (agents). Vibrational state of each individual oscillator depends on the states of neighbor oscillators (agents). (see Fig. 2.1a).[7] The energy of the whole system is constantly redistributed among the oscillators.

Next, let's to proceed with some physical model to deal with a key characteristics, i.e. entropy and information. For analogy, we can use a physical model based on the most basic laws of thermodynamics. First, let say we have a closed homogenous

---

[6]τ_0 can be interpreted as permeability for conduction of transversal wave energy. In the broader sense, it can be a function of location and time $\tau_0(x,t)$.

[7]A quantity which describes the oscillation state is written in the form $\omega t - \vec{k}\vec{r} + \phi$.

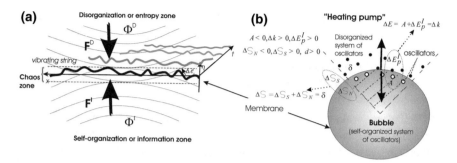

Fig. 2.1 One-dimensional physical model of stretchable string as analog for many lined up mutually coupled oscillators (agents). String is located between two opposing zones of self-organization and disorganization, see picture on the left (**a**). Picture on the right (**b**) depicts "Bubble's" model of the "Heating pump" aimed for entropy (disorder) filtering

macroscopic system of microscopic particles. According to the first law of thermodynamics: the total internal energy of the system is a conserved quantity, but energy can be transferred from one part of the system to another and different forms of energy may be converted into each other, but not all conversions are completely reversible.[8] Under such circumstances the closed system resides at adiabatic-isochoric equilibrium state (no heat exchanging with an environment and $V = $ const), distinguished by a maximum of entropy S_{max} and minimum of internal energy U. We should remember, though, that whole system is closed, but "Bubble's" (see Fig. 2.1) interior subsystem is open. Therefore, II law of thermodynamics is valid for a whole system.

Suppose further, after some time accidentally forms a small perturbation, i.e., self-organized "Bubble" (an analog for living entity or social agent). The "Bubble" has membrane with entropy (disorder) filtering properties, i.e., it works as a heat pump, which can be described as a thermodynamic machine operating according to the principle of a left-handed Carnot cycle: with expenditure of work it pumps heat from the colder ("colder" in the information sense means that inner oscillators have some vibrational frequencies filtered out compared to outside oscillators) inside to the warmer outside of the "Bubble". Efficiency of energy conversion is less than one (the ratio of the quantity of the work expended and the heat transferred to the hot system). This is irreversible process in a closed system; it leads to the increase of the whole entropy $\Delta S \geq 0$.[9] Meantime, in the subsystem of inner "Bubble", $\Delta S < 0$. This is possible only because of filtering properties of the "Bubble's" membrane. The whole "heating pump" process is following the law of energy conservation.[10]

---

[8]Thermal energy cannot be converted completely into other forms of energy, i.e. heat, internal energy and enthalpy contain fractions of energy, which cannot be converted at all.

[9]For isochoric process $\Delta S = C_V \ln(T_2/T_1)$.

[10]The change of the internal energy in an arbitrary, infinitesimal change of state is a total differential. The change of the internal energy depends only on the initial and final state, not on the path.

8 Claim *There are no closed processes in nature where the total entropy decreases, but this is locally possible in the open self-organized subsystems, which contain information how to make work the above described "heating pump". In this sense, self-organized living subsystems can be interpreted as entities capable to reduce inner entropy decreasing randomness of their vibrational states in expense of increased entropy in a whole environment.*

Oscillators inside the "Bubble" are self-organized, capable to sustain the inner order and membrane's entropy (disorder) filtering properties subject to the information they have. The dynamics of interplay between self-organized and disorganized states takes place in the membrane itself. In this way, by input of work A ($A < 0$ because work is done against prevailing opposite force from disorganized environment, see Fig. 2.1) membrane's oscillators get equivalent, but opposite sign potential energy $\Delta E_p^I > 0$, see Fig. 2.1b.[11] At the same time moment it changes the information content in the system relative to environment as part of existing uncertainty S (entropy $S > 0$) is decreased locally by $\Delta S^{\mathrm{bubble}} < 0$. System (the "Bubble") is freed from chaos, noise or more generally from uncertainty. It creates structured forms or complex arrangements increasing the information of the system. Sure, this process is far from trivial, because (i) in the closed system holds the second law of thermodynamics, (ii) every self-organized system has aproperty to split the flow of entropy into two opposite parts, i.e., entropy ΔS_S and negentropy ΔS_N [Brilloun-Schrodinger's negentropy (Brillouin 1953)] balancing with residual δ (δ—the overall increase of entropy in the transition process as required by II thermodynamic law in the closed system) between them

$$\Delta S_S + \Delta S_N = \delta, \quad \text{where} \quad \Delta S_N < 0, \Delta S_S > 0, \delta > 0. \tag{2.3}$$

Because of efficiency $\eta = (\Delta S_S + \delta)/\Delta S_N < 1$, the heating pump releases some additional irreversible heat $\delta > 0$ (increasing total entropy, i.e., disorder). In this way, we may also imply that order develops from chaos in the self-organized living systems. However, self-organized structures are far from a state of equilibrium.

Hence, the law of energy conservation and II thermodynamic law give us

$$A + \Delta E_p^I + \Delta \kappa = 0, \quad A < 0, \Delta \kappa > 0, \Delta E_p^I > 0, \tag{2.4}$$

where changes of all energy and entropy for the closed system can be described[12]

$$\begin{cases} \Delta E = A + \Delta E_p^I = \Delta \kappa \\ \Delta S = \Delta S_S + \Delta S_N = \delta \end{cases} \tag{2.5}$$

---

[11]Here holds energy conservation law. Superscript I in the ΔE_p^I term indicates information origin for this potential energy.

[12]Note that, in the process of information gain, increases kinetic energy $\Delta \kappa$ and closely related heat entropy δ yielding to $\Delta \kappa \simeq \delta$.

where ΔE—system's total energy, ΔS—increased entropy (heat) energy, $\Delta \kappa$—change of kinetic energy (see Eq. 2.1). Of note is that "Bubble's" self-organized membrane does a job of selective filtering of noise (uncertainty or entropy).[13]

In fact, energy conservation law is not violated, because the whole system's energy remains the same. Only some part of the inner heat (low order energy) is transferred locally from "Bubble's" inside to the outside. As if self-organized "Bubble" is filtering out some frequencies of white spectrum making only some inner oscillations to take strength.

9 Claim *Gained information (order) in a form of increased negentropy comes in the expense of the increase of heat (entropy related disorder) outside the boundaries of self-organized system.*

In the utmost case, absolute negentropy (order) can be achieved by filtering all possible oscillations from a white spectrum leaving a system idle of any oscillations (the point in the center of "Bubble" with zero potential energy). Self-organized systems have learned how to use surrounding chaos by filtering its random oscillations in this way transforming their oscillation spectra. Properties of membranes play crucial part in the process as they are multidimensional, i.e. segregate different energy states, entropy, and information (as well as various organizational) states.

10 Claim *From another perspective, we can also imply that white spectrum has all possible set of oscillations for any type of orders giving birth to any type of model representations for self-organized entities. Increasing complexity of the self-organized systems in informational sense means climbing up on the ladder of potential energy and getting closer to the realm of chaos, where more and more sophisticated patterns of oscillations occur. It leads to self-organized systems with chaotic, nonlinear, highly adaptive, and unpredictable behavior.*

Purely for the modeling purposes we imply that each economic agent can be interpreted as a unique spectrum of oscillations. According to the chosen approach, self organized oscillations play crucial role in the dynamics of emerging complexity and chaos structures on the border between disorder and self-organization.

Let's see how presented physical model can be related with the economic (financial) case. First, one can recall that continuous time economics always deals with time dependent relations as functions of time (in financial markets case we usually deal with stochastic dynamics). In every case, we have a set $\exists F$ of functions $\{f(t), p(x)\}$

$$F = \left\{ f(t), p(x) \middle| \begin{array}{l} f(t) \text{ time dependant functions describing properties of economic agents} \\ p(x) \text{ time independant parametric properties of economic agents} \end{array} \right\}.$$

(2.6)

---

[13]Unlike neo-classical economists, which assume the unphysical equivalent of a hypothetical economy made up of Maxwellian demonish like agents (who can systematically cheat the second law of thermodynamics), we do obey the fundamental laws here.

If the set of functions $f(t)$ satisfy the Dirichlet conditions, i.e., (a) the defined interval can be decomposed into a finite number of intervals where the function $f(t)$ is continuous and monotone, and (b) at every point of discontinuity of $f(t)$ the values f $(t + 0)$ and $f(t - 0)$ are defined, then the Fourier series of this function are convergent. At the points where $f(t)$ is continuous the sum is equal to $f(t)$, at the points of discontinuity the sum is equal to

$$(f(x - 0) + f(x + 0))/2.$$

Every function $f(t)$, satisfying the Dirichlet conditions, can be represented mathematically using Fourier series as a periodic function of period T (superposition of sine and cosine oscillations).

We can recall here that superposition of sine and cosine function is an oscillatory function itself. Arbitrary periodic phenomena may be represented as superposition of pure sine and cosine oscillations (Benenson et al. 2006). The result of Fourier analysis is represented by frequency–amplitude plot. We can use Fourier synthesis for construction of a complex time signal out of several sine and cosine functions of different frequencies and amplitudes.

In the case of nonperiodic functions on the interval $(-\infty, \infty)$ Fourier series is replaced by the Fourier transform and inverse Fourier transform. In sum, there is a way $\exists \Gamma$ to transform $\forall F$ set of only time dependant functions $\{f(t)\}$, which describe the behavior of economic agent, to a set $\exists \Psi$ of time–frequency dependant functions $\xi(A, \omega, t)$ and vice versa

$$\Gamma : F \Leftrightarrow \Psi, \quad \forall F|\{f(t)\}; \exists \Psi|\{\xi(A, \omega, t)\}. \tag{2.7}$$

In conceptual terms, this is a very important inference as it shows that our dynamic social model can be simulated using frequency terms. For the effective implementation of the proposed oscillations-based approach each economic agent is interpreted further as a source of unique composition of oscillations. A set of such agents represents PIF (pervasive information field) model, i.e., superposition of waves and distribution of associated energy accordingly.

11 Claim *All time dependant state functions of economic agents' can be mapped into the frequency domain. In this way, an agent becomes represented in terms of a unique composition of oscillations (spectra).*

Instead of time dependant functions Fourier transform gives an opportunity to use spectra for the same modeling purpose. We may also mention some other transformations like Gabor and wavelet, which can be employed too (Benenson et al. 2006).

Our aim is to find out how spectra can interact and carry out the information, which is the universal attribute we have for measuring the basic properties of information societies. In general sense, the most universal measures of information are entropy and

Fig. 2.2 The principle of quantization of energy-information states for social agents having an analogy to the rotational–vibrational states of diatomic molecular model (where k—the rotational quantum number and h—the vibrational quantum number)

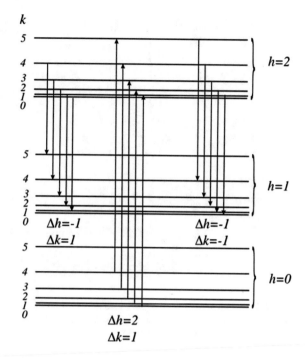

Brilloun-Schrodinger negentropy.[14] In the next section, we are going to see some hints for construction of PIF and wave-like interaction mechanism.

2.4   Pervasive Information Field (PIF) Concept

This section gives brief PIF outlines which are designed for (i) the construction of virtual media for information and resources (associated energy) exchange and (ii) the quantization of agents' internal states.

In a certain sense, we are proposing a universal energy-information state space model for the agents and their systems, see Fig. 2.2. All possible states are coded as

---

[14]Schrödinger gives the following example of the term elementary "negentropy" increase $N = -S = k \log \frac{1}{D}$ (where D—number of macroscopic states system is available to adopt, $1/D$—probability to be in one certain state) for living self-organized systems: living organism lives on food, which merely supplies our body with negentropy (for compensation of the loss of negentropy by the degradation of energy, i.e. transformation to heat). The energy contained in this food is not the crucial factor, because the energy is conserved. It is the negentropy that is the crucial factor, because the decrease of entropy, enabled by this negentropy, leads to an increase of the grade of energy (see Thompson, Kelvin grades of energy) and this is the reason living organisms can create complex structures.

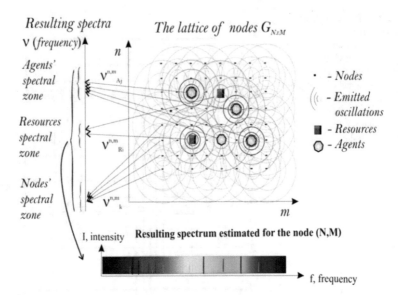

Fig. 2.3 PIF composed from agents, resources and nodes, which all are represented (see on the *left side*) by corresponding spectral bands. An example of resulting spectrum lines for the node (*N, M*) is given below the upper chart

unique oscillations in the frequency spectra.[15] In our proposed state space model, an arbitrary number of possible states for a given economic system can be coded and decoded.[16] Afterwards, with such a universal coding system we can create a set of conditions and rules for transitions between different energy-information levels.

We need a couple of quantum numbers to describe an infinitely large pool of states and transitions between them. We should bear in mind, however, that transitions between different energy states carry not only energy, but also information about quantum states, e.g., combination of quantum numbers like h and k as it is depicted in Fig. 2.2, bear exact information about the particular state.[17]

In the proposed PIF approach, fields are operating on the rectangular lattice $G_{N \times M}$ consisting from a set $\Omega_{N,M}$ of virtual nodes (we assume that nodes are distributed evenly on the lattice). Size of the lattice $N \times M$ is arbitrary. All resources $\{R_i\}$ and agents $\{A_j\}$ are distributed only on these nodes, see

Figure 2.3. Each particular node $\Omega_{N,M}$ represents a point on virtual lattice space ($n \in N, m \in M$), which functions are (1) for discrete time intervals to evaluate

---

[15]The agent's internal states are associated with possessed information, i.e. resources, behavioral patterns, internal parameters.

[16]Different agents can have concurrent oscillations (frequencies). It does not forbid an agent to have a unique set of own oscillations.

[17]Released or absorbed energy corresponds to the transitions between different energy levels.

incoming fields and produce corresponding spectra representations, (2) to oscillate at own fixed natural frequency $v_{n,m}$ emanating it to the surrounding PIF.

In order to reduce computations, resulting spectra are evaluated only for none-mpty nodes. Any node can be occupied either by agent from the set $\{A_j\}$ or by resource from the set $\{R_i\}$. Each agent and resource has own natural oscillation frequency $v_{n,m}^{A_j}$ and $v_{n,m}^{R_i}$ correspondingly, see Fig. 2.3.

Field intensities are decreasing according to the D^{-2} or similar negative power law, where D represents distance from the emanating source measured in virtual space using relative units.[18]

We will only discuss some ideas about the formal approach below. The most relevant model for PIF is adapted from physics using harmonic oscillators, where energy quantum for elastic wave

$$E = \hbar \cdot v, \tag{2.8}$$

where $\hbar$—quantum of action.

In physics, this energy quantum is called phonon.[19] Then the energy states E_k of the harmonic oscillator are quantized using phonons

$$E_k = \hbar \cdot v \cdot (k + 1/2), \quad k = 0, 1, 2, 3 \ldots, \tag{2.9}$$

where v stands as a base frequency. We may as well rewrite Eq. 2.9

$$E_k = \hbar \cdot (v \cdot k + v/2)$$

having two frequency terms $v \cdot k$ and $v/2$. The former term gives quantized frequency $v_k = v \cdot k$ as distribution in terms of phonons. The latter term gives the lowest possible frequency $v_0 = v/2$. In this way, corresponding frequency spectrum is $(v_0, 3v_0, 5v_0, \ldots v_0(1 + 2k))$, where $k = 0,1,2,3,\ldots$.

However, analogy with the physical model of phonons ends here. Strictly speaking, we cannot straightforwardly to adapt a set of quantized modes of vibrations occurring in a rigid crystal lattice for the economic systems. We have to design our own stylized set of quantized modes of vibrations suitable for simulations of processes taking place in the economic systems.

In our case, we want to have quantized spectrum, because we need fixed and separated frequencies allocated for different attributes (nodes, agents, resources),

---

[18]Virtual space axes are arbitrary chosen by the user. For example, in the investing MAS application, rectangular lattice could have two orthogonal axes for possessed capital and investing strategy.

[19]In physics, a phonon is a quantized mode of vibration occurring in a rigid crystal lattice, such as the atomic lattice of a solid. Phonons play a major role in many of the physical properties of solids, including a material's thermal and electrical conductivities. In particular, the properties of long-wavelength phonons give rise to sound in solids. In insulating solids, phonons are also the primary mechanism by which heat conduction takes place.

but instead of linear we are using natural logarithm to reduce high energy leaps at low k values, shortening available frequency interval at the same time. For the node $\Omega_{n,m}$ energy spectrum may look like

$$E_k^{n,m}(I^{n,m}, k^{n,m}, v^{n,m}) \sim \xi \cdot I^{n,m} \cdot v^{n,m} \cdot \ln(k+e),$$
$$I^{n,m} = A^2, \tag{2.10}$$

where ξ—chosen scaling factor, I—intensity (c.u.), A—amplitude of oscillation, $v^{n,m}$—first harmonic of the system of nodes, k—spectrum band number, and $e = 2.71$. Having Eq. 2.10, we may effectively characterize each system's spectrum in terms of distribution of k

$$D(E_k) \sim D(v_k) \Rightarrow D(k).$$

Actually, spectrum band number k, having virtual energy spectra described by Eq. 2.10, is the only thing we need to know in order to identify virtual nodes. Whereas, each nodes unique oscillations and corresponding spectral bands are derived by multiplying the first harmonic $v^{n,m}$ by the factor $\ln(k+e)$, which finally gives logarithmic scale for oscillations. Than $k = 0$, we obtain

$$E_0^{n,m}(v^{n,m}) = \xi \cdot I^{n,m} \cdot v^{n,m}. \tag{2.11}$$

which represents each node's $\Omega_{n,m}$ unique natural oscillation energy. Hence, different nodes have distinct frequencies $v^{n,m}$.

There is one major difference between representation of nodes and natural resources in the PIF model. Nodes are constant attributes of the system,[20] whereas, natural resources diminish or replenish themselves according to some time t dependant functions like[21]

$$I_{R_i} = f_t\left(E_{R_i}^{n,m}(t)\right). \tag{2.12}$$

For instance, in the economics domain, natural resources are productive economic sectors, regions, technologies, businesses, or securities like risky shares, government bonds, options, etc., which could generate capital gain or loss depending on initial model setup.

Finally, we have to discuss about agents' spectral energy representation in the PIF. Naturally, we will adapt the same approach as for resources. Let's assume, that agents $\{A_j\}$ radiate own group of frequencies. One unique frequency band is allocated for each agent, i.e., frequencies only are meant to identify a presents of the particular agent in the PIF.

---

[20]This is simplification as space metrics may be time dependant too, e.g. $\vartheta_{G_{N \times M}} = f_t^G(G_{N \times M}(t))$.
[21]We may employ whatever linear or nonlinear time dependant functions.

$$E_{A_j}^{n,m}(v_{A_j}^{n,m}) \Rightarrow \psi \cdot I_{A_j} \cdot v_{A_j}^{n,m} \cdot \ln(h+e), \quad h = 1,2,3\ldots, \qquad (2.13)$$

where $E_{A_j}^{n,m}$—energy of the agent A_j and characteristic frequency band $v_{A_j}^{n,m}$ (i.e., the first harmonic of agents'), ψ—scaling factor, I_{A_j}—emission intensity (represents quantity of the given resource) and h—spectrum band number. As for nodes and resources, h is the only thing we need to know in order to identify virtual agents. Each agent's unique oscillations and corresponding spectral bands are derived by multiplying the first harmonic $v_{A_j}^{n,m}$ by the factor $\ln(h + e)$.

We imply that $\min E_{A_j}^{n,m} > \max E_{R_i}^{n,m}$, i.e., available intervals of frequencies for natural resources and agents do not overlap. Hence, it means a limit for i (resource spectrum band number). We also assume that agents are located on the nodes only and one node could have only one agent.

In sum, PIF approach allocates unique frequencies for nodes, resources and agents. It also calculates resulting spectrum for each node. Such spectra include all bands coming from all resources, agents and other nodes. Bands do not overlap as they cover different spectral zones, see Fig. 2.3. In addition, we assume that all oscillations are transmitted instantaneously over the whole virtual space. Next, we will discuss wave-like interaction between agents.

2.5   Wave-like Interaction Mechanism (WIM)

In the proposed approach, nodes and resources are latent objects. They can only passively emit their own unique natural frequencies in the surrounding PIF. Agents passively emit their unique identifying frequencies too, but essentially they are proactive, i.e., agents, depending on their behavioral rules (internal production rules,

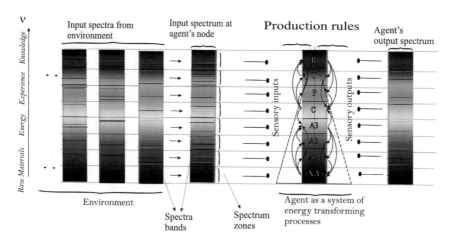

Fig. 2.4 An agent's production rules, which govern the transformation of internal states, represented as spectra bands

see Fig. 2.4), are capable of absorbing, transforming and emitting different frequencies. This interactive process is twofold: (i) automatic in the form of law, which holds whenever appropriate wave-like conditions are met and (ii) personalized, i.e., it depends on the agent's individual behavioral patterns governed by the oscillating agent model (OAM) rules, see Fig. 2.4. Now, let's take a look at some WIM (wavelike interaction mechanism) processes.

2.5.1   The Latent (Involuntary) Emission and Absorption Process

Each agent A_j has a set $\{\omega_{A_j}\}$ of natural frequencies. At these frequencies, even small incoming periodic oscillations can produce an agent's automatic response in the form of high-amplitude outgoing oscillations because an agent (like other physical systems) transforms and stores incoming energy in the form of natural or resonant frequencies.[22] If outside oscillations match an agent's natural frequency, then his natural stored energy is gradually released to the PIF. The general rule of thumb for the agent's involuntary resonance with surrounding PIF is

(a) if outside frequencies coincide with the agent's natural frequencies (then the change in the agent's internal energy for period T is negative $\Delta E_T < 0$, i.e., a responsive natural emission is taking place and agent is loosing stored energy)

(b) if outside frequencies do not coincide with the agent's natural frequencies (then the change in the agent's internal energy for period T is positive $\Delta E_T > 0$, i.e., latent absorption by agent is taking place)

$$\Delta E = \Delta E^f_{abs} - \Delta E^{\omega_{A_i}}_{emi} - = \begin{cases} <0, & \text{if } f = \omega_{A_i} \\ >0, & \text{if } f \neq \omega_{A_i} \end{cases}. \tag{2.14}$$

For the agent A_j absorbed energy is stored as free energy in the spectral range $\{\omega_{A_j}\}$ of the agent's natural frequencies

$$\Delta E \Rightarrow \Delta E^{\{\omega_{A_i}\}}, \tag{2.15}$$

---

[22]Resonant phenomena occur with all types of vibrations or waves: there is mechanical resonance, acoustic resonance, electromagnetic resonance, NMR, ESR and resonance of quantum wave functions. When damping is small, the natural frequency is approximately equal to the agent's resonance frequency.

then $\Delta E > 0$.

The agent's internal energy (see Eq. 2.13)

$$E_{A_j} = \sum_{h=0}^{H} E_{A_j}^{n,m}(v_{A_j}^{n,m})$$

$$= \sum_{h=0}^{H} \psi \cdot I_{A_j} \cdot v_{A_j}^{n,m} \cdot \ln(h+n), \quad h = 1, 2, 3 \ldots H \qquad (2.16)$$

2.5.2 The Active (Voluntary) Emission and Absorption Process

In the proposed approach, agents have a mechanism to look for useful information in the surrounding PIF. They inherit this property from a priori settled individual behavioral rules.[23]

Acquiring meaningful (successful) information means adopting a new behavioral pattern. This comes at a price, however. For instance, let us say agent A_1 wants to buy information P (where P—stands for the agent's A_2 behavioral parameters set) paying some amount of his own capital K (equivalent to the energy $E_K^{n,m}(v_K^{n,m})$) where $K \in R_i$ to the selling agent A_2

$$E_{A_1}^{emi}(v_K) \underset{K \text{ transfer}}{\Rightarrow} E_{A_2}^{abs}(v_K),$$

$$E_{A_1}^{abs}(v_P) \underset{P \text{ transfer}}{\Leftarrow} E_{A_2}^{emi}(v_P), \qquad (2.17)$$

where

$$\Delta E_{A_1}(P \to K) = E_{A_1}^{abs}(v_P) - E_{A_1}^{emi}(v_K) < 0,$$

$$\Delta E_{A_2}(K \to P) = E_{A_2}^{abs}(v_K) - E_{A_2}^{emi}(v_P) > 0, \qquad (2.18)$$

$$\Delta E_{A_1}(P \to K) + \Delta E_{A_2}(K \to P) = 0.$$

The agent's efficiency depends on his spectral coherence with the global criteria (e.g., market rules) which specify some spectrum measures like minimum internal energy (e.g., available capital, etc.) needed for survival. Extremely coherent agents (i.e., successful investment strategies in the current application case) can be rewarded, while extremely incoherent agents can be removed from the simulation process.

---

[23]The possession of meaningful information gives a certain advantage. For instance, acquiring information from PIF about successful investment strategy (i.e. a portfolio management rule) and adopting it gives a currently less successful agent a good prospect to becoming successful in terms of capital accumulation in the future.

In sum, this chapter presents an outline of a conceptually new approach for simulating local and global behavioral patterns observed in information-rich social networks. The main ideas cover the (i) pervasive information field (PIF), (ii) oscillating agent model (OAM), and (iii) the agents' wave-like interaction mechanism (WIM). The conceptual considerations provided can be interpreted as an initial first 'take' on the methods used for social phenomena simulation. More specific outlines and their explanatory sources are provided in the next chapters.

Hence, this chapter ended with the proposition of the stylized phonons' model, which quantifies agents' oscillatory energy and implements energy-information exchange mechanism. In the proposed model of harmonic oscillators, each agent absorbs incoming wave packets, linearly superposes them (producing unique spectra), and then transmits them to the environment. Putting mathematical notations aside, with the help of stylized production rules, we can obtain some quantitative means of modeling energy and information exchange mechanism between oscillating agents. The main drawback of such an approach, however, comes up from hypothetical and stylized modeling, which is based on the theoretical assumptions of the oscillatory nature of social agents' states. The next chapter makes some further considerations and assumptions in order to deepen and broaden the oscillations based approach.

Chapter 3
Oscillation-Based Social Simulation Paradigm: The Conceptual Premises

In this chapter, we start from a review of a wide range of theories across various disciplines in search of some common field-like fundamental principles of coordination and self-organization existing on the quantum, cellular, and social levels. Our findings outline some universal principles, which are further employed to formulate main premises and postulates for the proposed OSIMAS (oscillation-based multi-agent system) simulation paradigm. We elaborate on three conceptual models, whose design is based on neuroscience discoveries about the oscillating nature of the agents mind states and of the nonlocal field-like self-organization properties of modern information societies. To validate OSIMAS premises, we designed not only theoretical, but also a three level experimental and simulation research framework, which starts with modeling and benchmark estimates of the oscillatory nature of the human mind states, proceeds with estimates of the brain wave coherence measures for group-wide neurodynamics, and ends with review of some related case studies of virtual field-based modeling.

3.1 General Outlines

The relatively broad scope of our research inevitably touches upon many disciplines, like quantum mechanics, neuroscience, the cognitive and social sciences, multi-agent system research, etc. In this regard, we had to provide an extended introduction below, looking for fundamental coordinating laws that have the potential to unify all these areas of research.[1]

---

[1] Recent years have witnessed an explosion of interest and activity in the multidisciplinary areas bridging the gap between fragmented research in terms of disciplines, methods, and scale. The time has come to fully realize that our universe is not a static storehouse of separate microscopic, mesoscopic, or macroscopic objects, but rather a single organism of interconnected energy fields in a continuous state of becoming (Bohm 2002; Penrose 2007). According to Ervin Laszlo, science is laying the conceptual foundations of a transdisciplinary unified theory for the integrated science of an interconnected universe (Laszlo 1995).

© Springer International Publishing Switzerland 2016
D. Plikynas, *Introducing the Oscillations Based Paradigm*,
DOI 10.1007/978-3-319-39040-6_3

Let us start from some fundamental observations in the micro scale, i.e., quantum physics, where we can find a great number of theoretical and empirical studies suggesting that the peculiar properties of the micro scale occur not simply on the quantum level with subatomic or atomic particles, but also in the world of large molecules (Fröhlich 1968). For instance, some findings show that something as large as a molecule can become entangled or that collective bioelectromagnetic oscillations cause proteins and cells to coordinate their activities (Rossi et al. 2011) or that the Bose–Einstein condensate in living tissues produces the most organized light waves (i.e., biophotons) found in nature (Popp et al. 2002). All these suggests that, most probably, there are no separate laws of physics for the large (on the biological, sociological, or cosmological scale) and the small (on the atomic/subatomic scale), but rather, universal all-embracing laws for the self-organized, multifaceted information that permeates all living and nonliving states of energy-matter.

Admittedly, we are not referring here to some sort of Ervin Laszlo's "Akashic field" (Laszlo 1995), but rather to the unified field theory (UFT), which allows all that is usually thought of as fundamental forces and elementary particles to be written in terms of a single field. The term was coined by Einstein, who attempted to unify the general theory of relativity with electromagnetism. Actually, most physicists would agree that the final goal of physics is a UFT. However, at present, there is no accepted single UFT, and thus, it remains an open line of research. Because of the implicit involvement of mathematical content (hence information) in its description, the author prefers to describe this single field (UFT) in terms of the all-embracing laws of the multifaceted information that permeates all levels of energy-matter. In light of the UFT biological entanglement and nonlocal interactions are taken very seriously among top-notch scientist in the world.

Let me provide an analogous example concerning M-theory, an elegant theory by the Nobel laureate in Physics, Edward Witten. As we know, M-theory mathematically describes a framework for developing a unified theory of all of the fundamental forces of nature, where different elementary particles may be viewed as vibrating strings. Hence, unifying M-theory is one of the leading efforts toward understanding the fundamental, implicit laws of this universe, which govern the behavior of energy-matter in the space-time continuum. If physicists are right, and they finally succeed in identifying these most fundamental mathematical laws (only gravity has still not been unified with the other fundamental forces), then we will have some very serious arguments; that some sort of governing information in the form of these laws permeates and governs all levels of energy-matter in the universe. In this regard, my formulation of the UFT makes deep sense (see above). Consequently, following the leading M-theory of vibrating strings, our universe is of an oscillatory nature, beginning from its most fundamental level of self-organization. Thus, the proposed oscillations-based OSIMAS paradigm does

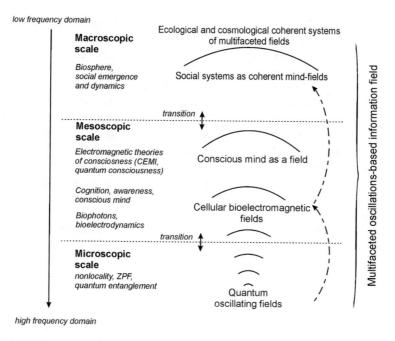

Fig. 3.1 Various research areas in the different spatial scales, united by the universal dimensionless concept of a multifaceted oscillation-based information field

not refer to some sort of supernatural or pseudoscientific laws, but rather, to the frontier research of fundamental physics.

The examples given earlier name just a few fundamental research results among many others that are forming a coherent transition from the quantum world of field-like reality to the mesoscopic world of field-like coordination in cellular biosystems. In short, the latest findings most probably indicate that different spatial scales (microscopic and mesoscopic) operate on the same spectral ladder, although in different spectral domains, see Fig. 3.5. For instance, the existence of biophotons and bioelectromagnetic dynamics (as contextual information distributed in fields) during vital metabolic processes expresses some field-based information locally perceived and communicated by cells (Zhou and Uesaka 2006).

On the other hand, some research results show how macroscale processes directly coordinate mesoscopic and even microscopic scale systems. For instance, DNA signals are sent out via externally activated surrounding water clusters (i.e., coherence domains) by 7.8-Hz naturally occurring Schumann waves on Earth

(Montagnier et al. 2011).[2] Such findings that processes taking place in the biosphere (macroscale) directly participate in the metabolism of cells.[3]

Hence, once we acknowledge the essential role of quantum bioelectromagnetic field-based metabolism in cellular systems, we automatically recognize the same wave-like metabolism in neural cells too, e.g., field-based coherent communication between neurons in the central nervous system (Libet 1994; Schwartz and Dossey 2010). Specifically, we may recall here some electromagnetic field-based theories of consciousness (or so-called quantum consciousness), e.g., holonomic brain theory (Pribram 1999). This theory describes processes occurring in visual neural webs (and in other sensory networks), where patches of local field potentials, described mathematically as windowed Fourier transforms or wavelets, change a space-time coordinate system into a spectral coordinate system within which the properties of our ordinary images are spread throughout the system.[4]

Some studies from the neurophysiological approach lead to unified field models of consciousness. For instance, some researchers study the large-scale dynamics of EEG (electroencephalography), precisely describing the interference patterns of standing waves of postsynaptic potentials that may be superimposed on neurons embedded in these potential fields. As these studies indicate, changes in long-range coherence between remote cortical regions of certain frequencies during cognitive tasks support the concept of globally dominated dynamics (Nunez et al. 2001).

Some MEG (magnetoelectric) studies show extensive cross-cerebral coherence, which led to the proposition that consciousness arises from the resonant coactivation of sensory-specific and nonsensory-specific systems that bind cerebral cortical sites to evoke a single cognitive experience (Llinas and Ribary 1998). At the same time, other studies are being coordinated to construct a field theory of consciousness (John 2002), which gave us an impetus to elaborate on the idea of unified field models of collective mind-fields of coherently convergent (congruent) human groups.

Here, we should also mention another well-known approach the conscious electromagnetic field theory (CEMI) (McFadden 2002). This electromagnetic field theory of consciousness is inherently attractive because of its natural solution to the

---

[2]Water nanostructures and their electromagnetic resonance can faithfully perpetuate DNA information.

[3]The fundamental Schumann resonance frequency is claimed to be extremely beneficial for the existence of the biological cycle phenomena of plants, animals, and humans. More intriguingly, human brains also produce electromagnetic waves in that range of extremely low EM frequencies. Prospective research may find out how external fields or our mind fields are involved in the physiological processes taking place in the whole body.

[4]Fourier transformations are routinely performed on electrical recordings from the brain such as EEG and local field potentials.

binding problem.[5] According to the CEMI theory, the brain generates an electromagnetic (EM) field that influences brain function through EM field-sensitive voltage-gated ion channels in neuronal membranes. The information in neurons is therefore pooled, integrated, and reflected back into neurons through the brains EM field and its influence on neuron firing patterns. The CEMI theory states:

The digital information within neurons is pooled and integrated to form an electromagnetic information field. Consciousness is that component of the brains electromagnetic information field that is downloaded to motor neurons and is thereby capable of communicating its state to the outside world.

In other words, the CEMI theory argues that we experience this field-level feedback processing as consciousness.[6] Its defining feature is the ability to handle irreducibly complex concepts such as a face, the self, identity, words, meaning, shape, and number as holistic units.[7] All conscious thinking involves the manipulation of such irreducibly complex concepts and must involve a physical system that can process complex information holistically. According to this theory, the only physical system that can perform this function in the brain is the CEMI field.[8] It is through this mechanism that humans acquired the capacity to become conscious agents able to influence the world (Malik 2002).

According to the CEMI theory and some recent experiments (Fries 2009), consciousness is usually associated with attention and awareness, which mostly are not correlated with a pattern of neural firing per se, but with neurons that fire in synchrony.[9] Similarly, we assume that self-organization and coherent behavior in economic (and broader social) systems is not so much correlated with the particular

---

[5]The binding problem of consciousness how our conscious mind integrates information distributed among billions of spatially separated neurons to generate the unity of conscious experience is one of the fundamental questions in the study of the mind (McFadden 2002).

[6]CEMI theory does not propose that consciousness is necessarily associated with the amplitude, phase, or frequency of the brains electromagnetic field. The defining feature is rather the informational content and its ability to be communicated to motor neurons.

[7]In other words, information that is encoded by widely distributed neurons in our brain is somehow bound together to form unified conscious percepts.

[8]The idea that our conscious minds are some kind of field goes back at least as far as the gestalt psychologists of the early twentieth century. They emphasized the holistic nature of perception, which they claimed was more akin to fields rather than particles. Later, Karl Popper proposed that consciousness was a manifestation of some kind of overarching force field in the brain that could integrate the diverse information held in distributed neurons. Only recently has an understanding emerged that this force field is actually generated by the bioelectromagnetic activity of neurons in a form of conscious mind as an electromagnetic field.

[9]Coherent oscillations are crucial to synchronize informational neural activity. Phase-locked gamma oscillations have been hypothesized to be the neural correlates of unified conscious awareness, e.g. during anesthesia (loss of consciousness) coherent oscillations significantly diminish across various parts of the brain. This suggests that consciousness is inherent in the synchronized state of the brain (John 2002).

patterns of agents actions, but with the coherence or even synchrony of their mental activity. Coherence of agents' mental activities leads to the social behavioral synchrony, which prevents otherwise dissipating and self-destructive activity patterns of the individual members of a society.

The proposition that the physical manifestation of consciousness is an electromagnetic field that exhibits wave mechanical dynamics has profound implications for our understanding of the phenomenon of consciousness and the nature of free will (McFadden 2002). We can acknowledge the viewpoint, at least theoretically, that societies can also be understood as global processes emerging from the coherent behavior of the brain and mind processes of the individual members of society. In this way, hypothetically, we can admit that some emergent social behavior could be at least partially induced by the coherent processes taking place between related individual bioelectromagnetic brain fields (some sort a collective consciousness) and therefore, can inherit some degree of wave-like excitation of mental states and their corresponding coherent behavior.

In fact, this is an important hypothesis as it offers a different worldview that opens new perspectives for modeling and simulating emergent social properties (Sawyer 2005) as collective mind-field effects. In that sense, we are finally getting to the main point of the OSIMAS (oscillations based multi-agent systems) research, i.e., the construction of a collective mind-field paradigm capable of simulating some (i) complex social cognitive (e.g., collective consciousness, memes spread, action at a distance, nonlocal communication, information broadcasting, common opinion formation, etc.) and (ii) behavioral phenomena, such as heard effects, economic cyclic activities, the network economy, social clustering, the predominance of intangible (informational) goods, generative micro–macro relations (Epstein 2006), etc.

Our study described in the following sections seeks to shed new light on the fact that there is a conceptually novel way of simulating the complex social processes taking place in an information (networking) economy. The main idea proposed assumes that all social processes can be reduced to the most fundamental oscillation-based representations (Plikynas 2010). In short, this interdisciplinary study is concerned with designing universal concepts and principles for modeling pervasive information-based social networks and context-aware intelligent agents.

In reality, social networks are highly heterogeneous with many links and complex interrelations. Uncoupled and indirect interactions among agents require the ability to affect and perceive a broadcasted information context (Mamei and Zambonelli 2006). In this regard, this research is looking for ways to model the information network as a virtual information field, where each network node receives pervasive (broadcasted) information field values. Such an approach is targeted to enforce indirect and uncoupled (contextual) interactions among agents in order to represent contextual broadcasted information in a form locally accessible and immediately usable by network agents.

As the following sections indicate, proposed paradigm does not start from ground zero, as a review of literature revealed striking similarities to some other approaches, such as the vibrating potential field (Yokoi et al. 1995), quantum computation as a model of consciousness (Hameroff and Penrose 2014), intra-and intercellular communication mechanisms (Rossi et al. 2011), the CEMI theory of consciousness (McFadden 2002), the neurophysics of consciousness (John 2001), field computations in natural and artificial intelligence (MacLennan 1999), field-based coordination mechanisms for multi-agent systems in a robotics domain (Mamei and Zambonelli 2006), organic computing for the emergent behavior of complex systems (Müller-Schloer and Sick 2006), amorphous or pervasive computing in contextual environments (Servat and Drogoul 2002), etc. It is beyond the purview of this monograph to provide a comprehensive review of the surge of publications on field-based paradigms.

A review of related similar research projects throughout the world shows some other international groups, such as the Self-organizing Systems Research Group of Harvard University in Cambridge (USA), the Pervasive Artificial Intelligence (PAI) Research Group of the University of Fribourg (Switzerland), the Centre for Computational Analysis of Social and Organizational Systems (CASOS) of Carnegie Mellon University, etc. An entire group of related projects is also being implemented under the EU research umbrella called FET (Future and Emerging Technologies) under the ICT program of information and communication technologies in the FP7 (Seventh Framework Program).

Next in this chapter, we further elaborate on the novel OSIMAS paradigm formulating main postulates. Afterwards we provide conceptual framework for the main pervasive information field (PIF) modeling principles, which are intended for designing field-based pervasive contextual environments in complex information-rich social networks. Further, we describe ways of measuring ordered structures in a field-like representation. We also provide an experimental research framework aimed at validation of the proposed paradigm.

3.2 The OSIMAS Paradigm: Basic Hypothesis, Assumptions, and Postulates

In this section, we introduce set of assumptions and postulates that lay the ground for the OSIMAS (oscillation-based multi-agent system) simulation paradigm. First, let us emphasize that, when formulating postulates, we are looking for universalities across different spatial scales and time horizons. In essence, we are searching for pervasive fundamental laws of self-organizing information unconstrained by space and time. If, for instance, some field-like fundamental principles work in the quantum world and in cellular biophysics, we must admit that the same principles manifest themselves in one way or another in the mesoscopic world of social systems, too. However, the form and expression of these fundamental principles vary across different scales.

Second, when formulating basic assumptions and postulates, we want to elaborate how field-based underlying reality can be applied in modeling pervasive contextual environments in complex information-rich social networks. In other words, we formulate the bases for modeling the emergent and self-organizing features of modern information-rich social networks, where not only intangible but also tangible natural resources and even social agents themselves can be simulated as oscillating processes immersed in an all pervasive contextual information field (PIF).

On the basis of a multidisciplinary research review (see above), the OSIMAS paradigm adapts and formulates five basic assumptions:

1. There are no separate laws for the large (on a biological, sociological, or cosmological scale) and for the physics of the small (on an atomic/subatomic scale), but rather universal all-embracing laws for the self-organized multifaceted information that integrally permeates all living and nonliving states of energy-matter.
2. In the mesoscopic scale, the most complex known form of self-organized information is the human mind. In electromagnetic (EM) field-based neurophysiological approaches, the human mind can be represented as a unified EM or other field-like model of consciousness.
3. Societies can be understood as global processes emerging from the collective behavior of the conscious and subconscious mind-fields of their individual members. In this way, some emergent social processes can be produced by a collective mind-field effects and inherit some degree of coherent (synchronized) field-like behavior.
4. Societies (the macro-world) can no longer be viewed as separate from the quantum effects taking place in the conscious mind-fields of society members. Self-organization and coherent behavior in social systems is not so much correlated with the particular patterns of agents behavioral actions, but with the coherence and synchrony of their mental activity.[10]
5. A core reason of social behavioral synchrony is the synergetic convergence of otherwise dissipating and self-destructive mental activity and, consequently, of the behavioral patterns of the individual members of a society. In this regard, the social coordination and organisation mechanism most probably can be also associated with the synchronicity mechanism between local self-organized information processes, supposedly taking place in our minds. Such synchronous mental activities may take an important part in the formation of self-organized social systems.

The OSIMAS paradigm is based on these key assumptions, which at least theoretically open up a new way for modeling and simulating emergent social properties as collective mind-field effects. To further clarify the OSIMAS paradigm, some underlying basic postulates are formulated below:

1st Postulate. Social systems can be modeled as complex informational processes comprised of semi-autonomous interdependent organizational layers, e.g., the individual, the group, and society.[11] Social information is coded and spread via

---

[10]We infer that mental activity patterns are explicitly or implicitly expressed in agents' behavior.

[11]In the book, there are mostly used broader terms like 'social agent' and 'social system' instead of more narrow terms like 'economic system' and 'economic agent'. However, the latter terms should always be in mind.

social network almost at the speed of light via broadcasting telecommunication networks. In the modern information societies information is propagated not primarily through peer-to-peer interactions between agents, but increasingly via virtual information fields transmitted through broadcasting information channels (Internet, GSM, radio, TV, etc.).

2nd Postulate. Like all complex systems, social systems are always on the verge of chaos. They are constantly balancing between order and disorder. In this way, social systems have the naturally inherited property of changing and adapting while searching for niches to survive. Hence, the main feature of social systems does not pertain so much to the ability to stay in internal and external states of equilibrium (which are constantly changing), but rather the ability to change and adapt while searching for internal and external equilibrium.

3rd Postulate. Uncoupled and indirect interactions among social agents require the ability to affect and perceive broadcasted contextual information. In this regard, a social information network can be modeled as a pervasive information field (PIF), where each network node receives pervasive (broadcasted) information field values.[12] Such a model provides an appropriate means of enforcing indirect and uncoupled (contextual) interactions among agents. It is expressive enough to represent contextual broadcasted information in a form locally accessible and immediately usable by network agents.

4th Postulate. The simulation results of social systems behavior do not adequately reflect observable reality unless simulated models acquire the features of open living systems, e.g., adaptability, self-organization, field-like inner coordination, and outer communication.

5th Postulate. The individual members of a society can be modeled as information-storing, -processing, and -communicating agents in an information network society.[13] From another perspective, information societies operate through agents, which are complex multifaceted self-organized information processes composed of mind-fields of quantum field-like processes originating in brains.[14]

---

[12]Following the broadcasting communication technology of telecommunication systems, we assume that similar principles should also be applied to simulating platforms in the social domain. In other words, communication between agents should operate not in the (peer-to-peer) vector-based multidimensional semantic space, but rather directly in the form of multimodal energy (spectra) emanated and absorbed over the social network. The flow of energy (and associated with it, information) in the form of fields, however, requires a somewhat different understanding of the agents role and their interaction mechanism.

[13]The modern information society can no longer be described by Adam Smiths classical theory as a statistically aggregated set of rational (mechanical) and static agents. The information society deals with a new type of economics where wealth is accumulated in procedural and declarative knowledge and the ability to use it effectively in the service and manufacturing sectors.

[14]We do not necessarily propose that collective wave-like processes originating from the set of human brains are associated with the amplitude, phase, or frequency of the set of the human brains electromagnetic fields. The essential feature is rather the informational content and its ability to be communicated through field-like virtual media.

6th Postulate. Agents, as complex multifaceted field-like information processes, can be abstracted using the physical analogy of multifaceted field-like energy, which is commonly expressed through spectra of oscillations. In this way, an agent becomes represented in terms of a unique composition of oscillations (individual spectrum).

7th Postulate. Agents inner states can be represented in terms of organized multifaceted information that expresses itself in the form of a preserved specific energy set. The latter can be modeled by means of a specific spectrum of oscillations. The distribution of oscillations over an individual spectrum, in contrast to a random distribution, carries information about the agents self-organizational features, i.e. negentropy (order). Hence, social agents are complex processes that dynamically change multifaceted inner information-energy states depending on the information received from a PIF.

8th Postulate. Artificial societies can be modeled as superimposed sets of individual spectra or, in other words, as part of the PIF. Hence, social order emerges as a coherent superposition of individual spectra (self-organizing information processes) and it can be modeled as coherent fields of information resulting from the superposition of the individual mind-fields of the members of a society.[15]

9th Postulate. Social order, i.e., self-organized and coherent behavior in social systems, is not so much correlated with the particular patterns of agents' actions, but with the synchrony of their mental activity. That multi-agent synchrony can be compared to the physical model of superposition of weakly coupled oscillators. Synchronicity is involved in the social-binding problem how information distributed among many agents generates a community. The social-binding process can be imagined as a global resonance state.

10th Postulate. The core reason for the emergence of social synchrony is related to the fundamental property of all self-organized systems, i.e., the preservation or increase of negentropy, which creates socially organized behavior.[16]

Of course, at the current stage of common understanding these postulates cannot be so evident as to be accepted as indisputably true. There is a long way to go before some of these postulates will find proper mathematical and experimental proof. Hence, putting mathematical notation aside, the above postulates define and delimit the realm of deductive analysis, serving as premises or a starting point for our reasoning.

---

[15]Emergent social processes are produced by the coherent behavior of individual mind-fields; hence, societies inherit some degree of wave-like behavior. Therefore, the investigation of emergent social processes is directly associated with collective (distributed) mind-field effects.

[16]In the process of self-organization, humans as well as other living organisms (and their societies) create (i) an inner organized metabolism and (ii) outer organized behavioral patterns. According to the second law of thermodynamics, in both cases entropy (disorder) has to decrease. Self-organized systems are open systems. This means that they produce inner negentropy (i.e. order as organized metabolism and outer organized behavioral patterns) at the expense of increased entropy (disorder) outside the system. The second law of thermodynamics holds for open systems.

On the basis of these postulates, we envisage that oscillating agents can be integrated into a common PIF spectrum as individual sets of oscillation bands (see Fig. 3.2), which can be described by the oscillating agent model (OAM). This latter model realizes the production rules for the transformation of internal energy (a set of natural oscillations), which can be defined a priori or induced from the agents behavioral strategy.

In the proposed paradigm, agents can communicate, voluntarily or involuntarily, their states of mind to each other. In this way, they form a distributed (collective) mind-field, where communication in technical terms is realized via a common medium, i.e., a PIF, and is managed by the wave-like communication mechanism (WIM). The next section briefly outlines the OSIMAS paradigm setup and specifically the PIF model.

3.3   The Pervasive Information Field (PIF): Extended Conceptual Framework

In the previous section, we formulated the major assumptions and postulates of the OSIMAS paradigm as guidelines for the prospective modeling framework. In short, OSIMAS is based on a conceptual trinity of models: the PIF (pervasive information field), the OAM (oscillating agent model), and the WIM (wave-like interaction mechanism), see Fig. 3.2.

These conceptual models describe an integral theoretical framework. In this section, we elaborate mostly on the PIF model. In essence, the PIF model serves as a means for contextual information (and associated energy) storage, dynamic distribution, and organization. According to the above assumptions, contextual information is distributed in virtual information fields, and fields although expressing some global information are locally perceived by agents, who are but a self-organizing part of the same PIF.

Fig. 3.2 The three major OSIMAS models: (i) the pervasive information field (PIF), (ii) the oscillating agent model (OAM) that generates a number of agents A_n, and (iii) the wave-like interaction mechanism (WIM), which realizes the interaction between those agents

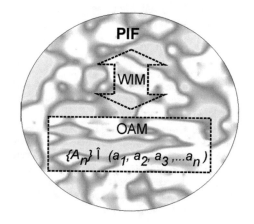

In this way, the PIF serves as a universal medium managing all kinds of multifaceted information in the form of self-organizing information fields. Hence, multifaceted information is conceptualized in the form of all-embracing virtual field, which can be realized as a programmable abstraction, where all phenomenalogically tangible and intangible observables are represented as a set of oscillations (energy equivalents). However, for the effective implementation of spectra as a universal energy-information warehouse, we first have to transform all tangible objects-resources into their energy equivalents and then interrelate different types of energy as intangible information stored in the form of corresponding sets of spectral bands (natural frequencies).

The reasoning behind this is based on the principle of reductionism and universality as we are looking for the most fundamental means of representing a multiplicity of phenomenological forms in a single informational medium. In the OSIMAS paradigm, this medium is modeled by means of a system of oscillations. In other words, all tangible and intangible system resources are represented by means of sets of spectra of oscillations. Consequently, the PIF represents a grand total of all individual spectra.

Hence, the PIF model is constructed to follow basic assumptions and postulates of the OSIMAS paradigm. Following these principles, we further summarize the PIF model framework:

1. Social systems constitute yet another layer of self-organizing information, where the same scale-free and field-like universal laws of self-organization apply.
2. Social systems behave in a coherent way because they are integral holistic units, where each part is inseparable from all the rest. Each part is the sum of the influences of all the other parts. Likewise, each part directly or indirectly influences all the other parts.
3. Such holistic multi-agent systems consist from the interdependent parts, i.e., agents. In this way, we can simulate agents as local processes of self-organized information in a global all-embracing multifaceted and pervasive information field (PIF).[17]
4. In the PIF, self-organizing information processes, i.e., agents, can be modeled by using various models, e.g., standing waves[18] or, in other words, resonant

---

[17]Such an approach greatly resembles what the coherent photon field does within a living system, where it is responsible for intra-and intercellular communication and the regulation of biological functions (Popp et al. 2002).

[18]Standing waves are generated by the superposition of two waves of equal frequency, amplitude, and phase, but with opposite directions of propagation. Systems that are able to store the energy of standing waves are able to fix the phases of particular oscillations, and doing so is necessary for the synchronization of inner and outer processes. In fact, the specific energetic fingerprint for each agent can be determined not only by frequency, amplitude, and phase but also by some dynamic characteristics such as the rhythms occurring in chaotic phase synchronization cases, etc. (Pikovsky et al. 2003).

frequencies.[19] At resonant frequencies the agent stores free energy, i.e., self-organized information (see previous chapter). This information is used to enhance internal processes and external behavioral patterns.

5. A homeostatic agent can be represented in terms of the local energy spectral density (LESD) distribution, which describes how the internal energy (or its variance in time series) is distributed with frequency.[20] Meanwhile, system-wide distribution of LESD provides the global energy spectral density (GESD) distribution, which uniquely describes the state of the PIF at each moment.

6. The main way for the agent to increase negentropy (negative entropy or information) is by adopting some set of resonant frequencies, which may yield to the beneficial behavioral patterns in a dynamic environment.[21] The adoption of new resonant frequencies (information) changes the agents LESD and GESD distributions accordingly.[22]

7. Homeostatic agents as self-organized information processes are usually proactive. They search for ways to sustain and increase self-organized order by increasing internal negentropy.[23] There are many ways of reaching the same level of negentropy by employing different LESD distributions.

8. In general, order can be represented by levels of synchronization and of coherence locally within an agent as system of oscillations and globally within interacting populations of agents.[24] These deviations constitute local and global negentropy or self-organizing information. In such a way, above average virtual field potentials in the PIF environment indicate the degree of coherence among agents.

9. Coordination between agents can be modeled through coherent convergence, i.e., the synchronization of oscillation phases. Such technical approach to social binding involves the synchronous oscillations of agents as self-organized oscillating processes.[25] In such a way, social order emerges as a consequence of

---

[19]A system can have as many resonant frequencies as it has degrees of freedom, where each degree of freedom can vibrate as a harmonic oscillator (Benenson et al. 2006). In the same way, simulated agents can have as many resonant frequencies as they have degrees of freedom.

[20]Homeostasis is the property of an open or closed system to regulate its internal environment, allowing an organism to function effectively in a broad range of environmental conditions.

[21]The Brilloun–Schrodinger term negentropy is reciprocal to entropy (Arndt 2012). Schrodinger gives the following example of the term negentropy: increase $N = -S = k \log 1/D$ (where D the number of macroscopic states the system is able to adopt, $1/D$ the probability to be in one certain state) for living self-organized systems.

[22]The covariance of energy spectra among regions of agents across frequencies mediates information transactions.

[23]As homeostatic systems, agents regulate their internal environment by maintaining the stable, constant condition of most important properties such as the inner level of negentropy.

[24]In essence, coherence, i.e. deviation from the random distribution of oscillations, indicates order.

[25]For instance, social information signals such as various mass media news spread within multiple social agents by means of a significantly nonrandom synchronization mechanism. Critical information causes a massive reaction and excited (coherent) state in social systems.

the coherent resonance of self-organized local field potentials (agents as systems of oscillations).

10. Synchronization as a process locally involves searching for beneficial information and globally means minimizing systems entropy. In that sense, both local and global processes are homeostatic and self-organized to maintain or increase negentropy (an analog of order). The dynamics of synchronous oscillations supposedly creates self-organized social systems.

Hence, the PIF computation is a theoretical model of information processing operations that take place in natural systems. The PIF can be treated mathematically as a multifaceted function Ψ over a bounded spatial set Ω. The value of the function Ψ is restricted to some bounded subset of real numbers $\Psi: \Omega \rightarrow K$ for a K-valued field. Thus, for the time-varying field we have $\Psi(k, t)$, where $k \in K$. In general, we assume that Ψ for each moment and space location are uniformly continuous, square-integrable, finite energy $\|\Psi\|^2 = \langle \Psi | \Psi \rangle < \infty$, Hilbert space functions (MacLennan 1999).[26]

In general terms, the time-varying field $\Psi(t)$ can be defined by field transformations and differential equations. A linear field transformation can be described by using integral operators of the type $\Phi(\Psi(k)) = \int_{t_1}^{t_2} K(k,t)\,\Psi(t)\mathrm{d}t$, where Ψ-input field, Φ-output (transformed) field, K—kernel function, or nucleus of the transform. One important class of linear field transformations, as continuous mapping functions,[27] consists of integral operators of the Hilbert–Schmidt type $\int_\Omega K_{ku}\,\Psi_u \mathrm{d}u$, which map input field Ψ into output field Φ over Ω.

In the presence of multiple stimuli, multilinear integral operators can be applied (MacLennan 1999), which map one or more input fields Ψ into one or more output fields Φ

$$\Phi_k = \int_{\Omega_n} \cdots \int_{\Omega_2} \int_{\Omega_1} K_{ku_1 u_2 \ldots u_n}\,\Psi_1(u_1)\Psi_2(u_2)\ldots\Psi_n(u_n)\mathrm{d}u_1\,\mathrm{d}u_2 \ldots \mathrm{d}u_n. \quad (3.1)$$

Such mapping represents interference from all the stimuli,[28] i.e., incoming fields. Let us take an example of two simple fields represented by two linear harmonic

---

[26]Hilbert spaces are widely used as models of continuous knowledge representation, but not all elements of Hilbert space are physically realizable.

[27]In brief, an integral transform maps an equation from its original domain into a target domain where manipulating and solving the equation can be much easier than in the original domain. The solution is then mapped back to the original domain with the inverse of the integral transform.

[28]In a more restricted sense, interference means the superposition of coherent waves. Linear waves are coherent if their phase differences do not vary with time. In the superposition of incoherent waves, there is no interference; the intensities of the individual waves are simply added up (Schrödinger 1955).

waves with the same frequency and amplitude $y_1(x, t) = 2A\cos(\omega t - kx + \varphi_1)$ and $y_2(x, t) = 2A\cos(\omega t - kx + \varphi_2)$. The resulting wave

$$y_{\text{res}}(x,t) = y_1(x,t) + y_2(x,t) = 2A\cos\left(\omega t - kx + \frac{\varphi_1}{2\varphi_2}\right)\cos(\Delta\varphi/2) \qquad (3.2)$$

has a doubled amplitude, the same frequency, and a changed phase. Actually, the phase difference $\Delta\varphi$ determines the resulting waves amplitude. In essence, the phase difference shapes the outcome, i.e., the magnitude of the resulting oscillations. This dependence holds for other types of interference, too.

In fact, synchronization phenomena are directly related to the law of wave interference and consequently to the frequency and phase management mechanism, which provides a key for modeling coherent systems and self-organization processes. Indeed, according to the phase synchronization theory of chaotic systems,[29] dynamic coherent behavior emerges as a consequence of nonlinear synchronization in complex networks. In the framework of such a frequency and phase approach, it is quite natural that synchronization processes in various systems of different nature have close similarities and can be studied by using universal oscillation-based tools (Osipov et al. 2007).

The holistic nature of social systems is rich in connections, interactions, and communications of many different kinds and complexities. In this regard, very often observed synchronization is the most fundamental phenomenon as it is a direct and widespread consequence of the interaction of multifaceted complex oscillating systems of agents. There is a great deal of material on different aspects and effects of synchronization (Balanov et al. 2008).

Fortunately, we can employ the essential contribution made by some earlier research (Pikovsky et al. 2003), that has provided a contemporary view of synchronization as a universal phenomenon that manifests itself in the entrainment of rhythms in interacting self-sustained systems. In fact, one of the practical tasks in the next stage of the further development of OSIMAS is to create a model for the frequency and phase synchronization mechanism (FPSM) and, on the basis of the PIF framework, to test it for the simulation of some complex self-organizing social processes such as aggregated cycles of activity, the clustering of social agents, the spread of particular behavioral patterns, etc.

The next research phase requires the quantification and measurement of order in spectral patterns, as it represents coherence and the social-binding effects observed in real social oscillatory networks. In the following section, we shall discuss how synchronization contributes to the observed order in oscillatory networks and how coherence can be employed as a measure of order.

---

[29]There are some relatively new aspects of synchronization such as phase multistability, dephasing, self-modulation, etc. (Pikovsky et al. 2003).

3.4 Information Versus Order in the Case of Coherent Oscillations

The terms information and order are found quite often in the literature about complex self-organizing systems, but often they are provided without an adequate reference framework. In this sense, we have to make a bit clearer what we mean when using these terms in the OSIMAS paradigm and particularly in the PIF conceptual model. Only afterwards we can discuss the issue of quantification of order in terms of spectral coherence and synchronicity for agents and their systems.

First, we will begin with a brief review of the most common classical approaches. The use of the mathematical formalism of information theory may seem preferable in the social domain, but we have to be cautious, as information theory is severely limited in its scope. It was originally developed for practical needs by telecommunications engineers to investigate how the characteristics of closed systems, i.e., information channels, influence the amount of information transmitted from the source to the receiver in a given time. Consequently, the amount of transmitted information is usually measured by employing entropy S, which according to Shannons theory (Shannon 2001) expresses a logarithmic measure of the density of possible states[30]

$$S = -k \sum_{i=1}^{N} p_i \ln(p_i), \tag{3.3}$$

where p the number of elementary complexions (the ratio of observed states to the possible number of states), k the scaling factor, p_i the probability of events i. Hence, the information content of an event is defined not by what has actually happened, but only with respect to what might have happened instead (Sheldrake 2009).

We argue that the traditional interpretation of entropy as an information measure in the case of biological or social domains has two basic limitations:

(1) it neglects the fundamental fact that organisms are not closed systems their organizational structures embrace many horizontal and vertical channels of internal and external communication, extending beyond the physical boundaries of the organism or social agent itself;

(2) it does not account for the proper meaning and multifaceted aspects of information manifested in ordered structures and behavioral patterns.

Bearing in mind these limitations of classical information theory, let us investigate how classical thermodynamics and statistical mechanics deal with the concepts of entropy and information, respectively. In the former case, the concept of entropy is defined phenomenologically by the second law of thermodynamics,

---

[30]Shannon's information theory assigns a quantity of information to an ensemble of possible messages. All messages in the ensemble being equally probable, this quantity is the number of bits needed to count all possibilities.

which states that the entropy of an isolated system always increases or remains constant. This means that the total entropy of any system will not decrease unless the entropy of some other close system is increased or, in other words, higher order in one system means less order in another (Plikynas 2010). Hence, in a system isolated from its environment, the entropy of that system will tend to increase.

In this way, by applying the universal law of conservation of energy we could infer that there is a law of conservation of information in an isolated system, too. According to this analogy, we can imply that the total amount of information in an isolated system should remain constant or decrease over time, as information can neither be created nor destroyed, but it can be transformed from one form to another. According to the second law of thermodynamics, the entropy of an isolated system always increases or remains constant. Thus, entropy measures the process of constant or diminishing information (order) in an isolated system. We infer that this apparent contradiction can be resolved by admitting that the term isolated system is only a useful abstraction as all systems in terms of pervasive information are open, i.e., holistic in their essential nature.

Meanwhile, in statistical mechanics entropy is a measure of the number of ways in which a system can be arranged, often taken to be a measure of disorder (the higher the entropy, the higher the disorder). This definition describes entropy as being proportional to the logarithm of the number of possible microscopic configurations of the individual atoms or molecules in the system (microstates), which can give rise to the observed macroscopic state (macrostate) of the system.

Hence, as we see, according to thermodynamics and statistical mechanics, the most general interpretation of entropy results in a measure of uncertainty about a system or, in other words, disorder. In fact, such a measure has the opposite meaning to the order observed in the inner structures and behavior of living systems as well as societies. This provides clear incentives for living systems to employ another measure, which is called negentropy. This term was first used by Schrödinger. He introduced the concept of negative entropy for a living system as entropy that it exports to keep its own entropy low.[31] In this way, negentropy can be understood as a measure of the distance D of the entropy state S to the white noise state S_{max}

$$D(p_x) = S(r_x) - S(p_x) = S_{max} - S(p_x) = \int p_x(u)\log p_x(u)du, \qquad (3.4)$$

where $S(r_x)$-the entropy of the Gaussian white noise distribution r_x with the same mean and variance as of the investigated systems distribution p_x;[32] $S(p_x)$-the entropy of the investigated system. When an investigated systems state differs from the Gaussian white noise distribution, then negentropy $d(p_x) > 0$, and when it is equal

---

[31]Negentropy for the dynamically ordered subsystem can be redefined as the specific entropy deficit relative to the surrounding chaos.

[32]It is assumed that, if the signal is random, then the signal has a normal (Gaussian) distribution.

to a random distribution, then negentropy $d(p_x) = 0$. In the first case, we have some degree of order, and in the second case, there is no order at all.

This makes perfect sense as a random variable with a Gaussian white noise distribution would need the maximum length of data to be exactly described. If p_x is less random, then something about it is known beforehand, i.e., it contains less unknown information, and accordingly it needs less length of data to be described. In other words, negentropy measures something which is known about a systems state.[33] In this way, negentropy serves as a measure of order.

It is apparent that unlike engineering (where negentropy takes the form of digital information and is quantized in bits), social, and biological systems are much more complex self-organizing processes and require a more sophisticated approach. Consequently, two important questions arise:

(i)   what can be known about the ordered states of social systems?
(ii) what kind of quantification method can be applied to measure order in such ordered states?

These two fundamental questions are directly related to the qualitative and quantitative assessments of order respectively. Looking for the answer to the first question, we should first recall our basic OSIMAS and PIF assumptions (see previous sections):

– *social order, modeled as a coherent field of information, results from the superposition of the individual mind-fields (self-organizing information processes) of the members of a society, represented as sets of natural (resonant) oscillations;*
– *agents, as complex multifaceted field-like information processes, resemble a physical analogy of potential field-like energy, stored in the form of natural oscillations, which is commonly expressed through spectra of oscillations; in this way, an agent becomes represented in terms of a unique composition of resonant oscillations, i.e., an individual spectrum;*
– *societies, however, can be modeled as a superimposition of agent-based individual spectra; consequently, social order emerges as a coherent global field based upon superimposed sets of resonant individual oscillations.*

In brief, the distribution pattern of single-agent A_j natural resonant oscillations (an individual spectrum) for the time moment t, described as the spectral density $\Phi^{Aj}(\omega)$ of its energy distribution over a range of frequencies,[34] in contrast to a random distribution $\Phi^R(\omega)$, carries (i) quantitative knowledge about the spectral

---

[33]Gaussian white noise refers to the probability distribution with respect to the average value, in this context the probability of the signal reaching certain amplitude, while the term white refers to the flat power spectral density distribution. In general, Gaussian noise is not necessarily white noise, yet neither property implies the other.

[34]An agents energy distribution over a range of frequencies is the square of the magnitude of the discrete Fourier transform.

energy distribution patterns leading to particular behavioral patterns $\Phi^{Aj}(\omega) \rightarrow B^{Aj}$, and (ii) formative (qualitative) knowledge about ordered patterns as such.

In accordance with OSIMAS assumptions, social behavior emerges, and then individual spectral density patterns interfere, producing coherent behavior in social systems. This process has some specific synchronicity effects basically related to:

1. the superposition of an agents individual resonant (natural) oscillations, which form common resonant patterns in the social mind-field;
2. the feedback loop; then the social mind-field influences and adjusts individual mind-fields by creating the social-binding process (that is, individual mind-fields accommodate themselves to the social mind-field, which generates a coherent social community).

We infer that the driving force behind both synchronization effects is programmed into the individual and social optimization (binding) processes. For instance, the individual and social optimization processes can be optimized to search for (i) synergy effects, (ii) internal energy minimization effects,[35] (iii) emerging self-organization and complexity effects.

All these effects, in one way or another, are related to negentropy and indicate various aspects of the order observed in social systems. In quantitative terms, self-organizing order (negentropy) can be quantized as the difference D_n between the non-infinite flat white noise power distribution spectra described by the maximum entropy S_{max} and the given power distribution spectra described by the entropy S divided by the norm S_{max}

$$D_n = \frac{D}{S_{max}} = \frac{S_{max} - S}{S_{max}}, \tag{3.5}$$

where D_n denotes normalized negentropy. In two extreme cases in which $S \rightarrow S_{max}$, we get $D_n = 0$, which indicates zero quanta of order in a given system, and in another extreme in which $S \rightarrow 0$, we get $D_n = 1$, which indicates the maximum possible quanta of order in a given system. D_n serves as a relative measure of order, which can be used to compare different systems. Other, more sophisticated measures of order can be applied, too, e.g., chaotic invariants, phase space reconstruction, etc. (Pikovsky et al. 2003).

Let us recall the intriguing fact that uniform noise power distribution spectra expressed by the maximum entropy S_{max} need the maximum length of data to be

---

[35]In thermodynamics, the free energy (free energy is a state function) is the internal energy of a system minus the amount of energy that cannot be used to perform work (it is given by the entropy multiplied by the temperature of the system). Likewise, for social systems the qualitative aspect of free energy can be understood as the internal energy (described by the spectral density in the PIF model) of a social system minus the amount of energy that cannot be used to perform social work (background white noise spectral density). The definition of social work depends on the specific simulation case, when an individual agent's behavior (the spectral pattern described by the OAM) is bonded (resonating) with some social activities (global spectral patterns).

exactly described.[36] This means that in terms of information content random spectra hold the most information, although we do not observe any ordered structures in them. From another angle, white noise power distribution spectra are uniform without any apparent structures or disturbances—a fact which is typical for perfect equilibrium an ideal order. Of course, such thinking substantially differs from our usual understanding. However, it is important to note that from the perspective of the equilibrium state of uniform white noise, so-called order is a consequence of imperfect and temporal, states of disequilibrium.[37] These ordered states constantly need running the process of self-organization in order to sustain their existence.

To make this fact clearer, let us give a simple example: assume that we have a cup in a given space; by removing this cup from the given space, we do not remove the space itself. In a similar manner ordered structures emerging in initially uniform noise power distribution spectra neither create nor destroy information, which is dimensionless and attributeless.[38] This example shows the qualitative difference between information (attributes: objective, potential, nominal) and order (attributes: subjective, actualized, phenomenal).

We could also cite Claude E. Shannon's assertion that information has two basic forms, i.e., latent and manifested (Shannon 2001). The manifested form of information we call order in the OSIMAS. In contrast, there is little to be said about unmanifested (latent) information as our minds are used to grasping ordered structures and have difficulty interpreting background noise in a meaningful manner.

To return to ordered systems, it is important to comprehend that they arise because of characteristic processes which support their existence. For instance, in the case of living systems, there are complexes of continuous self-organizing processes $P_{S \rightarrow D}$ (where $P_{S \rightarrow D} \in (p_1, p_2, \ldots p_n)$), which are able to process inward and outward entropy and produce order, i.e., negentropy D

$$[P_{S \rightarrow D}(S)]_t = [P_{S \rightarrow D}(S_{in} + S_{out})]_t \rightarrow D_t = [D_{st} + D_{bh}]_t, \qquad (3.6)$$

where D sustains vital internal structures D_{st} and behavioral patterns D_{bh}, respectively.[39] This can be achieved only through the constant transformation of entropy into negentropy. Otherwise, after some time the second thermodynamic law

---

[36]In the usual sense, randomness suggests a non-order or noncoherence in a sequence of values, such that there is no intelligible pattern or combination which, expressed in an algorithmic way, would reduce the amount of information needed to describe such randomness.

[37]In this way, order is not absolute, but only relative to the observer, depending on the frame of reference he is employing for measurement.

[38]It is well known that energy is similar to information (e.g. it can be transported and stored in several physical quantities; however, information is not additive).

[39]In reference to Stuart A. Kauffman, propagating the organization of processes in biological systems takes place by building classical boundary conditions that are constraints on the release of energy that does work to build further constraints on the release of energy and so forth (Kauffman 1993).

of increasing entropy transforms negentropy back to entropy $[P_{D \to S}]_t$. In general, the dynamics of a local ordered state can be described in the following way:

$$\text{if } \frac{d(P_{S \to D})_t}{dt} + \frac{d(P_{D \to S})_t}{dt} = \begin{cases} > 0, \text{then order increases } x < 0, \\ 0, \text{then order persist } x = 0, \\ < 0, \text{then order increases } x > 0. \end{cases} \qquad (3.7)$$

The increase of negentropy in a local system does not violate the second law of thermodynamics as it is compensated for by the equivalent increase of entropy in the surrounding open system, i.e., $P_{S \to D} + P_{D \to S} \to 0$.

Actually, this relationship confirms the earlier mentioned assumption that ordered structures emerging in initially uniform noise power distribution spectra neither create nor destroy information, which is omnipresent, dimensionless, and attributeless, see Fig. 3.3. We can understand such information as fundamental laws, which are governing energy and matter.

The self-organization (or actualization) of ordered structures takes place on the plane of space-time continuum, where energy and matter are actualized (E, S + T; see Fig. 3.3). The term "actualization" has similar sense as a collapse of the wave function in quantum theory.

Let us see, how quantum mechanics describes a quantum system evolving through time. Depending on the situation, the most general form is the time-dependent Schrodinger equation

$$i\hbar \frac{\partial}{\partial t} \psi = \hat{H}\psi; \quad \hat{E} = i\hbar \frac{\partial}{\partial t}, \qquad (3.8)$$

where i-the imaginary unit, $\hbar$-the reduced Planck constant, $\hat{H}$—the Hamiltonian operator (whose form depends on the system), $\hat{E}$-the energy operator, and ψ-the wave function, which in the case of the superposition of many plane waves of

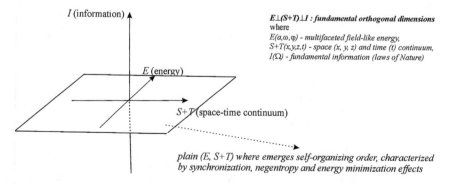

Fig. 3.3 Illustration of orthogonality between dimensionless information and space-time continuum, where energy and ordered structures evolve

neighboring frequencies for one-dimensional motion along an x-direction produces the wave packet $\psi(x, t)$ (Benenson et al. 2006).[40]

$$\psi(x, t) = \frac{1}{2\pi} \int_{-\infty}^{+\infty} f(k) e^{j[kx - \omega(k)t]} \, dk, \tag{3.9}$$

where the spectral function $f(k)$ determines the weight distribution of the plane waves of various frequencies. Following John von Neumanns notation (Neumann 1955), Eq. 3.9 can be rewritten as

$$|\psi\rangle = \sum_i c_i |\varphi_i\rangle, \tag{3.10}$$

here c_i-complex probability amplitudes corresponding to each eigen basis $|\psi\rangle$, where $|\psi_1\rangle$, $|\psi_2\rangle$, $|\psi_3\rangle$...$|\psi_n\rangle$—specify a particular quantum state (the different quantum alternatives available). Technically, they form an orthonormal eigenvector basis. The moduli square of c_i, that is $|C_i|^2 = c^*_i c_i$ (c^*_i-denotes a complex conjugate), then $\sum_i |c_i|^2 = 1$ gives the probability of measuring the systems observable position, momentum, time, energy, spin, orbital, total angular momenta, etc. in the state $|\phi_i\rangle$. The wave function at the moment of measuring collapses to a particular quantum state $|\psi\rangle \rightarrow |\phi_i\rangle$.

This is the process of reducing potential possibilities into a single manifested (physically observed) possibility as seen by an observer. In other words, the superposition of several different possible eigenstates composes a wave packet which appears to be reduced to a single eigenstate after interaction with an observer. In that sense, the information constitutes the plentitude of possibilities, i.e., the potentiality. Underlying potentiality, according to quantum physics, emerges in actualized form as one or another sort of manifested observable (Bohm 2002).[41]

Unfortunately, we cannot directly observe underlying potentiality. We study it indirectly through actualized energetic footprints. For instance, those simplified global and individual (agent-based) spectral density distributions in the PIF model are meant to virtually simulate actualized (multifaceted) energetic footprints. However, the degree and depth to which PIF and OAM conceptual models reflect

---

[40]Wave functions are complex valued functions, i.e. convenient mathematical constructs first introduced by Erwin Schrodinger in 1926. They cannot be observed but still continue to describe the nature of wave-like reality. Wave function is used for predicting the probability of measurement results.

[41]To put it very briefly, information, being attributeless, cannot be observed directly, only through its self-actualized energy and matter forms. In our frame of reference, the potential form of information is not available for us to observe; instead, we observe a self-actualized form of order which is expressed in terms of energy spectral density distributions. In this way, information as the third basic quantity beside matter and energy becomes of central importance. According to Albert Einstein's theory of relativity, mass is a form of energy, and energy is a form of information. So everything depends on the relative system of reference.

actualized states of observed reality essentially depends on the chosen actualization principles, i.e., the operations in an oscillating field-like environment.

Here, we come close to one of the most fundamental philosophical questions: who is the observer who actualizes the potentiality to be observed? This kind of question goes beyond the accepted limits of objective science, which deliberately avoids inquiries about the nature of subjectivity. In fact, everything we observe, even our thoughts, is in the realm of actualized potentiality, which is the object of research for objective science. But the problem is that the available methods of objective science are in principle not able to cross the line into subjectivity, which is somehow involved in the actualization of objects in the process of observation. Indeed, the question we have posed leads us to the limits of rational thinking. Not surprisingly, nobody has been able to deliver a rational, solid-proof answer. Neither can we.

We can only conjecture that the observer does not originate from the physical plane $(E, S + T)$, see Fig. 3.3, because this is the plane where the observer actualizes the potentiality to be observed.[42] Logically, the objects of observation cannot create the observer who actualized them. It does seem that the observer originates from the potentiality (information dimension) itself and, being multifaceted in nature, it chooses (in a process of self-exploration) which potential forms of itself to realize for observation.

Leaving aside the metaphysics of potential information actualization, let us recall here again Claude E. Shannons words that information has two basic forms, i.e., latent (e.g., natural laws) and manifested. The manifested form of information is actualized and takes an ordered form in one or another energetic footprint (in the case of OSIMAS—energy distribution in a form of spectra). In our research, we are mostly interested in those energetic footprints (agents) that have the properties of self-organization through adaptation, creation, maintenance, and the propagation of order. Thus, we use order as explained above, see Eqs. 3.4–3.7.

Now, let us discuss in greater detail how negentropy is embodied in the OS IMAS framework. First of all, we should recall that in the OSIMAS paradigm every simulated resource, e.g., agent A_j, is represented by the system of natural resonant oscillations, which are coded in the form of individual spectra and described for the time moment t as spectral density $\Phi^{Aj}(\omega)$. Thus, negentropy takes the form of a coherent distribution of natural frequencies relative to flat white noise $\Phi^R(\omega)$ spectra.[43]

In order to find a quantification measure for negentropy in the form of a coherent distribution of natural frequencies, we have to specify Eq. 3.5, which estimates negentropy as the difference D_n between non-infinite flat white noise energy distribution spectra described by the maximum entropy S_{max} and given energy distribution spectra described by entropy S divided by the norm S_{max}. In our case,

---

[42]One of the major postulates of quantum mechanics is that all observables are represented by operators which act on the wave function, and the eigenvalues of the operator are the values the observable takes (Bohm 2002).

[43]The term "white" refers to the flat power spectral density distribution, which collects all possible spectra from the whole system.

when we deal with the frequency domain, $S_{\max}$ and S take an equivalent form of energy spectral density distributions represented as flat white noise energy spectral density $\Phi^R(\omega)$ and $\Phi^{R+Aj}(\omega)$, respectively.

Energy spectral density (ESD) is a positive real function $\Phi(\omega)$ of a frequency variable associated with a stationary stochastic process which is commonly expressed in energy per hertz (Benenson et al. 2006). ESD is often simply called the spectrum, and it is expressed as a square of the magnitude of the continuous Fourier transform of the time-dependent signal $f(t)$

$$\Phi(\omega) = \left\| \frac{1}{2\pi} \int\limits_{-\infty}^{+\infty} f(t)e^{-i\omega t}dt \right\|^2. \tag{3.11}$$

Hence, according to Eqs. 3.5 and 3.11 we can specify negentropy D_n by integrating the normalized difference between the ESD of flat white noise $\Phi^R(\omega)$ and the ordered process (i.e., agent) $\Phi^{R+Aj}(\omega)$ as follows:

$$D_n^{\mathrm{ESD}} = \int \frac{\Phi^R(\omega) - \Phi^{R+A_j}(\omega)}{\Phi^R(\omega)} d\omega. \tag{3.12}$$

In Eq. 3.12, we integrated the distance in energetic terms between flat white noise and the agent-based individual (ordered) spectra bands $\Phi^{Aj}(\omega)$. The latter spectra bands, i.e., the distribution of the agents natural frequencies, are embodied in the flat white noise $\Phi^R(\omega)$

$$\Phi^{R+A_j}(\omega) = \Phi^R(\omega) + \Phi^{A_j}(\omega), \tag{3.13}$$

here, $\Phi^R(\omega)$-means background white noise and $\Phi^{Aj}(\omega)$-means individual A_j agent-based spectrum bands. In fact, Eq. 3.13 denotes the individual spectrum embodied in the background white noise spectrum. In Eq. 3.12 we integrate the normalized individual spectra bands which are left after subtracting the flat white noise. In this way, we obtain the measurement of negentropy $D_n^{\mathrm{ESD}_{A_j}}$ for the individual ordered process, e.g., agent A_j, which gives

$$D_n^{\mathrm{ESD}_{A_j}} \to 0, \text{ then } \Phi^R(\omega) - \Phi^{R+A_j}(\omega) \to 0, \tag{3.14}$$

i.e. the individual distribution of natural frequencies, which diminishes and merges with the background noise. This actually happens, depending on the measurement of distance d to the agent location, see Fig. 3.4.

As in the electromagnetic field case, individual field intensities can decrease according to the d^2 or similar negative power law, where d represents the distance from the emanating source

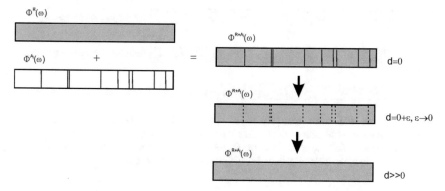

Fig. 3.4 Stylized illustration of the background white noise spectrum $\Phi^R(\omega)$, agent-based local spectrum $\Phi^{Aj}(\omega)$, and the common spectrum Φ^{R+Aj} (ω) dependence on the distance d from the agent's location

$$D_n^{\mathrm{ESD}_{Aj}}(d) \cong kD_n^{\mathrm{ESD}_{Aj}}\bigg/ d^2, \qquad (3.15)$$

where the k coefficient is adjusted to the measurement scale. A somewhat opposite effect in appearance and in nature takes place when some agents spectral density patterns interfere, producing common spectral bands which act further as attraction centers for the rest of the population. In this way, the social-binding process emerges, i.e., stronger common spectral bands work as attractors for individual agents. In this process, individual agents are transformed so that they can better fit into the common spectral patterns, see Fig. 3.5.

In this way, the transformation function T_Ω may be interpreted as a social harmonization function that makes a social system more coherent. This is the case when, through harmonization, a system's global state influences the microstates on the agents level. On the systems level, however, the harmonization function T_Ω can be counterbalanced by the deharmonization function T_R, which is produced by the fluctuating microstates on the agents' level.[44] The interaction of these two functions produces complex, constantly self-organizing social behavior, which can be characterized by the negentropy

$$T_\Omega + T_R \to D_n. \qquad (3.16)$$

When these two levels (micro and macro) interact with each other, they are restricted only by the boundary conditions of self-preservation, which were hard-wired during biological evolution so that micro systems (agents) do not destroy

---

[44]A good example is the increasing effect of heat taking place on the cellular level, i.e. after reaching some critical point heat destroys cellular structures and molecular patterns (negentropy decays to zero).

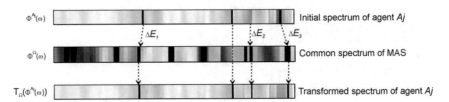

Fig. 3.5 An illustration of the process of social binding, when a multi-agent system's (MAS) common spectral pattern $\Phi^{\Omega}(\omega)$ (see the *middle spectrum*) works as an attractor for the individual (agent-based) spectral pattern $\Phi^{Aj}(\omega)$ (see the *upper spectrum*), producing the transformed individual spectral pattern $T_{\Omega}(\Phi^{Aj}(\omega))$ (see the *lower spectrum*)

the macro system and the macro system does not destroy the underlying micro systems (Fröhlich 1968). This cooperation mediated by coherence and synchronization constantly searches for new niches of coexistence in order to increase overall negentropy. As we have mentioned earlier, the social-binding process of coherent behavior for self-organizing systems is driven by the fundamental law of increasing negentropy, which is realized here for a spectrally represented population of agents through the principles of spectral coherence and synchronicity.

The idea of spectrally represented agent stems from the OSIMAS paradigm, where the concept of oscillating agent model (OAM) is used in the context of the pervasive information field (PIF), where all individual spectra submerge, see Fig. 3.6.

Robust theories tend to have an empirical basis. In this regard, in the next chapters we also provide empirical validation (or falsification) framework. In the section below we briefly mention some closely related other field-based research approaches.

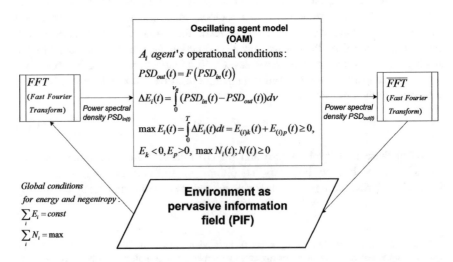

Fig. 3.6 Oscillating agent and pervasive information field in the spectral representation

3.5 Related Other Approaches

This section briefly discusses research issues and results from other related approaches in which contextual information is distributed in virtual fields (or field-like representations), and virtual fields although expressing some global information are locally perceived by agents. In this way, agents' decisions depend not only on local neighborhood states but also on contextual information from a distance. To our knowledge, the major area of virtual field-based social simulations and applications is related to emerging research in social-networking, agent-oriented or multi-agent systems (MAS). As a matter of fact, field-based modeling is usually applied in the nature inspired studies while searching for the better simulation results of very complex, large and highly dynamic physical, biological, and even social networks. Unfortunately, these studies mostly employ field-based coordination approach as technical solution without deeper conceptual understanding of the fundamental nature of such an approach. In this sense, OSIMAS provides at least some fundamental background and reasoning for the field-based coordination applications in the social domain.

For instance, in social-networking research, because of its large scale and complexity, often attempts are being made to simulate social networks using wave propagation processes. Some of these applications deal with message-broadcasting and rumor-spreading problems (Wang et al. 2012), etc. Other applications deal with behaviors spread in dynamic social networks (Centola 2010; Zhang and Wu 2011) or with the diffusion of innovations (Valente 1996), etc. In this latter approach the authors capture the effect of clusters and long links on the expected number of final adopters. They found that the expected number of final adopters in networks with highly clustered subcommunities and short-range links can be less effective than in networks with a smaller degree of clustering and with long links (nonlocal interaction). Basically, all these social-networking approaches employ graphs theory using nodes and connections to represent links between agents and social networks in general. Hence, in social-networking research nonlocal interactions are mostly realized through random connections between pairs of distant agents. In fact, this is an intermediate solution toward a virtual field-type representation of information diffusion.

As was mentioned in the previous chapter, a number of other field-based approaches have been introduced, like gradient routing (GRAD), Directed Diffusion, Co-Fields at the TOTA Programming Model (Mamei and Zambonelli 2006), CONRO (Shen et al. 2002), etc. In fact, almost all of these methods are employed for various technological or robotic applications,[45] and very few of them, like MMASS, Agent-Based Computational Demography (ABCD), or Agent-Based

---

[45]Field-based approaches are used in modular robots, routing in mobile ad hoc and sensory networks, navigation in sensor networks, situated multi-agent ecologies, coordination of robot teams, artificial worlds, etc.

Computational Economics (Tesfatsion and Judd 2006),[46] are suitable for pro-grammable simulations of social phenomena.

Let us take a closer look at one particularly interesting proposal in that direction, i.e., the multilayered multi-agent situated system (MMASS), which defines a formal and computational framework by relying on a layered environmental abstraction (De Paoli and Vizzari 2003). MMASS is related to the simulation of artificial societies and social phenomena for which the physical layers of the environment are also virtual spatial abstractions. In essence, MMASS specifies and manages a field emission-diffusion-perception mechanism, i.e., asynchronous and at-a-distance interaction among agents. In fact, different forms of interaction are possible within MMASS: a synchronous reaction between spatially adjacent agents and an asyn-chronous and at-a-distance interaction through a field emission-diffusion-perception mechanism. Fields are emitted by agents according to their type and state, and they are propagated throughout the spatial structure of the environment according to their diffusion function, reaching and being eventually perceived by other spatially distant agents (Bandini et al. 2006). Differences in sensitivity to fields, in capa-bilities, and in behaviors characterize agents of different types.[47] The MMASS simulation platform has been applied in various modeling cases like crowd behavior, adaptive pedestrian behavior for the preservation of group cohesion, websites as context-aware agents environments, awareness in collaborative ubiq-uitous environments, etc. (Bandini et al. 2007; Vizzari et al. 2013). In fact, MMASS corresponds well to the OSIMAS paradigm in terms of its information diffusion mechanism. However, it uses graph theory and does not seek to model the deeper, i.e., oscillatory, nature of agents themselves, see Fig. 3.7.

Space-dependent forms of communication (at-a-distance interaction) comparable to MMASS are pheromone-based models such as those adopted by Swarm (and other simulation platforms that are based on it, like Ascape, Repast, and Mason). Swarm-based platforms generally provide an explicit representation of the envi-ronment in which agents are placed and of the mechanisms for the diffusion of signals (Krishnanand and Ghose 2008). However, this diffusion mechanism is not well documented, and even though it allows a certain degree of configurability (e.g., through the definition of the constants regulating signal diffusion and evaporation), it does not allow the definition of specific diffusion functions. Swarm and other similar approachesmay thus represent a possible solution for specific field-based simulations, but it would require a huge effort to design and implement more

---

[46]For instance, ABCE Trading World application: the simulation of economic phenomena as complex dynamic systems using large numbers of economic agents involved in distributed local interactions.

[47]The MMASS simulation platform also supports the implementation of applications based on the situated cellular agents (SCA) model (Bandini et al. 2004), which is a particular class of MMASS characterized by a single-layered agent environment and specific constraints on field definition. In fact, our CA approach, described in the previous section, also employs a similar approach.

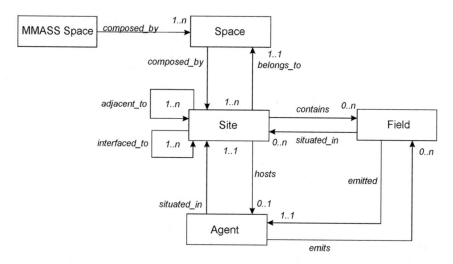

Fig. 3.7 MMASS model elements and relationships among them (De Paoli and Vizzari 2003)

general spatial structures and diffusion mechanisms (Bandini et al. 2007; Bandini et al. 2004).[48]

Recently, the Co-Fields model has been proposed within the area of agent coordination, and it provides a novel interaction method for agents through an explicit description of agent context (Mamei and Zambonelli 2006). In essence, Co-Fields propose an interaction model inspired by the way masses and particles in our universe move and self-organize according to contextual information represented by gravitational and electromagnetic fields. The key idea is to have actions driven by computational force fields, generated by the components themselves or by some infrastructures, and propagated across the environment. Hence, agents are simply driven by abstract computational force fields generated either by agents or by the environment. Agents, driven in their activities by such fields, create globally coordinated behaviors, for instance, in the case of urban traffic control with Co-Fields (Camurri et al. 2007). Although this model still does not offer a complete engineering methodology, it can provide a unifying abstraction for self-organizing intelligent systems. Despite this drawback, the content-based information access in TOTA middleware, which implements the Co-Fields approach in distributed environments, represents an interesting and strong support for the implementation of field-based distributed applications.

---

[48]For instance, stigmergy, as a form of self-organization, is a mechanism of indirect coordination between agents or actions. The principle is that the trace left in an environment by an action stimulates the performance of the next action by the same or a different agent. In that way, subsequent actions tend to reinforce and build on one another, leading to the spontaneous emergence of coherent, apparently systematic activity. This result produces complex, seemingly intelligent structures without the need for any planning, control, or even direct communication between the agents.

Another potential area of application is in the emerging domain of pervasive computing (Hansmann et al. 2003), e.g., in the cases of amorphous (Nagpal and Mamei 2004) and ubiquitous (Poslad 2009) computing. There is fast-growing empirical research about the gradual development of context-aware pervasive computing environments, and it will create yet another area for virtual field-based communication approaches.

In short, as following chapters disclose in more detail, comparing with the other related research, our approach presents novelty related to the connection between the proposed oscillating agent model (OAM) and what recently neuroscience has proven about the coherently oscillating nature of human mind states (i.e., EEG-recorded mind-fields). In other words, to validate the OSIMAS premises, we designed not only a theoretical but also an experimental validation framework.

Following this line of thought, we have also proposed to interpret social order in terms of oscillatory processes emerging from the collective coherent behavior of the conscious and subconscious mind-fields of individual members of a society. In this way, emergent social processes can be interpreted as collective mind-fields that inherit some degree of coherent (synchronized) field-like behavior. On the basis of such reasoning, a social information network can be understood as a virtual information field, where each network node (agent) receives pervasive (broadcasted) information field values. Such an approach is targeted to enforce indirect and uncoupled (contextual) interactions among agents in order to represent contextual broadcasted information in a form locally accessible and immediately usable by network agents.

This chapter also provides some universal scale-free and field-like principles valid in diverse complex systems. On the basis of these principles, we can formulate a set of claims which form the core of the proposed OSIMAS (oscillations-based multi-agent system) paradigm. Following the proposed assumptions, agents are complex multifaceted field-like information processes that resemble the physical analogy of potential field-like energy stored in the form of natural oscillations, which is commonly expressed through the individual spectra of oscillations. In this way, an agent becomes represented in terms of a unique composition of resonant oscillations, i.e., an individual spectrum of brainwaves (some experimental findings based on EEG recordings are presented at our website http://osimas-eeg.vva.lt/), while society is understood as a superimposition of agent-based individual spectra. Consequently, social order emerges as a coherent global field composed of resonant individual oscillations.

We have to admit, however, that because of the immense complexity, we have left many unanswered questions open for discussion and further research, such as synchronization principles, the model for an oscillating agent, the rules for energy transformation and communication, how the opposing processes of coherence and decoherence produce order, etc. However, this chapter provides some guidelines, along with explanatory sources, for further conceptual, empirical and simulation studies of the OSIMAS models in the next chapters.

Chapter 4
From Baseline Individual to Social Neurodynamics: Experimental Framework

This chapter is looking for the biologically inspired oscillating agent modeling. In other words, it searches for the experimental neuroscience based evidences of the oscillatory nature of social agents as approximations of real humans. In this regard, we noticed from the neuroscience domain, that basic mind states, which directly influence human behavior, can be characterized by the specific brainwave oscillations.

For the experimental validation (or disproof) of the biologically inspired OSIMAS paradigm we have designed a framework of EEG (electroencephalography)-based experiments. Initial baseline individual tests of spectral cross-correlations of EEG-recorded brainwave patterns for some mental states have been provided in this chapter. Preliminary experimental results do not refute the main OSIMAS postulates.

4.1 Overview of the Related Research Scope

This chapter addresses the previously mentioned conceptual problem—the lack of a multidisciplinary connecting paradigm, which could link fragmented research in the fields of neuroscience, artificial intelligence (AI), multi-agent systems (MAS), and social network domains. The need for a common multidisciplinary research framework arises essentially because these fields share a common object of investigation and simulation, i.e., individual and collective behavior. Although the fields of research mentioned above all approach it from different perspectives, their main object of investigation remains the same. Let us briefly review latest developments in each of the above-mentioned domains.

Hence, new opportunities for multidisciplinary integration have emerged following technical neuroscience advancements in the research area of brain activity mapping. This has enabled a qualitatively new level of cognitive, behavioral, and computational neuroscience (Bunge and Kahn 2009; Haan and Gunnar 2011). With the increase of computing power neuroscience methods have crossed the borders of

© Springer International Publishing Switzerland 2016
D. Plikynas, *Introducing the Oscillations Based Paradigm*,
DOI 10.1007/978-3-319-39040-6_4

individual brain and mind states research. Hardware and software, used for mapping and analyzing electromagnetic brain activations, have enabled measurements of brain states across groups in real-time (Grinberg-Zylberbaum and Ramos 1987; Lindenberger et al. 2009; Newandee and Reisman 1996; Nummenmaa et al. 2012; Stevens et al. 2012). This research frontier has made room for emerging multidisciplinary research areas like field-theoretic modeling of consciousness (Libet 1994; McFadden 2002; Pessa and Vitiello 2004; Pribram 1991; Thaheld 2005; Travis and Arenander 2006; Travis and Orme-Johnson 1989; Vitiello 2001), social neuroscience, neuroeconomics, and group neurodynamics (Cacioppo and Decety 2011; Loewenstein et al. 2008).

From the other side, some perspicacious biologically–inspired simulation approaches have emerged in the areas of computational (artificial) intelligence, agent-based, and multi-agent systems research (Nagpal and Mamei 2004; Raudys 2004). In turn, these advances have laid the foundations for simulation methods oriented towards intelligent, ubiquitous, pervasive, amorphous, organic computing (Hansmann et al. 2003; Poslad 2009; Servat and Drogoul 2002), and field-based coordination research (Bandini et al. 2007; De Paoli and Vizzari 2003; Mamei and Zambonelli 2006).

A closer look at applied social networks research also reveals some related approaches, which deal, in one way or another, with simulations of field-like information spreading in social networks. For instance, behaviors spread in dynamic social networks (Zhang and Wu 2011), spread of behavior in online social networks (Centola 2010), urban traffic control with coordinating fields (Camurri et al. 2007), mining social networks using wave propagation (Wang et al. 2012), network models of the diffusion of innovations (Valente 1996), etc.

In sum, research trends in neuroscience, AI/MAS, and social networks are leading to increasingly complex approaches, pointing toward oscillations-based or field-theoretic representations of individual and collective mental and behavioral phenomena as well (Haven and Khrennikov 2013). In this regard, the major insights of this chapter are derived from the novel oscillation-based multi-agent system (OSIMAS) social simulation paradigm shortly described below, which links previously mentioned emerging research domains via neurodynamic oscillation-based representation of individual human mind and society (as coherent collective mind) states as well.[1]

Major conceptual implications of the OSIMAS paradigm, which are presented in the previous two chapters and briefly summarized in the next section, essentially are oriented to field-theoretic ways of modeling and simulating individual and collective mind states. Whereas, major prospective practical applications of the

---

[1]Some electroencephalographic (EEG) experimental evidences provided in the sections below led to an idea of interpreting human basic behavioral patterns in terms of mind states, which can be characterized by unique electromagnetic power spectral density distributions in EEG channels (McFadden 2002).

OSIMAS paradigm are targeted to the simulations of real social phenomena.[2] Major methodical tools suitable for the appropriate simulations are within the area of agent-based and multi-agent systems (in short ABS and MAS, respectively) research.

The current peer-to-peer-based ABS and MAS direct communication approaches have been unable to incorporate huge amount of indirect (contextual) information. This is due to the associated complexity and informal information intangibility, and the lack of a foundational theory that could create a conceptual framework for the incorporation of implicit information in a more natural way. Hence, there is a need to expand prevailing ABS/MAS conceptual frameworks in such a way that nonlocal (contextual) interaction and exchange of information could be incorporated.

Hence, our idea is to incorporate implicit information in the form of nonlocal (contextual) information, which, as in the case of natural laws (e.g., the laws of gravity, entropy, symmetry, energy conservation, etc.), would affect an entire system of social agents at once. Following such an analogy with natural laws, we assume that explicit local activities of social agents can be influenced by implicit (contextual or nonlocal) information. This could influence entire systems of agents at once in a form of global laws (Reimers 2011).[3] Each agent would respond to this contextual (nonlocal) information in a different way depending on individual characteristics.

Our assumptions naturally follow from real life observations, where agents interact locally (interchanging information with neighbors), but also are affected by the nonlocal states of the whole system (e.g., traditions, cultures, fashions, national mentalities, political situations, propaganda, economical/financial situations, etc.). Here the term 'nonlocality', which we borrowed from quantum physics, could have many social interpretations, but we prefer to understand it as Jung's archetypes of the collective unconscious, which can be thought of as laws of nature in terms of structures of consciousness.

If we admit that laws of nature can be applied in terms of common structures of consciousness, then from the very ground up our mindfields patterns would be cohesively adjusted. This cohesion can be interwoven so deeply in our own reasoning nature, that we individually could not directly perceive it.[4] In this sense, common mind-field would work as an intangible medium through which the individual cognitive processes are activated and mutually coordinated. We assume, though, that the aftermath of this very subtle and intangible common mind-field could be experimentally observed and explored in terms of neuroscientifically measured synchronicities and coherence between individual brainwave activations.

---

[2]For instance, contextual (implicit) information spread in social media (like propaganda, political campaigns, information wars, etc), network models of the diffusion of innovations, models of self-excitatory wave propagation in social media, etc.

[3]We assume that some information can be broadcasted (shared as in the modern mass media communication channels) by agent to the entire system of other agents.

[4]Similarly, like we cannot see, smell or hear an air (as medium), through which other things and people can be seen, heard or smelled. Nerveless, this medium exists, despite the fact that our senses do not perceive it directly.

Social neuroscience has just started exploration of this very perspective area of research. In the next sections, we will elaborate on various conceptual and experimental aspects of the presented ideas in more detail.

4.2   The OSIMAS Paradigm: Conceptual and Experimental Framework

In the OSIMAS paradigm, we propose field-based approach for modeling pervasive contextual environments in complex information-rich social networks. In other words, we formulate the bases for modeling the emergent and self-organizing features of modern information-rich social networks, where not only intangible but also tangible natural resources and even social agents themselves can be simulated as oscillating processes immersed in an all-pervasive contextual information field,[5] which is called pervasive information field (PIF) in the OSIMAS paradigm (Plikynas et al. 2014).

On the basis of the above presented multidisciplinary research review (and earlier mentioned claims and postulates), the OSIMAS paradigm formulates five basic assumptions:

1. There are no separate laws for the large (on a biological, sociological, or cosmological scale) and for the physics of the small (on an atomic/subatomic scale), but rather universal all-embracing laws for the self-organized multifaceted information that integrally permeates all living and nonliving states of energy-matter.
2. In the mesoscopic scale, the most complex known form of self-organized information is the human mind. In electromagnetic (EM) field-based neurophysiologic approaches, the human mind can be represented as a unified EM or other field-like model of consciousness.
3. Societies can be understood as global processes emerging from the collective (distributed) behavior of the conscious and subconscious mind-fields of their individual members. In this way, emergent social processes are produced by a collective mind-field and inherit some degree of coherent (synchronized) field-like behavior.
4. Societies (the macro-world) can no longer be viewed as separate from the quantum effects taking place in the conscious mind-fields of their members. Self-organization and coherent behavior in social systems is not so much correlated with the particular patterns of agent's behavioral actions, but with the synchrony of their mental activity (inferring that mental activity patterns are explicitly or implicitly expressed in agent's behavior).

---

[5]In the early 90s were widely discussed similar all-pervasive contextual information field ideas referred then as nonlocal 'mind field' along the lines that (Grinberg-Zylberbaum and Ramos 1987) and (Orme-Johnson et al. 1982) (Orme-Johnson 1981) found some experimental evidences for existence of experimental evidences of interbrain direct communication at a distance.

5. A core reason of social behavioral synchrony is the convergence of otherwise dissipating and self-destructive mental activity of the individual members of a society. In this regard, the social coordination mechanism, or so-called social binding, involves the synchronicity mechanism between local self-organized information processes, i.e., agents. The dynamics of synchronous oscillations creates self-organized social systems.

The OSIMAS paradigm is based on these key assumptions, which open up a new way for modeling and simulating emergent social properties as collective mind-field effects. In short, in the proposed OSIMAS framework, all agents are represented in the common PIF spectrum as individual sets of oscillation bands, which are managed by the oscillating agent model (OAM). This latter model realizes the production rules for the transformation of agents' internal energy states (a set of active oscillations), where agents communicate the information they possess according to their behavioral strategy. Communication takes place through a common medium, i.e., a PIF, but it is managed by the wave-like communication mechanism (WIM). To further clarify the OSIMAS paradigm we refer to the next chapters.

Concerning the experimental aspect, some prominent studies indicate brainwave spatial and spectral coherence and synchronicities for the simultaneous measurements of neurodynamic activations in groups of people (Newandee and Reisman 1996; Nummenmaa et al. 2012; Orme-Johnson et al. 1982; Stevens et al. 2012). Below we will briefly outline the reasoning of the chosen OSIMAS experimental validation framework. First, while formulating our experimental framework, we faced some challenging fundamental questions—like how to bring human interaction, occurring in a complex social environment, under the scrutiny of laboratory testing and how to identify human interaction itself, i.e., what are the most basic social communication artifacts to measure? Obviously, human external behavior shows only the tip of the iceberg and can only vaguely represent the states of individual and collective mindfields. Hence, we had to look for more fundamental and unambiguous artifacts, i.e., brainwaves activation patterns, which uniquely identify brain states. It is truism in the cognitive neurodynamics that certain brain states lead to the appropriate thought patterns and consequently to the associated decisions and behavior (Nunez and Srinivasan 2005).

In this regard, our next step was to test whether basic mind states have similar cross-correlated brainwave activation patterns in different individuals and can be distinguished by their characteristic power spectral distributions. However, our main goal was not to improve well-known experimental methods of measuring brain states. Rather, we were investigating whether experimentally measured spectral patterns of different mind states can be unambiguously associated with the mind-field concept theoretically deduced from the OSIMAS paradigm.

Hence, we identified and cross-correlated various mental states of temporally separated individuals in terms of their characteristic brainwave patterns (delta [1–4 Hz], theta [4–8 Hz], alpha [8–12 Hz], beta [13–30 Hz], and gamma [30–70 Hz] frequency ranges), which were recorded using EEG (electroencephalography) methods. The next section provides concrete description of the

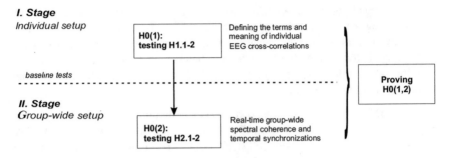

Fig. 4.1 Experimental EEG research stages and series of tests

experimental setup and some initial findings which were designed for defining
(a) mind-field states in terms of the EEG spectra and (b) the terms and meaning of
cross-correlations for the mind-field states between individuals.

The main purpose of the proposed experimental EEG framework is to find out
whether brainwave patterns, i.e., EEG-recorded mindfields, can demonstrate
mutually correlated brainwaves' patterns depending solely on the states of people's
minds. Hence, the main experimental (null) hypothesis (see Fig. 4.1) consists of
two parts—H0(1) and H0(2):

H0(1): temporally separated people doing the same mental activities demonstrate
an increase of cross-correlations in their brainwave patterns;

H0(2): a real-time spatially close, but mutually separated group of people
demonstrates a statistically significant increase of the mutual synchronization in
their brainwave patterns during collectively coordinated or even uncoordinated, but
mutually induced states.

In the current state of research, we are dealing with the hypothesis H0(1), which
has these series of tests:[6]

H1.1: EEG recordings of different mind states for different individuals have the
fewest cross-correlations and thus have less similar spectral patterns.

H1.2: Temporally separate EEG recordings of the same mind states for different
individuals are cross-correlated in terms of similar spectral patterns.

At this particular stage the goal was to find out the cross-correlations of brainwave signals for separate individuals in various mind states. In the next section, we
will demonstrate whether temporally separated people doing the same mental
activities have a significant increase of cross-correlations in their EEG-recorded
brainwave patterns. These individual baseline tests are designed to define the terms
and meaning of cross-correlations between the same mind states. The experimental
observations—that similar mind states can be uniquely characterized by the specific

---

[6]Experimental testing of the H0(2) hypothesis is foreseen in the prospective research.

oscillations-based electromagnetic power spectra distributions in the brain—are essential for proving the basic conceptual OSIMAS assumption about the oscillatory nature of agents.

4.3   EEG Experimental Setup: Cross-Correlation of Individual Mind States

The experimental EEG cases presented below illustrate observed cross-correlations of individual mind states. However, the main purpose of this section is not to present thorough EEG-based statistical results. This will be carried out in the prospective research. At this stage, however, our main focus is to:

– present the experimental method used to prove the basic conceptual OSIMAS assumption about the oscillatory nature of agents; and
– based on the individual tests presented below, to present a group-wide EEG experimental framework, which could test cross-correlations of EEG signals between individuals within groups.

Our experimental findings are based on the EEG signals recorded using the BioSemi ActiveTwo Mk2 system with 64 channels. We employed standard positions of 64 channels in the BioSemi ActiveTwo Mk2 EEG measuring system. This system obtains and records high-quality (resolution LSB = 31.25 nV) and low-noise (total input noise for Z_e < 10 kOhm is 0.8 μV_{RMS}) electric local field potentials from the surface of the cranium.

The time-density (sampling frequency) of the recorded signals was 1024 points per second for all 64 channels from the head surface of participants. We denote the set of recorded signals $u^i(t) = 1,..., 64$ for every channel. We have used spectral activation differences for the EEG-recorded brainwave signal visualization and analyses. Below we present the methodological setup.

The results of our findings are based on the EEG signals recorded from 10 healthy volunteers (five women and five men; four of them were experienced yoga practitioners aged between 30 and 50, and the remainder were beginners) who were all asked to stay in five different mental states, each for a duration of 30 min.[7] Between targeted states there were 5-min pauses for relaxation. For further reference, these states are denoted with numbers as follows: (1) *Meditation*—each volunteer was asked to meditate in a transcendental state with eyes closed in the complete silence; (2) *Meditation with acoustic sound signals*—each volunteer was asked to meditate for 30 min with four different short sounds—noise, beep, standard, and oddball played in random sequence—a total of 750 sounds played during

---

[7]Recordings took place in the laboratory premises of the Meditation Research Institute (MRI) in SRSG (Rishikesh, India). Recorded signals were analyzed with the EEGLab software package, which was used to remove artifacts, filter data and analyze it.

the session; (3) *Counting of sound signals*—each volunteer was asked to count one
of four different sound types; (4) *Thinking*—each volunteer was asked to think,
contemplating the chosen idea or subject; (5) *Thinking with acoustic sound signals*—
each volunteer was asked to think while random sounds were playing.

Hence, we calculated the spectral density of the EEG signals, using the fast
Fourier transformation and denoting it by $f^i(w)$. For all of the results presented here
we have used 600-second intervals. We applied frequency filters to obtain useful
signals in the spectral range [1–38] Hz. In this way, electric power lines noise
(50 Hz frequency) was eliminated. In order to compare the EEG signal changes for
different persons, we normalized the spectral density to unity so that

$$\frac{1}{C^i} \int_{1\,\text{Hz}}^{38\,\text{Hz}} f^i(w)dw = 1, \tag{4.1}$$

where $1/C^i$ is the normalization constant. Afterwards, the separation of different
brainwaves was done for the delta, theta, alpha and beta frequencies: Δ = 1–4 Hz,
Θ = 4–8 Hz, α = 8–13 Hz, and β = 13–38 Hz.

Consequently, we obtained normalized brainwaves values for every person in
every mental state. For example, the α signal of the ith channel for the first person in
the second mental state

$$\alpha_{12}^i = \frac{1}{C^i} \int_{1\,\text{Hz}}^{4\,\text{Hz}} f^i(w)dw. \tag{4.2}$$

We did it for all persons, their mental states and brainwaves spectral ranges (like
Δ, Θ, α, β) accordingly. We also carried out a comparison between the different
mental states by calculating the total differences (summing the activations in all
EEG channels) for different brainwaves. For example, the difference between
meditation (indicated by the number 1) and thinking (indicated by the number 4) for
the second person in the delta frequency range is expressed as follows:

$$\Delta_{21} - \Delta_{24} = \sum_{i=1}^{64} |\Delta_{21}^i - \Delta_{24}^i|. \tag{4.3}$$

The above formula is used for all brainwave frequency ranges and for all persons
in all five different mental states.

Hence, the obtained results showed a modulus of aggregated differences summed up for all EEG channels between brainwaves in different mind states.
However, to illustrate clearly the observed differences, we also used a colored
head-map representation in which the differences between the separate channels are
not added up but are averaged out using spline interpolation over the 64 channels
for a long time interval, see Fig. 4.2. This approach helps greatly to distinguish
similar brain activations when both activation patterns are very different and tend to
have rapidly changing dynamics. In Fig. 4.2 we illustrated an example of the EEG

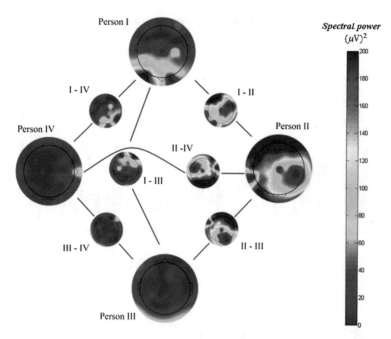

Fig. 4.2 Spectral power activation (alpha range) of brainwave dynamics illustrated for 4 persons (experienced meditators) in thinking state. Smaller diagrams and connecting lines indicate spectral power differences between respective persons

head-maps of four experienced yoga practitioners, who were able to produce distinguishable differences in the EEG activation patterns in the meditation and thinking states (less experienced practitioners were not so successful to produce the distinguishable meditative state). One more reason to use only two or four head-maps for the visual analyses of EEG activation differences is due to the unresolved technical, organizational, and synchronization problems, which are still common during the simultaneous group-wide experiments. The habitual EEG analyses usually explore differences of EEG activations for two head-maps only (Radin 2004; Standish et al. 2004; Travis and Arenander 2006; Wackermann et al. 2003). Hence, purely for illustrative reasons we depicted a few examples of activation differences for two or four head-maps in some figures below. Whereas, aggregated data for 10 persons is depicted in the tables and some graphs.

Such a visualization of brainwave dynamics is very helpful for the recognition of group synchronization patterns for different persons, mind states, and frequency ranges. For the sake of clarity, we have added smaller head-maps that show direct differences in brainwave activation patterns (for the chosen mind state) between pairs of people. If brainwave activations for the pair of people are dissimilar, the corresponding differences between activations produce a color and intensity-rich activation pattern in the difference head-map. In the opposite case, if the brainwave activations for the pair of people chosen are similar, the corresponding difference

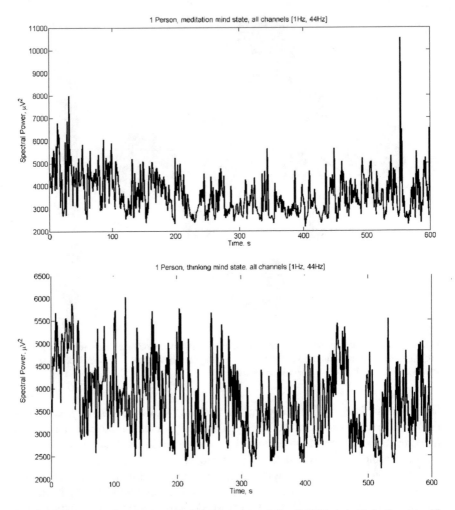

Fig. 4.3 Illustration of spectral power dynamics summed for all EEG channels in the states #1 (meditation) and #4 (thinking) for Person #1

map tends to be less rich with a low intensity in color. This approach helps greatly to distinguish similar brain activations when both activation patterns are very different and tend to have rapidly changing dynamics. Our proposed group-wide mind-field visualization approach extends well-known brain imaging techniques (BIT). Having this in mind, we named it the group-wide mind-field imaging method—GMIM, see more at http://osimas-eeg.vva.lt/.

However, the highly fluctuating nature of the EEG signals makes it difficult to compare them for the different individuals and mind states. Our experiments showed that in order to have meaningful comparison of EEG activations between different persons the baseline individual tests are needed. Hence, below we provide

Table 4.1 Statistical estimates of average total spectral power (SP) fluctuations measured in c.u. for 10 persons over the period 600 s in each mind state	State no.[a]	Statistical estimates of total spectral power (SP) dynamics, c.u.	
		Mean SP	Standard deviation
	1	4053.8	969.2
	2	3908.8	1157.1
	3	4294.4	810.5
	4	4078.4	1011.5
	5	3944.8	1257.1

[a]Labeling of the states is explained in the beginning of this section

few baseline tests. For instance, in Fig. 4.3 we illustrated an example of spectral power dynamics summed for all EEG channels for person #1 (this participant was randomly chosen from other four experienced meditators, which were capable to stay in the meditative state).

In Fig. 4.3, we depicted only one person's case, however, similar spectral power dynamics were observed for other persons too, i.e., we observed volatile dynamics of the summed spectral power activations. Averaged spectral power dynamics for all 10 persons is provided in Table 4.1.

High variance of total spectral power has been reported in other experiments too (Nummenmaa et al. 2012; Standish et al. 2004; Travis and Arenander 2006). Observation of spectral power dynamics over the EEG channels in various spectral ranges clearly indicate spatial location as playing important role, see Fig. 4.4 for an illustration. These measurements also clearly show that some locations (represented by the specific EEG channels) play a major role for determining differences between the mind states across all spectral ranges, see Fig. 4.4.[8]

In the prospective research, thorough examination of the set of the specific channels and there dominant influence for the determination of the mind states will considerably reduce amount of data and computer power for analyses. In our earlier research, we found evidence that spatially close brain regions well correlate in terms of spectral power distributions in Δ, Θ, α, and β frequency ranges. However, this is not valid for the distant brain areas (Nummenmaa et al. 2012; Travis and Arenander 2006).

It is worth noting, that EEG measures spectral power radiated from the surface of the brain, but not from the deeper brain layers, which take a very active part in the energy exchange with the surface layers of the brain (Nunez and Srinivasan 2005). In the perspective research, energy flows from the deeper layers of the brain have to be explored in order to get a complete picture of the spatial spectral power dynamics.

---

[8]Figure 4.4 illustrates data of only 3 of the 10 participants. They were randomly chosen. However, similar tendencies were observed for other participants too. In this way, we outlined a direction for further thorough investigation based on the specific spatial locations.

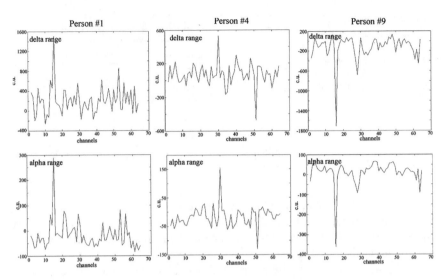

Fig. 4.4 Illustration of power spectra differences between states #1 and #4, recorded for all channels in delta and alpha frequency ranges for persons #1, 4, and 9

We also found, that spectral power dynamics takes place not only in the spatial sense, but also in the sense of redistribution of brainwaves energy over spectral regions. This is especially obvious during transitions between mind states, when redistribution process of spectral power takes place between spectral ranges Δ, Θ, α, β, and γ, see http://osimas-eeg.vva.lt/.

Intrinsic spatial and spectral dynamics show very complex processes involved, which cannot be grasped by the average estimates. However, for the baseline testing we made average measurements of the spectral power distributions over the Δ, Θ, α, β, and γ frequency ranges for all 10 persons, see Fig. 4.5 for an illustration.[9]

Despite high standard deviation for 0.05 significance level, results clearly demonstrated characteristic spectral power distributions in different states. Hence, regardless of highly dynamical nature, mind states can be distinguished in terms of brainwaves activations. Although, we admit that thorough further investigation is needed to provide more solid statistical estimates.

Let us recall, that proposed individual EEG baseline tests were designed to test the basic conceptual OSIMAS assumption, that different states of social agents can be represented in terms of coherent oscillations. In fact, EEG experimental results revealed how theoretically deduced idea of agent as system of coherent oscillations can find solid empirical backup in terms of coherent Δ, Θ, α, β, and γ brainwave oscillations, which uniquely describe agent's mind states. Hence, following Eqs. 4.1–4.3 and methodic developed in our previous research (Plikynas et al. 2014), statistical estimates for the average spectral density differences (summed up

---

[9]We included γ frequency range in order to embrace whole brainwaves region.

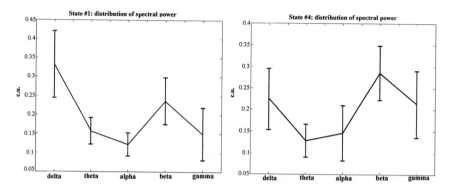

Fig. 4.5 Spectral power distribution results over Δ, Θ, α, β and γ frequency ranges for the states #1 and 4 (averaged for all channels and all 10 participants)

for all EEG channels through all frequency ranges) between various mind states were depicted in the Table 4.2.[10]

Variability of differences of obtained power spectral density distributions is very high due to the intrinsic dynamical nature of human mind. However, the results in the Table 4.2 clearly indicate nonzero and for some states even very high power spectral density distribution differences. For instance, thinking and thinking with noise states substantially differ from meditation and meditation with noise.

Hence, further in our study, we distinguished meditation and thinking states. For instance, refer to Fig. 4.6 for an illustration of differences in EEG-recorded electric field potentials (for 64 channels) between the meditation and thinking states for two experienced meditators in the Δ, Θ, α, and β frequency ranges, respectively. For the sake of clarity, we have normalized the differences over all frequency ranges. So, adding up the activation differences via all frequency ranges yields a total of zero; refer to Fig. 4.6 for diagrams A, B, C, and D.

Our proposed visualization of activation differences can be used for recognizing brain areas that are responding differently (in terms of active brainwave frequencies) depending on mental states, but it does not provide absolute activation values. In this regard, we have provided absolute activation values (in conditional units) for the corresponding mental states and frequency ranges; refer to Fig. 4.6 for diagrams E, F, G, and H. In this way, we can minimize the possibility of misinterpreting these

---

[10]We carried this out for a comparison between the different mental states by calculating the total differences (summing the activations in all EEG channels) for different brainwaves, see Eq. 4.3. However, absolute values of power spectral density distributions for the different states and persons varied significantly, see Figs. 4.3 and 4.4. Therefore, we converted absolute measures to conditional units [c.u.] so that estimates of differences would fit in the normalized scale [0–10]. We assumed that a significant difference is beyond 3.3 c.u., because an average standard deviation $\pm 2\sigma$ for the 95 % confidence level reaches ± 1.1 c.u. (see Table 4.2), which is one third of the face value. Beyond this level, relatively the values make 95 % confidence level meaningless as the signal is dissipated in the noise, i.e., noise and signal ratio is too high. Some related details are provided in the other article (Plikynas et al. 2014).

Table 4.2 Power spectral density distribution differences (in c.u.; scale [0–10]) and their standard deviation $\pm 2\sigma$ for the 95 % confidence level are depicted between various mind states, summed for all channels through all frequency ranges. Results are provided for 10 voluntaries

Person 1

State	I	II	III	IV	V
I	0.0	3.6 ± 1.3	5.5 ± 1.8	5.9 ± 1.9	5.6 ± 1.7
II	3.6 ± 1.3	0.0	3.1 ± 1.2	4.8 ± 1.4	3.2 ± 1.4
III	5.5 ± 1.8	3.1 ± 1.2	0.0	2.9 ± 1.0	3.0 ± 1.1
IV	5.9 ± 1.9	4.8 ± 1.4	2.9 ± 1.0	0.0	2.2 ± 0.8
V	5.6 ± 1.7	3.2 ± 1.4	3.0 ± 1.1	2.2 ± 0.8	0.0

Person 2

State	I	II	III	IV	V
I	0.0	3.0 ± 1.0	5.1 ± 1.7	4.9 ± 1.9	5.7 ± 1.8
II	3.0 ± 1.0	0.0	2.1 ± 1.2	4.8 ± 1.2	3.7 ± 1.2
III	5.1 ± 1.7	2.1 ± 1.2	0.0	2.9 ± 1.0	3.3 ± 1.2
IV	4.9 ± 1.9	4.8 ± 1.2	2.9 ± 1.0	0.0	2.0 ± 0.7
V	5.7 ± 1.8	3.7 ± 1.2	3.3 ± 1.2	2.0 ± 0.7	0.0

Person 3

State	I	II	III	IV	V
I	0.0	2.6 ± 1.0	3.7 ± 1.1	4.2 ± 1.2	4.4 ± 1.5
II	2.6 ± 1.0	0.0	3.3 ± 1.3	3.8 ± 1.3	3.3 ± 1.3
III	3.7 ± 1.1	3.3 ± 1.3	0.0	1.9 ± 0.5	1.5 ± 0.3
IV	4.2 ± 1.2	3.8 ± 1.3	1.9 ± 0.5	0.0	1.3 ± 0.6
V	4.4 ± 1.5	3.3 ± 1.3	1.5 ± 0.3	1.3 ± 0.6	0.0

Person 4

State	I	II	III	IV	V
I	0.0	3.5 ± 1.4	3.0 ± 1.0	3.6 ± 1.2	4.4 ± 1.5
II	3.5 ± 1.4	0.0	3.1 ± 1.4	3.3 ± 1.0	2.3 ± 1.1
III	3.0 ± 1.0	3.1 ± 1.4	0.0	2.7 ± 0.8	1.9 ± 0.4
IV	3.6 ± 1.2	3.3 ± 1.0	2.7 ± 0.8	0.0	1.0 ± 0.4
V	4.4 ± 1.5	2.3 ± 1.1	1.9 ± 0.4	1.0 ± 0.4	0.0

Person 5

State	I	II	III	IV	V
I	0.0	1.7 ± 0.5	3.5 ± 1.1	4.6 ± 1.4	4.4 ± 1.5
II	1.7 ± 0.5	0.0	2.1 ± 0.6	1.3 ± 0.3	2.2 ± 1.0
III	3.5 ± 1.1	2.1 ± 0.6	0.0	1.7 ± 0.5	1.0 ± 0.3
IV	4.6 ± 1.4	1.3 ± 0.3	1.7 ± 0.5	0.0	1.5 ± 0.4
V	4.4 ± 1.5	2.2 ± 1.0	1.0 ± 0.3	1.5 ± 0.4	0.0

Person 6

State	I	II	III	IV	V
I	0.0	3.7 ± 1.3	4.7 ± 1.4	5.2 ± 1.8	4.9 ± 1.9
II	3.7 ± 1.3	0.0	3.4 ± 1.3	3.7 ± 1.5	3.9 ± 1.5
III	4.7 ± 1.4	3.4 ± 1.3	0.0	2.9 ± 0.6	2.5 ± 0.5
IV	5.2 ± 1.8	3.7 ± 1.5	2.9 ± 0.6	0.0	2.3 ± 0.6
V	4.9 ± 1.9	3.9 ± 1.5	2.5 ± 0.5	2.3 ± 0.6	0.0

(continued)

Table 4.2 (continued)

Person 7

State	I	II	III	IV	V
I	0.0	2.7 ± 1.0	**5.7 ± 1.9**	**5.9 ± 1.8**	**5.0 ± 1.8**
II	2.7 ± 1.0	0.0	**4.1 ± 1.2**	**3.8 ± 1.3**	**3.0 ± 1.2**
III	**5.7 ± 1.9**	**4.1 ± 1.2**	0.0	3.2 ± 1.1	**3.0 ± 1.0**
IV	**5.9 ± 1.8**	**3.8 ± 1.3**	3.2 ± 1.1	0.0	2.7 ± 0.8
V	**5.0 ± 1.8**	**3.0 ± 1.2**	**3.0 ± 1.0**	2.7 ± 0.8	0.0

Person 8

State	I	II	III	IV	V
I	0.0	3.1 ± 1.3	2.7 ± 1.4	**3.9 ± 1.8**	2.5 ± 1.0
II	3.1 ± 1.3	0.0	3.1 ± 1.0	**3.4 ± 1.0**	2.1 ± 0.8
III	2.7 ± 1.4	3.1 ± 1.0	0.0	3.2 ± 1.1	2.0 ± 0.7
IV	**3.9 ± 1.8**	**3.4 ± 1.0**	3.2 ± 1.1	0.0	2.2 ± 0.5
V	2.5 ± 1.0	2.1 ± 0.8	2.0 ± 0.7	2.2 ± 0.5	0.0

Person 9

State	I	II	III	IV	V
I	0.0	3.1 ± 1.1	**3.7 ± 1.2**	**4.2 ± 1.8**	**4.0 ± 1.4**
II	3.1 ± 1.1	0.0	**3.3 ± 1.0**	**3.9 ± 1.1**	3.1 ± 0.7
III	**3.7 ± 1.2**	**3.3 ± 1.0**	0.0	3.0 ± 1.0	2.4 ± 0.6
IV	**4.2 ± 1.8**	**3.9 ± 1.1**	3.0 ± 1.0	0.0	1.2 ± 0.4
V	**4.0 ± 1.4**	3.1 ± 0.7	2.4 ± 0.6	1.2 ± 0.4	0.0

Person 10

State	I	II	III	IV	V
I	0.0	**3.5 ± 1.0**	2.1 ± 1.0	**3.4 ± 1.2**	3.0 ± 1.1
II	**3.5 ± 1.0**	0.0	2.1 ± 1.2	**5.0 ± 1.4**	**4.7 ± 1.3**
III	2.1 ± 1.0	2.1 ± 1.2	0.0	3.1 ± 1.0	2.3 ± 0.6
IV	**3.4 ± 1.2**	**5.0 ± 1.4**	3.1 ± 1.0	0.0	1.0 ± 0.2
V	3.0 ± 1.1	**4.7 ± 1.3**	2.3 ± 0.6	1.0 ± 0.2	0.0

Large differences between appropriate states are marked in bold

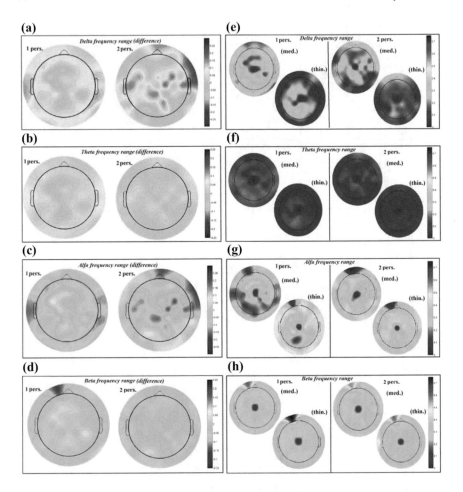

Fig. 4.6 Differences in the EEG-recorded electric field potentials between meditation and thinking states for two experienced meditators in the Δ, Θ, α, and β frequency ranges (diagrams A, B, C, and D, respectively). EEG diagrams E, F, G, and H show actual activations of the corresponding mental states for the chosen frequency ranges. All power spectra are normalized (c.u.)

results, which are primarily based on the analysis of activation differences (see diagrams A, B, C, and D).

In short, from diagrams like E, F, G, and H we have drawn the following conclusions:

– in the thinking state Δ and Θ the frequency ranges are least activated (refer to dark diagrams E and F), but most active are the α and to some degree the β frequency ranges, whereas in the meditation state the Δ waves are relatively more activated, and the α waves less activated;

- the Θ frequency range is not active for the meditation and thinking states (see diagram F);
- the left hemisphere of the frontal lobe activates in the β frequency range, which corresponds with the well-known activity of logical reasoning, i.e., thinking (see diagram H).

Thus, a more detailed analysis of the EEG-recorded differences (see diagrams A, B, C, and D) in electric field potentials between the meditation and thinking states has revealed even more intriguing results for the OSIMAS project:

1. In the meditation state, the Δ (delta) frequency range dominates. These findings confirm other research results like those of (Newandee and Reisman 1996; Travis and Arenander 2006).
2. In the thinking state the α frequency range is dominant. Hence, the thinking state can be distinguished by the appearance of characteristic α waves—a fact also confirmed in the results of other related research.
3. In the meditating and thinking states, the Θ and β frequency ranges are activated considerably less than the Δ and α ranges, respectively.
4. The topological distribution of activated brain zones in different states is almost identical in the Δ, Θ, and α ranges.

The latter inference leads to some important observations:

(i)   *Electric field potentials are activated in the similar spatial distribution, but in the different spectral ranges for different states. For instance, the meditation state is dominated by Δ and the thinking state by α waves while the most active EEG channels remain the same. This observation suggests that both states are similar in nature—a conclusion that is not surprising, after all, since the transcendental meditation method is about the contemplation of pure consciousness itself dissociated from all other objects of contemplation, which is still a sort of deep thinking (Orme-Johnson et al. 1982).*
(ii) *People have specific and unique spatial distributions of activated electric field potentials in each mind state, but independently of the person each mind state has common features that can be recognized in terms of their dominant frequency ranges. For instance, the transition from the meditation to the thinking state is characterized by the transition of the dominant brainwaves from Δ to α frequency (i.e., from lower to higher frequencies).*

These observations can be inferred from the OSIMAS paradigm, where different mental states are perceived as part of one and the same multivariate mental field composed of a set of dominant frequencies which are called natural resonant frequencies. According to the OSIMAS paradigm brain is able to store free energy in these superimposed spectral bands. We argue that theoretically inferred agent's coherent oscillations can be associated with the dominant frequencies of the mind-field, i.e., with the experimentally observed Δ, Θ, α, and β brainwaves.

From OSIMAS paradigm follows that various combinations of activated dominant frequencies in the common mind-field yield unique mental states. As a matter of fact, our experimental data for the meditation and thinking states validate this assumption. We and some other results observed spectral energy redistribution over frequency ranges for opposing mental states like meditation versus thinking (Travis and Arenander 2006; Travis and Orme-Johnson 1989).

The experimental study described above uses EEG-based brain imaging techniques that are targeted for mapping activated areas of electric local field potentials on the human cranium, but this study says little about numerical estimates of the total activations (added up over 64 channels) in particular frequency ranges. Hence, we have made numerical analyses, refer to Fig. 4.7 for an illustration of results,[11] where we can see the absolute activations (in c.u.) for both states and their differences in the Δ, Θ, α, and β frequency ranges (for more results, see http://osimas-eeg.vva.lt/).

We will limit ourselves to the conclusion that each specific mental state has its own characteristic spectral mind-field pattern in terms of numerical estimates of the total distribution of activations in the Δ, Θ, α, and β frequency ranges (the total power of activations in these frequency ranges is normalized to 1). Once we have this specific set of spectral mind-field patterns for a particular person, we can recognize how deeply and in which state this person is in a particular time period. We can do so by making a relative comparison of the spectral mind-field pattern being investigated with the specific set of spectral mind-field patterns already recorded for this person. Then, after this relative comparison, we can obtain differences in the same way as we show in our charts; refer to Fig. 4.7 for an illustration of the charts. The set of these differences obtained for various specific states can precisely identify the state being investigated.

Although each individual has a unique distribution of spatial-temporal activations in different mind states, the above-mentioned spectral mind-field patterns follow the same characteristic distribution of spectral activation patterns in the Δ, Θ, α, and β frequency ranges for every normal individual. Our initial findings and other research show that this is a universal rule.

In order to find social evidence of coherent collective mind-field effects, we looked also for experimental estimates of the coherence and synchronicities simultaneously measured for group-wide (collective) neurodynamics (Plikynas et al. 2014). There are some related other studies that indicate brainwave spatial and spectral coherence and synchronicities for the simultaneous measurements of neurodynamic activations in groups of people (Newandee and Reisman 1996; Nummenmaa et al. 2012; Stevens et al. 2012). Usually there are applied intra- and

---

[11]Figure 4.7 illustrates analyses of the differences in brainwave activations for the appropriate frequency ranges in Med.-Thin. states of two experienced meditators, who were depicted due to the longest meditation practice. For the baseline testing we needed the most coherent participants in terms of ability to stay in the meditative state. As we have mentioned before, the habitual EEG analyses usually explores differences of EEG activations for two head-maps only (Radin 2004; Standish et al. 2004b; Travis and Arenander 2006a; Wackermann et al. 2003). Hence, these two participants were chosen purely for illustrative reasons to visualize results of our measuring methodic (see Eqs. 4.1–4.3).

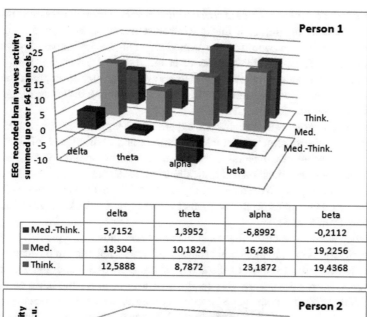

	delta	theta	alpha	beta
■ Med.-Think.	5,7152	1,3952	-6,8992	-0,2112
▨ Med.	18,304	10,1824	16,288	19,2256
■ Think.	12,5888	8,7872	23,1872	19,4368

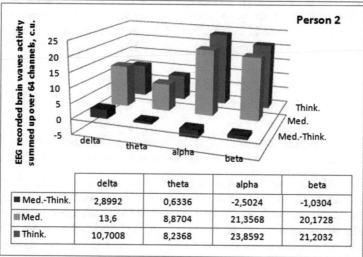

	delta	theta	alpha	beta
■ Med.-Think.	2,8992	0,6336	-2,5024	-1,0304
▨ Med.	13,6	8,8704	21,3568	20,1728
■ Think.	10,7008	8,2368	23,8592	21,2032

Fig. 4.7 EEG-recorded brainwave activity (c.u.) added up in all frequency ranges for 64 channels in the meditation and thinking states. The difference in activations for the appropriate frequency ranges is denoted as Med.-Thin

interbrain synchronization measures like phase-locking index (PLI), partial directed coherence (PDC) or interbrain phase coherence (IPC). The latter measures employ full multivariate spectral estimates based on the concept of Granger causality. They indicate significant increase in phase synchronization within and between the brains during periods of coordinated actions (Orme-Johnson et al. 1982).

These and many other studies of social cognition reported evidences indicating that brain oscillations synchronize within and between the brains when people engage in various forms of action coordination (Achterberg et al. 2005; Grinberg-Zylberbaum and Ramos 1987; Radin 2004; Standish et al. 2004; Wackermann et al. 2003). Group-wide EEG studies are giving birth to new interdisciplinary research areas like social neuroscience, team neurodynamics, neuroeconomics, etc (Lindenberger et al. 2009; Newandee and Reisman 1996).

In sum, our goal was to show an experimental way to test major OSIMAS assumptions. We constructed testable hypothesis and showed few EEG cases as examples. However, it is beyond the scope of the current book to provide in depth EEG studies in order to validate statistically solid evidences. Further studies should follow in this regard. We can only say that at this early stage our and other above-mentioned studies do not contradict to the basic OSIMAS assumptions (see postulates from 2 to 5).

In other words, presented few experimental cases illustrate typical examples of cross-correlations of individual mind states, but have no intention to be interpreted as final statistically solid proves or disproves of the OSIMAS paradigm. These individual cross-correlations show (i) experimental way to prove the basic conceptual OSIMAS assumption about agents' oscillatory nature, (ii) a way for the construction of prospective group-wide EEG experimental framework to test cross-correlations of EEG signals between individuals in the groups (Plikynas et al. 2014).

4.4   Further Prospects

In short, main contribution of the OSIMAS paradigm with respect to the conceptually related other research approaches can be summarized in a following way:

(a) OSIMAS provides a multidisciplinary connecting framework, which links fragmented research in the domains of above-mentioned neuroscience, artificial intelligence (AI), multi-agent systems (MAS), and social research,

(b) following dynamical causal modeling (DCM), OSIMAS provides a missing link between fundamental field-theoretic approaches and experimental neuroscience findings of individual and group-wide coherent brainwaves' oscillations,

(c) proposed unique oscillating agent and pervasive information field conceptual models (OAM and PIF respectively) can extend and considerably deepen above-mentioned other field-based approaches, providing necessary means for the simulation of coherent nonlocal and contextual interactions taking place in various social mediums.

However, we have to name few drawbacks and limitations of the proposed OSIMAS paradigm:

- in technical sense it is hard to implement group-wide EEG experimental validation setup (there are some organizational, equipment synchronization, and
  methodological issues),
- lack of ready to use field-based methods and simulation platforms tailored for
  the agent-based and MAS modeling,
- lack of direct relation between observed social phenomena and simulated social
  modeling results.

In general, major limitations stem from the conceptual, experimental, and
methodical constraints of the currently available field-based approaches. We challenged these limitations, providing not only some insights, but also novel experimental and simulation approaches.

The obtained experimental observations, that mind states can be uniquely characterized by the specific oscillations-based electromagnetic power spectra distributions in the brain, provide solid proof for the validity of the basic OSIMAS
assumption, that agents can be represented in terms of coherent oscillations, which
identify their state and, consequently, behavior. These individual baseline tests outline
prospective investigation of group-wide neurodynamic processes of the temporal and
spatial synchronizations and cross-correlations of EEG-recorded brainwave patterns.

Based on our conceptual and experimental findings, we provided group-wide
modeling framework and means for construction of simulation models, which are
presented in the online virtual lab simulation platform (7 original simulation models
are created at http://vlab.vva.lt/; Guest: guest555).

Let us briefly discuss some further research prospectives. Hence, in neurodynamics, there are two classes of effects: dynamic effects and structural effects. The
duration and form of the resulting dynamic effect depends on the dynamical stability of the system to perturbations of its states (i.e., how the system's trajectories
change with the state). Structural effects depend on structural stability (i.e., how the
system's trajectories change with the parameters). Systematic changes in the
parameters can produce systematic changes in the response, even in the absence of
input (David 2007).

OAM should address both classes of neurodynamic effects, i.e., dynamic and
structural. In our prospective research (see next chapters) we use the quantum
mechanical representation of self-organized mind states.[12] In other words, the OAM
simulates human basic mind states (BMS) dynamics and structural effects
employing stylized oscillations-based representations of experimentally observed
characteristic EEG power spectral density (PSD) distributions of brainwaves
(Buzsaki 2011; Nunez and Srinivasan 2005) in delta, theta, alpha, beta, and gamma

---

[12]Emmanuel Haven's and Andrei Khrennikov's recent monograph "Quantum Social Science"
forms one of the very first contributions in a very novel area of related research, where information
processing in social systems is formalized with the mathematical apparatus of quantum mechanics
(Haven and Khrennikov 2013).

spectral ranges.[13] We infer, that PSD patterns can objectively identify BMS, see Figs. 4.5, 4.6 and 4.7.

We elaborate further, that the wave-like nature of coherent human mind-field states (BMS) can be approximated using the wave mechanics approach, i.e. wave function (also named as state function)[14] and linear operators (Buzsaki 2011; Haven and Khrennikov 2013). This assumption is based not only on the recent theories like Pribram's and Bohm's the holonomic brain theory (Pribram 1999; Pribram 1991) or Vitello's the dissipative quantum model of brain (Pessa and Vitiello 2004; Vitiello 2001), but also on recent evidences, which show "warm quantum coherence" in plant photosynthesis, bird brain navigation, our sense of smell, and brain neurons' microtubules (Engel et al. 2007; O'Reilly and Olaya-Castro 2014). These evidences well corroborate with 20-year-old 'Orch OR' (orchestrated objective reduction) theory of consciousness proposed by Stuart Hameroff and Sir Roger Penrose (Hameroff and Penrose 2014). According to this theory and recent experimental evidences, EEG rhythms (brainwaves) derive from deeper level microtubule (protein polymers inside brain neurons) vibrations in the megahertz frequency range,[15] which govern neuronal and synaptic function, and also connect brain processes to self-organizing processes in the fine scale (Georgiev and Glazebrook 2014). Hence, according to 'Orch OR' theory, consciousness depends on biologically 'orchestrated' coherent quantum processes in collections of microtubules within brain neurons. These quantum processes correlate with, and

---

[13]We chose these basic mind states (BMS)—sleeping, wakefulness, thinking, and resting. We make use of the fact that each BMS has characteristic brainwave pattern, which can be identified using power spectral density (PSD) distribution analyses (Müller et al. 2008). At the level of neural ensembles, synchronized activity of large numbers of neurons give rise to macroscopic oscillations, i.e., brainwaves, which can be observed in the electroencephalogram (EEG). For instance: (i) delta range (frontal high amplitude waves; frequency range up to 4 Hz) is mostly associated with deep sleep and deep meditative states; (ii) theta range (found in various locations; frequency range 4–8 Hz) is mostly associated with relaxed, meditative, and creative states (Lutz et al. 2004; Travis and Arenander 2006b); (iii) alpha range (both sides posterior regions of head; frequency range 8–12 Hz) is mostly associated with reflective, sensory and motor activities; (iv) beta range (both sides, symmetrical distribution, most evident frontally, low-amplitude waves; frequency range 12–30 Hz) is mostly associated with active thinking, focus, hi alert, anxious states; (v) gamma range (frequency range approximately 30–100 Hz) is mostly associated with brain binding into a coherent system for the purpose of complex cognitive or motor functions.

[14]In quantum mechanics, the wave function describes the quantum state of a particle and how it behaves in terms of its wave-like nature. Although human mind consists from billions of particles, we assume that in BMS prevails some sort of coherence mechanism, which binds these particles and makes them oscillate coherently (in unison) as one. Therefore, in a most simplified way we assume that wave function in principle can be used as approximation for the representation of BMS.

[15]Despite a century of clinical use, the underlying origins of EEG rhythms have remained a mystery. However, microtubule quantum vibrations (e.g. in the megahertz frequency range) appear to interfere and produce much slower EEG "beat frequencies" in the range 4–70 Hz. Clinical trials of brief brain stimulation—aimed at microtubule resonances with megahertz mechanical vibrations using transcranial ultrasound—have shown reported improvements in people mood (Hameroff and Penrose 2014b).

regulate, neuronal synaptic, and membrane activity. Continuous Schrodinger evolution of each such process terminates in accordance with the specific Diósi–Penrose (DP) scheme of 'objective reduction' ('OR') of the quantum state (Hameroff and Penrose 2014). This orchestrated OR activity ('Orch OR') is taken to result in moments of conscious awareness and/or choice. The DP form of OR is related to the fundamentals of quantum mechanics.

Following OSIMAS paradigm and earlier cited related other research, we want to employ OAM as a building block for the construction of multi-agent systems (MAS), which could lead to the simulation of collective coherent mind states. However, it requires further empirical research so that oscillations-based conclusive and experimentally proven social simulation theory could further evolve.

Chapter 5
Oscillations-Based Simulation of Human Brain EEG Signal Dynamics Using Coupled Oscillators Energy Exchange Model (COEEM)

For the sake of clarity we would like to note, that the COEEM model here presented was created in the context of a larger scheme of multi-agent systems (MAS) simulation research. Within the framework of the OSIMAS paradigm we investigate opportunities to make use of a biologically inspired approach, where basic human mind states can be represented in the form of the EEG oscillations. Based on the EEG experimental findings, we were looking for ways to model (i) human (social agent) mind states in terms of distributions of characteristic oscillations and (ii) transitions between basic mind states in terms of redistributions of characteristic oscillations.

In order to establish relationship between experimentally measured EEG signal oscillations and conceptually described oscillating agents (OAM) in the OSIMAS paradigm, we have created the coupled oscillations-based COEEM model. Investigation and further refinements of this biologically inspired experimental model helped to define features of our artificially constructed OAM in the OSIMAS paradigm. Hence, oscillation-based modeling of human brain EEG signals oscillations, using a refined Kuramoto model (Pessa and Vitiello 2004), not only helps to specify the OAM but also significantly contributes to EEG prognostication research, which is the major topic of this particular neuroscience research direction.

Hence, this chapter introduces the coupled oscillator energy exchange model (COEEM) which simulates experimentally observed human brain EEG signal dynamics. This model is in some ways similar to the Kuramoto model, but essentially differs in that the Kuramoto model oscillator amplitude is constant, while the COEEM model is dependent on the phase of the oscillators. The reasoning behind the COEEM model construction is based on an energy exchange and synchronization simulation in a localized brain area using (i) the coupled oscillators approach and (ii) experimental (nonfiltered and filtered) EEG observations. For this purpose, we proposed (1) a novel coupled oscillators' phase-locking mechanism (PLM) and (2) a unique and very narrow spectral band prognostication and

© Springer International Publishing Switzerland 2016
D. Plikynas, *Introducing the Oscillations Based Paradigm*,
DOI 10.1007/978-3-319-39040-6_5

superposition method of just 0.01–0.1 Hz. It has been shown that the COEEM model, which is based on the PLM, is suitable not only for accurate short term prognoses of human brain EEG signal dynamics (several ms) but also for the long term (several seconds). In the latter case, for the chosen mind states and EEG channel prognostication, we created and effectively applied a method for the superposition of prognostication results for very narrow spectral bandwidths. In short, based on the promising prognostication results for the real EEG signals, we infer that the oscillatory model presented here well simulates phase locking and energy exchange features characteristic to localized human brain activations dynamics.

5.1 Neuroscience Context

Recent years have witnessed an explosion of interest and activity in the area of human brain research. For instance, the Human Brain Project in the EU and the Brain Mapping Project (BRAIN initiative) in the US are just two examples of large scale research programs (Alivisatos et al. 2012; Zador et al. 2012), which are dedicated to brain simulation, neuroinformatics, high performance computing, medical informatics, brain imaging and mapping, neuromorphic computing, neurorobotics, etc. These advances in theoretical and experimental methods and techniques (Gobbini et al. 2007) not only help to reveal brain disease states but also broaden conceptual knowledge of processes taking place in the brain. However, the degree and depth of fundamental understanding of the underlying processes lags far behind any advances in brain activity measuring techniques (Zacks 2008).

The complexity of the human brain and mind requires us to search for new modeling approaches. Consequently, this chapter addresses an attempt to narrow the gap between theory and experiments. In this way, we expand on the work of other researchers (Friston and Price 2001; Matsuoka 2011; Onton and Makeig 2006; Pessa and Vitiello 2004; Quiroga et al. 2001; Rubinov and Sporns 2010) in the area of human brain activations simulations.

First, let us recall that historically, Berger was the first to quantify the EEG (electroencephalogram) using Fourier analysis (Berger 1929). Berger was also the first to examine the effects of maturation on the EEG. Hence, the EEG consists of complex electrical field oscillations that are assumed to indicate underlying neural network activation dynamics (Anokhin et al. 2006). Traditionally, neural networks are understood as networks of neurons interacting through excitatory and inhibitory connections, where neurons are used for only one kind of activity at a time (Düzel et al. 2010).

Analyses of complex neural networks, based largely on graph theory, have led to studies of brain network organization. Structural brain networks can be described as graphs that are composed of nodes (vertices) denoting neural elements (neurons or brain regions) that are linked by edges representing physical connections (synapses or axonal projections) (Bullmore and Sporns 2009).

Mathematical frameworks for oscillatory network dynamics in neuroscience use tools of weakly coupled phase oscillator theory. It have had a profound impact on the neuroscience community, providing insight into a variety of network behaviors ranging from central pattern generation to synchronisation, as well as predicting novel network states such as chimeras (Ashwin et al. 2016).

Historically, linear regression models were mostly used for the forecasting of EEG time series. For instance, autoregressive (AR), moving average (MA), and autoregressive moving average (ARMA) models are still quite often employed (Blinowska and Malinowski 1991; Hoon et al. 1996; Kim et al. 2013). In most cases, these models are applied for the forecasting of epilepsy or other mental diseases from the EEG time series analyses.[1] Linear regression models, however, have certain limitations for the forecasting of highly nonlinear data (Kim et al. 2013).

Nonlinear modeling of brain activation dynamics, based on the physiological signals of brain activations, has been introduced in recent years, including, for example, studies of: Fourier spectra, chaotic dynamics, wavelet analysis, scaling properties, multifractal properties, correlation integrals, $1/f$ spectra, and synchronization properties (Acharya et al. 2010; Übeyli 2009). Some research employs a time–frequency analysis of the EEG data analyses, using wavelet packets and cost function (Blanco et al. 1995), while other research classifies EEG signals using wavelet coefficients (Hazarika et al. 1997) or cognitive correlates (Kahana 2006).

Despite such efforts, one of the most difficult problems remains—the explanation of the dynamic relationships between different brain regions (Ho et al. 2005). Another problem is the explanation of the partial coherence of EEG signals (Pessa and Vitiello 2004). In this regard, some research focuses on analyses of the relationship among neural signals, using partial directed coherence (Schelter et al. 2006). One of the more promising areas of research close to our work is the study of the brain oscillation control of one single-neuron activity, which attempts to uncover the temporal relationship between brain oscillations and single-neuron activity (Jacobs et al. 2007).

After careful revision of the above-mentioned prognostication methods and models, we have elaborated a completely different approach, i.e., oscillation-based modeling and prognostication of short (several ms) and long time series (several seconds) of real EEG signals. We created and employed an iterative COEEM scheme for the short time prognostication using fourth-order Runge–Kuta algorithms (RK4) with the nonfiltered spectral range 1–512 Hz (including noise) EEG data. For the long time prognostication we used (i) filtered EEG data of 1–38 Hz (without power noise) and (ii) superposition of prognostication results for very narrow spectral bands of 0.01–0.1 Hz.

The COEEM that we propose here is not intended for modeling a complex brain network from the point of detailed mapping of neural networks activations in

---

[1]Epilepsy EEG data has special characteristics including nonlinearity, nonnormality, and nonperiodicity.

various states. Instead, it attempts to address both of the above-mentioned fundamental neuroscience problems by modeling connectivity of strictly localized brain
area activations (in the area of one EEG channel) for the chosen mind state, via the
general electromagnetic field representation using the COEEM approach. Hence,
the COEEM is designed to analyze the temporal evolution of the experimentally
recorded EEG data of one localized EEG channel area. It employs a synchronization (partial coupling) mechanism between simulated artificial oscillators (as
substitutes of biological neurons) to simulate and prognosticate the observed EEG
signal temporal dynamics.

Proposed COEEM model is similar to the Kuramoto model (Kuramoto 2012).
The latter investigates the temporal evolution of coupled phase oscillators, i.e.,
collective synchronization, where the oscillators are weakly coupled and nearly
limit-cycle (Gong et al. 2007; Moioli et al. 2012). It should be noted that in the
Kuramoto model the amplitude of the oscillators is constant, i.e., not dependent on
time, while in the COEEM model the amplitude of the oscillators varies in time.
This fact makes a fundamental difference in comparison to the classical Kuramoto
model, as the COEEM realizes, additionally, an energy exchange between oscillators. In the COEEM, the energy exchange mechanism between two oscillators is
described by the phase difference between them. In the following sections, we
introduce COEEM, PLM and the superposition of very narrow spectral bands
prognostication in detail.

The EEG research results that we obtained provided us with experimental evidence that basic human mind states can be characterized and modeled employing
specific set of electromagnetic (delta, theta, alfa, beta, and gama) oscillations
(Bonita et al. 2013; Cantero et al. 2004; Travis and Arenander 2006).

This chapter is organized as follows. First we describe the COEEM setup and
PLM. Next, there are described the COEEM model and the very narrow spectral
band prognostication technique. Finally, we present the experimental setup and our
findings.

5.2 The COEEM Setup and Phase-Locking Mechanism

The COEEM estimates energy fluctuations in the localized brain area. In this regard, the
COEEM model is essentially a coupled oscillator energy exchange model. Equations
using the COEEM model (see Eq. 5.1) describe the evolution of complex amplitudes of
oscillators. The energy of each oscillator is proportional to the modulus squared of the
complex amplitude. In general, energy exchange takes place between all oscillators,
despite internal or external division. In the COEEM model, though, we model energy
exchange between the inner and outer system of oscillators. Hence, the EEG signal is
modeled as the total energy change of an internal oscillator system over time.

In the COEEM model, each neuron is modeled as an oscillator, which has a
phase and complex amplitude. Thus, modeling the energy exchange between
groups of neurons is replaced by the model of the energy exchange between groups

of oscillators.[2] These oscillators are coupled with each other in a relationship function that can be freely chosen. This model is similar to the Kuramoto model, where the same coupling equations of oscillators are used. As in the Kuramoto model, we use the sinus coupling function of oscillators. In short, the COEEM model investigates systems of coupled oscillators, where each oscillator is characterized by an angular frequency w_i, see Eq. 5.1. This frequency is also emitted in a form of the field, which enables the energy exchange between the oscillators. In this regard, coupling constants were chosen equal to w_j, see Eq. 5.1. The COEEM model equations are as follows:

$$\frac{\partial \theta_i}{\partial t} = w_i + \sum_{j=1, j \neq i}^{N} w_j \sin(\theta_i - \theta_j),$$

$$\frac{\partial A_i}{\partial t} = A_i w_i + \sum_{j=1, j \neq i}^{N} A_j w_j \sin(\theta_i - \theta_j), \tag{5.1}$$

where A_i is the complex amplitude of the i-th oscillator (the amplitude A_i is itself a complex number), θ_i—the phase of the i-th oscillator, w_i—the angular frequency of the i-th oscillator, and t—the temporal coordinate.

Here we have $2N$ linear equations for the phases and complex amplitudes of the oscillators. The system Eq. 5.1 of $2N$ linear equations describes the temporal evolutions of the phases and the complex amplitudes of the oscillators. The system in Eq. 5.1 consists of two phases: the oscillator phase θ_i and the radiated field complex amplitude phase arg (A_i). In this case, the energy of an oscillator is proportional to the modulus squared of the complex amplitude A_i (A_i describes the EEG signal). We are modeling this signal as a radiated field (A_i is proportional to radiated field) using the system of internal oscillators. In this case, oscillators are coupled via the phase differences between them. The phase differences between phases of the radiated fields (sin(arg (A_i) − arg (A_j))) are not used for the coupling. In this way, we can get the temporal evolution of the energy of each oscillator. Thus, we can evaluate the energy redistribution between oscillator groups after a certain time.

The phase difference between two oscillators describes the energy exchange mechanism between them: the energy exchange process takes place between oscillators where the phase difference is not equal to zero (if the phase difference is zero, then energy exchange is absent). In short, by using the COEEM model, we seek to characterize the dynamic mechanism of the relationships between different

---

[2]We asssume that the brain's neural activation dynamics, characterized by the EEG signals, can be interpreted as an energy exchange mechanism between different neurons (or groups of neurons). Thus, a model of the energy exchange between groups of neurons can be replaced by a model of the energy exchange between groups of oscillators. Hence we claim that our model reveals energy exchange features present in the form of neurons interacting as coupled oscillators.

brain regions (Ho et al. 2005) and explain the partial coherence of EEG signals (Pessa and Vitiello 2004).

Hence, we study the temporal EEG signal dynamics recorded from a single channel. Since the surface area of one EEG channel electrode is relatively small when compared to the total surface area of the head, we can say that the analyzed EEG signal is generated by only a very small part of all neurons. In this regard, we explore two different size groups of oscillators, which form a single closed oscillator group, i.e., a closed system U consisting of two open systems W and w; see Fig. 5.1:

$$U = W + w,$$
$$N_U = N_W + N_w, \tag{5.2}$$

where U is a closed global system, W—an external open system, w—an internal open system, N_U—the number of global oscillators, N_W—the number of external oscillators.

Note that for a closed system, the energy conservation law is valid and that for each of the two open systems the energy conservation law is not valid:

$$E_U = E_W + E_w, \tag{5.3}$$

where E_W denotes the energy of the external system, E_w denotes the energy of the internal system, and E_u denotes the energy of the global system.

One of the two open oscillator groups we call the internal system, and the other—the external system; see Fig. 5.1. Note, that the number of external oscillators is much higher. In this case, we calculate the energy exchange between the internal and the external oscillators, where each oscillator exchanges energy with any other remaining oscillator.

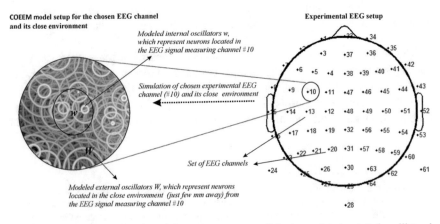

Fig. 5.1 The experimental and modeling setup: systems of the internal and external oscillators'

Actually, Fig. 5.1 reflects the fact that we were modeling only a small brain area, very close to the chosen EEG channel. COEEM has not been performed to model the connectivity distant brain areas. In fact, COEEM model calculations are localized at a chosen EEG channel area around 1 cm in diameter. Therefore, there is no sense in calculating signal delays. In principle, the COEEM model is not designed for the accurate topological imaging of brain activity across large areas of the cranium. Instead, it provides an approximate model, which allows us to prognosticate the temporal evolution of a real EEG signal at the chosen localized area in the brain.

The time-evolution of the simulated EEG data is modeled as the total energy fluctuations of an internal oscillator (or their system). The dependence of energy on time is derived as follows: the modulus squared of the EEG data is integrated over the intervals of a certain short time. For the internal oscillators, the dependence of energy on time is obtained by summing up the energies of all the internal oscillators at certain moments. This procedure follows an iterative scheme (see Fig. 5.2) in which the COEEM model uses experimentally measured EEG data and finds the time dependences of the amplitudes and phases of the external and internal oscillators. The iterative scheme is as follows:

Step 1   The phases and amplitudes of all the oscillators are assigned random values. The distribution functions of random values are similar to the uniform distribution. That is, all random numbers are generated with approximately equal probability.

Step 2   The amplitudes of all the oscillators are normalized to the total energy of all the internal oscillators. Hence, we then get the amplitudes and phases of all the oscillators at $t = t_1$, see Fig. 5.2.

Step 3   We solve the COEEM model equations by using the RK4 method (see the next section). The solution gives the amplitudes and phases of all the oscillators after time-step Δt at $t = t_2$.

Step 4   The amplitudes of all the internal oscillators are replaced in such a way that the total energy of the internal oscillators is equal to the value of the EEG power at $t = t_2$:

$$A'_k = \sqrt{\frac{E_{\text{EEG}}(t_2)}{N_w}} \exp[i \, \arg(A_k(t_2))] \tag{5.4}$$

Step 5   We solve the COEEM model equations by using the RK4 method (back propagation: $\Delta t := -\Delta t$). After this step, we have the modified amplitudes and phases of all the oscillators at $t = t_1$.

Step 6   We reset these modified amplitudes of the internal oscillators to the initial values at $t = t_1$[3] see Fig. 5.2.

---

[3]We cannot replace the amplitudes of internal oscillators because the initial conditions of internal oscillators must remain the same in the first layer.

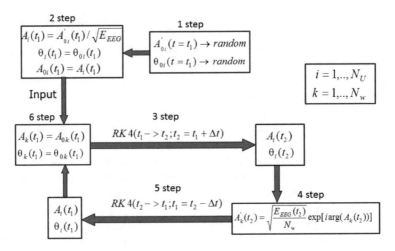

Fig. 5.2 The iterative scheme: E_{EEG} is the energy of the EEG signal, RK4—the fourth-order Runge–Kuta algorithms, A and θ denote complex amplitude and phase, respectively

The third, fourth, fifth, and sixth steps are repeated until the amplitudes of the internal oscillators come close to the initial values at $t = t_1$ and close to the values at $t = t_2$ (where the squared sum of the modulus is equal to the experimentally measured EEG signal power values). After this iterative algorithm, we have the initial conditions (initial amplitudes and phases) for the external oscillators at $t = t_1$. After these conditions are met, the sum of the energy of the internal oscillators is close to that of the EEG energy at $t = t_2$.

In this way, this iterative algorithm is repeated for the other initial conditions of the external oscillators found at $t = t_2, t_3, \ldots, t_n$ (second layer, third layer, etc.). Thus, this iterative scheme finds the time dependences of the amplitudes and phases of the external oscillators for the chosen EEG channel data. In the next section, we discuss how to improve COEEM model for prognostication of long periods. Whereas, our simulation results are presented in the Sect. 5.4.

5.3   COEEM Refind with Narrow Spectral Bands Superposition Approach

COEEM model originally was designed for prognostication of EEG signals of broad spectral ranges but short periods (several miliseconds). In this section, though, we use the COEEM model for prognostication of EEG signals with close to constant frequencies and maximums of amplitudes for long periods (several seconds). Actually, EEG signals' amplitudes and frequencies have small fluctuations only for very narrow spectral widths (bands). In this sense, our idea is to apply COEEM prognostication model for broad spectral ranges, using superposition of narrow spectral bands prognostication results.

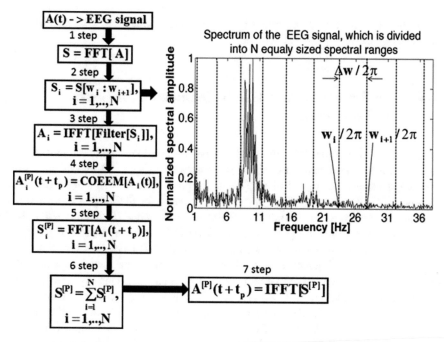

Fig. 5.3 Principal scheme of narrow spectral bands superposition approach, which is used for long time prognostication

As our simulations showed, narrow spectral bands' prognostication yields very good results (see next section). Consequently, superposition of such results for all spectral bands also yields very promising prognostication results for even much longer periods. In this way, we succeeded to get much better EEG prognostication results for longer periods. Below, we present the applied methodology in more details.

This methodology consists of the following steps, see Fig. 5.3:

Step 1 Original EEG signal is transformed to the spectral representation using fast Fourier transformation (FFT).

Step 2 The obtained spectral representation is divided into N equally sized narrow spectral bands.

Step 3 After filtration and inverse fast Fourier transformation (IFFT) for each narrow spectral band, we obtain N temporal EEG signals.

Step 4 Each temporal EEG signal (N) is prognosticated employing COEEM.

Step 5 Fast Fourier transformation is applied for each prognosticated temporal EEG signal in order to get N spectral representations.

Step 6 Superposition is applied for the obtained N spectral representations.

Step 7 After IFFT of the superposed spectral representation, we obtain prognosticated temporal EEG signal for the wide spectral range.

In the section below, we present COEEM model applications with and without the above-described principal scheme.

5.4   The Results of the COEEM Simulation for Short and Long Periods

The study was performed in two stages. First, we employed an iterative COEEM scheme for the short term EEG signal prognostication (up to 92 ms) using a non-filtered spectral range of 1–512 Hz (including power line noises) and fourth-order Runge–Kuta algorithms (RK4). In the second stage, for the long term prognostication we employed an iterative COEEM scheme using (i) filtered EEG data (without power noise) in the range of 1–38 Hz and (ii) superposition of prognostication results for very narrow spectral bands of 0.01–0.1 Hz, (see Fig. 5.3).

Our EEG experimental results, discussed below, were obtained using the BioSemi ActiveTwo Mk2 system with 64 channels. This system records high-quality (resolution LSB = 31.25 nV) and low-noise (total input noise for Ze < 10 kΩ is 0.8 µVRMS) electric local field potentials from the surface of the cranium (Metting et al. 1990). The time-density of the recorded signals from the head surface of participants was 1024 points per second for all 64 channels.

The EEG data was recorded for four people, who voluntary attended the experimental sessions.[4] The results of our findings were based on the EEG signals recorded from these people, who were all asked to stay in five different mental states for 30 min (Plikynas et al. 2014).[5]

Prior to our analyses, the EEG signals had been treated to eliminate any artifacts like eye blinking, swallowing, etc. We denoted the set of recorded signals $u^i(t) = 1$, … ,64 for every channel. Afterwards, we calculated the spectral density of the EEG signals, using the FFT and denoting it by $f^i(w)$. The highest value of the frequency was taken to be 38 Hz, in order to eliminate the intervention of electric power lines on 50 Hz and multiple higher frequencies. In order to compare the EEG signal changes in different persons, we normalized the spectral density to unity so that

$$ {}^1\!/_{C^i} \int_{1\,\text{Hz}}^{38\,\text{Hz}} f^i(w)\mathrm{d}w = 1, $$

---

[4]Our EEG measurements were not conducted as medical experiments. According to local law, prior consent from volunteers was therefore not required. All the volunteers were healthy 30–40 years old men.

[5]EEG was measured for the states: meditation, meditation with noise, counting with noise, thinking, with noise. Each state lasted for 30 min with 5 min brake between different states. Experiments were conducted with closed eyes in a soundproof room.

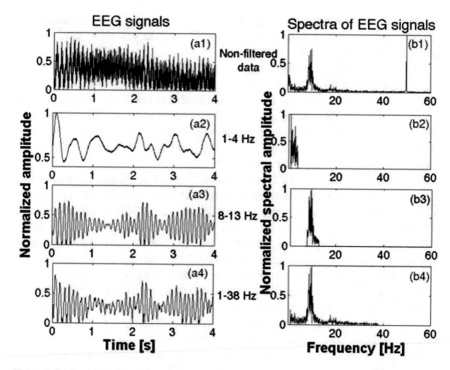

Fig. 5.4 Temporal (*left* side) and spectral (*right* side) representation of the EEG signal for a nonfiltered EEG signal (A). Data represents EEG oscillations for Person #1 in the state of thinking

where $1/C^i$ is the normalization constant. Afterwards, the separation of different brain waves was conducted for the following frequencies: $\Delta = 1$–4 Hz, $\Theta = 4$–8 Hz, $\alpha = 8$–13 Hz, and $\beta = 13$–38 Hz.

For the illustration of obtained results below, we picked few experimental EEG signal sets: (A), (B), (C), and (D).[6] Signal (A) corresponds to the state of thinking (channel #10; analyses started 300 s after the beginning of the session). EEG signal (B) corresponds to the state of meditation (channel #30; analyses started after 420 s from the beginning of the session). The EEG signal (C) corresponds to the state of meditation and oddball count (channel #10; analyses started 140 s after the beginning of the session). The EEG signal (D) corresponds to the state of thinking and oddball count (channel #15; analyses started 50 s after the beginning of the session).

Hence, in Fig. 5.4, we illustrate obtained temporal and spectral representations of the signal (A): for nonfiltered representation graphs (a1) and (b1), for frequency range 1–4 Hz graphs (a2) and (b2), for frequency range 8–13 Hz graphs (a3) and (b3), for frequency range 1–38 Hz graphs (a4) and (b4).

---

[6](A), (B) and (C) signals were recorded for Person #1 and (D) signal for Person #2.

Below, we offer a step-by-step description of the optimization procedure and the results of the short time prognostication (up to 92 ms) using COEEM model for the nonfiltered EEG signal (A).

At the initial time moment, $t = t_0$, we do not know anything about the amplitudes and phases of the external oscillators. Their values are obtained with the help of the experimentally observed EEG curve. For that purpose, we used an iterative scheme of the COEEM model, see Fig. 5.2. This iterative algorithm generates the initial conditions (initial amplitudes and phases) of external oscillators at the initial moment in time. In this way, the temporal dynamics of total power for internal oscillators is calibrated by the experimental EEG curve.

In our initial experimental setup, the number of internal oscillators N_W is equal to 1, the number of external oscillators N_W is equal to 15, and the time-step Δt is equal to 0.25 ms.[7] Calibration was performed for a different number of iterations: the number of iterations is equal to 3 in case (a), the number of iterations is equal to 6 in case (b), and the number of iterations is equal to 16 in case (c), see Fig. 5.5. When the number of iterations increases, the calibrated curve comes closer to the original curve. Hence, the phases/amplitudes of external oscillators converge towards fixed values. For our further estimates, we used a setup with the number of iterations 16 because under such conditions the COEEM curve is sufficiently calibrated to the experimental EEG curve, see Fig. 5.5c. Note that for the simpler COEEM calculations we used only one internal oscillator, which corresponds to one EEG channel, see Fig. 5.1.

After proper calibration, we can state that the COEEM simulated curve, which is generated by one internal oscillator,[8] well matches with the experimental EEG curve, see Fig. 5.5. Our findings revealed that each external uncalibrated oscillator could radically change the COEEM generated signal. In this sense, the whole system of oscillators has to be well calibrated. Hence, it looks like each external oscillator potentially contains information about all the rest of the system. It should be noted, that (i) phase differences of all the external oscillators' does not change with time, (ii) all external oscillators' are coupled and their phases are locked. The phase locking of all external oscillators is determined by the COEEM model and its parameters. In short, these results practically illustrate a coupled oscillator (PLM).

Hence, we calibrated all external oscillators using the chosen EEG data range [0, 1388] ms. After proper calibration,[9] the COEEM simulated curve was fitted to the experimental curve in the chosen data range. At the end of the calibration step we depicted the data range [268, 1000] ms for validation of the prognostication

---

[7]A chosen number of external oscillators (15) were calibrated, but during the optimization step we found that only a few of them were actually needed in order to get the best prognostication results. As a matter of fact, before the optimization process was complete, we did not know how many of the calibrated external oscillators should be used for the prognostication.

[8]The number of internal oscillators can be chosen freely. For reasons of simplicity we chose just one internal oscillator.

[9]The calibration process can be associated with a learning process, i.e. external oscillators are learning to oscillate according to the experimental EEG pattern.

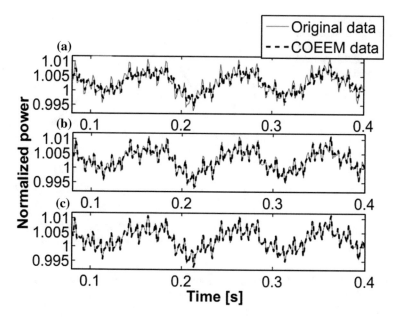

Fig. 5.5 The COEEM curve calibrated to the experimental EEG curve: the number of iterations is equal to 3 in case (**a**), the number of iterations is equal to 6 in case (**b**), and the number of iterations is equal to 16 in case (**c**). EEG data is nonfiltered

results. Using the test and trial method, we observed better prognostication results applying this particular data range, though it can be freely chosen.

It should also be noted that not all calibrated external oscillators were used for the prognostication. For this reason, we applied another optimization step to find an optimal number of calibrated external oscillators. We used time interval [1388, 1398] ms to find an optimal number of calibrated oscillators. For this particular EEG data set it equaled 5. Below, we present prognostication optimization depending on the number of calibrated oscillators, see Fig. 5.6.

As we can observe from the Table 5.1, the best prognostication estimates in terms of Pearson correlation coefficient and MSE were obtained when the number of the calibrated external oscillators equals to 5.[10] As we can see from Table 5.1, the best obtained correlation coefficient for the prognostication period 35 ms equals 0.80 and for period 92 ms equals 0.76.

---

[10]MSE is often used to measure prognostication errors (Blinowska and Malinowski 1991; Hoon et al. 1996; Kim et al. 2013), but this is an absolute measure, which makes it of little use for straightforward comparisons with estimates from other data. Instead, a phase-locking factor (PLF) is often used to measure synchronization for EEG data comparisons (Bonita et al. 2013). However, according to our estimates, PLF changes little for highly correlated data. That is, we got PLF values for our data close to 1 (ranging from 0.9998 to 1.0000). Hence, it does not work well as a discriminating factor. Therefore, we preferred to use the dimensionless and relative Pearson correlation coefficient.

Fig. 5.6 Dependency of
prognostication correlation
coefficient from the number of
calibrated oscillators. EEG
data is nonfiltered

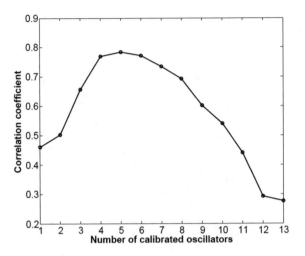

Table 5.1 Values of correlation coefficient and mean squared error for the chosen periods of prognostication (35 and 92 ms). EEG data is nonfiltered

	(a)	(b)	(c)
Prognostication of 35 ms (Fig. 5.7)			
Mean squared error (MSE)	0.43E−06	2.36E−06	1.52E−06
Correlation coefficient	0.80	0.12	−0.30
Prognostication of 92 ms			
Mean squared error (MSE)	0.70E–06	2.39E−06	1.72E−06
Correlation coefficient	0.76	0.19	−0.17

Next, we present results of the prognostication for the experimental EEG data using the optimized (calibrated) COEEM model. Prognostication results were obtained starting from the 1398 ms time mark, see Fig. 5.7.

As we already mentioned, in order to effectively prognosticate EEG signals for long periods of time, we (i) filtered out noise (leaving 1–38 Hz spectral range), (ii) applied COEEM for the narrow spectral bandwidths (0.01–0.1 Hz) prognostication, (iii) performed superposition of the obtained (prognosticated) narrow bandwidth spectra, (iv) used an inverse Fourier transformation to obtain prognostication of the temporal EEG signal in the spectral range (1–38 Hz), see Fig. 5.3.

Below we illustrate the intermediate results of the EEG signal (A) prognostication for one narrow spectral bandwidth, see Fig. 5.8. Here the filtrated narrow bandwidth is 0.1 Hz. In this regard, the EEG signal is very close to a harmonic sinus function (monochromatic wave).

The prognostication correlation coefficient for this narrow bandwidth equals 0.9944. In this way, we got good prognostication results for just one narrow spectral bandwidth. For instance, in the case of the 1–38 Hz spectral range, we would have 370 such narrow spectral bands. All these narrow spectral bands have to be

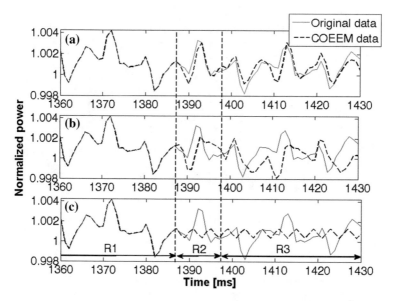

Fig. 5.7 Calibration (R1), optimization (R2), and prognostication (R3) of the COEEM curve versus the original nonfiltered EEG data. Number of the calibrated external oscillators equals to 5 in case (**a**), to 15 in case (**b**), and to 10 in case (**c**)

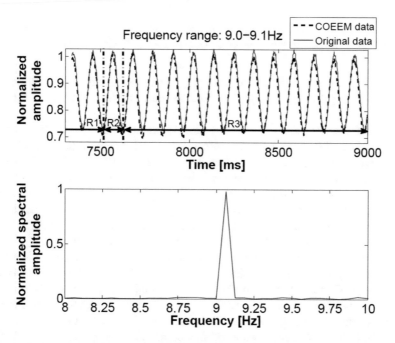

Fig. 5.8 Prognostication of the filtered EEG signal (picture on the *top*) for the spectral range from 9.0 to 9.1 Hz (picture on the *bottom*). Calibration range R1 equals to 7500 ms, optimization range R2 equals to 100 ms and prognostication range R3 equals to 1400 ms

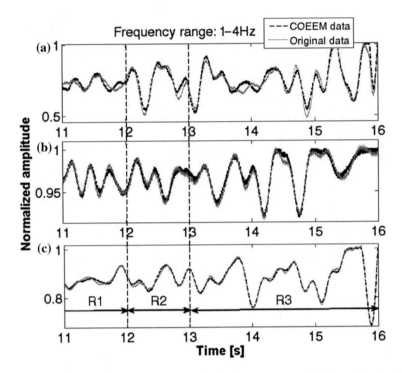

Fig. 5.9 Filtered EEG signals (A), (B), and (C) in the frequency range 1–4 Hz: calibration (R1), optimization (R2), and prognostication (R3) range (prognostication period 3 s)

prognosticated using COEEM. Then the prognostication results are superpositioned (see Fig. 5.3) and with the help of the Fourier transform inversed back to the time scale.

We did this procedure for longer, i.e., 3 s duration filtered (1–38 Hz) EEG signals. We used R1 = 12 s for calibration, R2 = 1 s for optimization and R3 = 3 s for prognostication. The whole spectral range (1–38 Hz) was divided into the 0.07 Hz almost monochromatic spectral bandwidths. Then we made prognostications for each narrow band and superpositioned these results to get a wide (1–38 Hz) spectral range. Below the presented prognostication results illustrate different brain waves and mind states: for the delta waves (1–4 Hz), see Fig. 5.9; for the alfa waves, see Fig. 5.10; for the whole spectral range, see Fig. 5.11.

Correlation coefficients between the prognosticated curve and the original EEG signal at frequency range 1–4 Hz for cases (A), (B) and (C) are, respectively, 0.9590, 0.9840, and 0.9961, see Fig. 5.9. Very similar values of correlation coefficients are obtained for the same EEG signals at the other frequency ranges: from 8 to 13 Hz and from 1 to 38 Hz, see Table 5.2, Figs. 5.10 and 5.11. As we can see from Table 5.2, the largest value of the prognostication correlation coefficient equals 0.9961 at the frequency range 1–4 Hz and the smallest value equals 0.9297 at the frequency range 8–13 Hz.

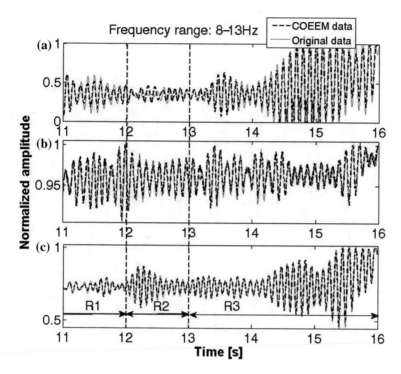

Fig. 5.10 Filtered EEG signals (A), (B), and (C) in the frequency range 8–13 Hz: calibration (R1), optimization (R2), and prognostication (R3) range (prognostication period 3 s)

In short, high values of the correlation coefficients show that prognoses of a 3 s time period at different ranges of frequencies for all three EEG signals (A), (B), and (C) is quite accurate, i.e., we obtain good matching between the COEEM prognosticated data and the original EEG data, see Figs. 5.9, 5.10, 5.11 and Table 5.2.

For the effective implementation of the proposed approach, we also investigated how prognostication results depend on the chosen narrow spectral bandwidth, which is used to get almost monochromatic waves for the narrow band prognostication, see Fig. 5.12.

Figure 5.12 indicates that the correlation coefficient acquires values between 0.92 and 0.99, when the spectral bandwidth is less than 0.1 Hz (see bottom graph). It therefore seems obvious, that for better prognostication results we have to choose a narrow bandwidth, which in this illustrated example ranges from 0.05 to 0.07 Hz. Contrary, in the case of a 0.7 Hz bandwidth, prognostication correlation may be lower than 0.2. As we can see from the same figure, very similar tendencies hold for all cases (A), (B) and (C).

Next, we present a few prognostication results, applying the same procedure of superposition to filtered (1–38 Hz) very long, i.e., 28 s time periods. For instance,

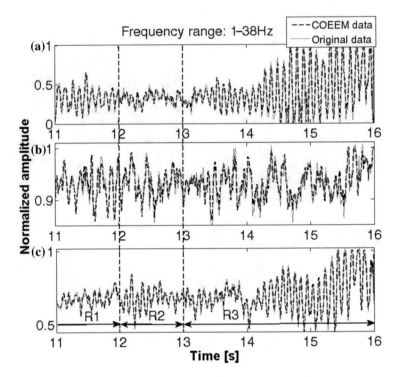

Fig. 5.11 Filtered EEG signals (A), (B), and (C) in the frequency range 1–38 Hz: calibration (R1), optimization (R2) and prognostication (R3) range (prognostication period 3 s)

Table 5.2 Correlation coefficients for various states and frequency ranges (prognostication period 3 s)

States	Correlation coefficients		
	1–4 Hz	8–13 Hz	1–38 Hz
(A)	0.959	0.9589	0.9821
(B)	0.984	0.9297	0.9562
(C)	0.9961	0.9952	0.9893

Fig. 5.13 illustrates prognostication correlation 0.9744 between the COEEM and real EEG data curves.

Figure 5.14 illustrates prognostication correlation 0.9829 between the COEEM and real EEG data curves. The bottom graph in Fig. 5.14 illustrates almost perfect matching between the prognosticated and real EEG signal at the end of the prognostication period.

Figure 5.15 illustrates prognostication correlation 0.9640 between the COEEM and real EEG data curves.

The above examples illustrate that the prognostication correlation coefficient is very high (0.96–0.98) even for quite long prognostication periods. Regarding the

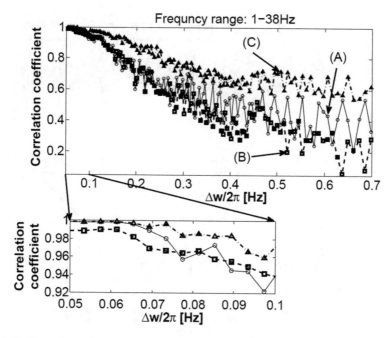

Fig. 5.12 Dependence of prognostication correlation coefficient from the size of spectral bandwidth for states (A), (B), and (C) at frequency range 1–38 Hz and prognostication time period 3 s

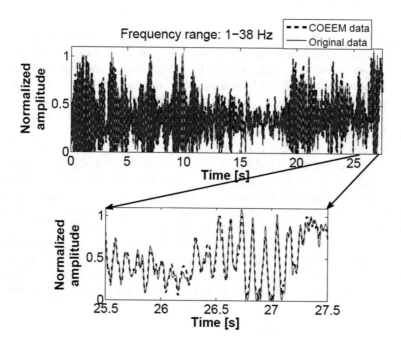

Fig. 5.13 Prognostication of the EEG signal (A) at frequency range 1–38 Hz for R3 = 28 s time period (R1 = 36 s, R2 = 1 s)

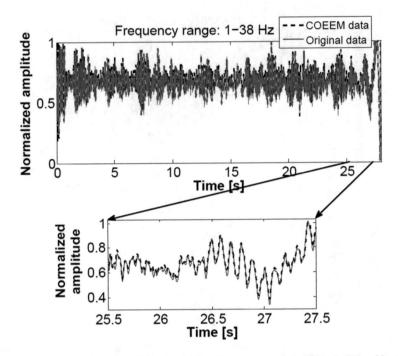

Fig. 5.14 Prognostication of the EEG signal (C) at frequency range 1–38 Hz for R3 = 28 s time period (R1 = 36 s, R2 = 1 s)

relation between the prognostication period and the correlation coefficient, we conducted an additional investigation, which is presented in Fig. 5.16. Here we observe counterintuitive results. That is, the prognostication correlation (i) for the short periods is much more volatile than for the long periods of time, (ii) remains high and stable for quite long periods of prognostication. Regarding these results, it is evident that the proposed approach for longer prognostication periods is able to smooth out prognostication errors.

In another set of experiments, we revealed the relation between the size of the chosen narrow spectral bandwidths and the superpositioned prognostications results, see Fig. 5.17. Our findings show a clear tendency: the smaller the bandwidth, the better the superpositioned prognostication results. For instance, for the EEG signal (D), we obtained correlation coefficients 0.9534, 0.6407, and 0.2495 using bandwidths 0.065, 0.167, and 0.500 Hz accordingly, see Fig. 5.17.

In sum, based on the results provided in this section, the authors argue that the proposed prognostication approach generates very promising results. Our online virtual lab for the interactive testing and modeling of the proposed COEEM approach is available at http://vlab.vva.lt/ (login as Guest, password: guest555).

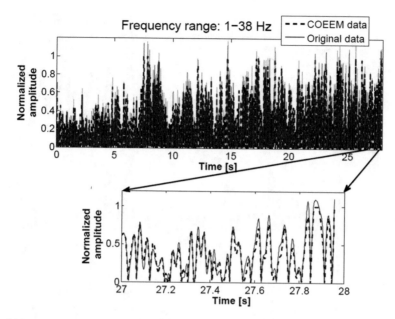

Fig. 5.15 Prognostication of the EEG signal (D) at frequency range 1–38 Hz for R3 = 28 s time period (R1 = 36 s, R2 = 1 s)

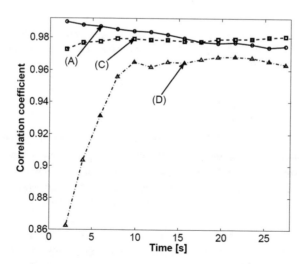

Fig. 5.16 Relation between prognostication period and correlation coefficient for EEG signals (A), (C), and (D) at frequency range 1–38 Hz

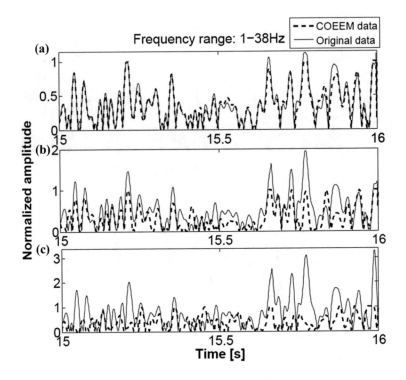

Fig. 5.17 Relation between superpositioned prognostication correlation and chosen narrow spectral bandwidth for the EEG signal (D). Chosen narrow spectral bandwidths: **a** 0.065 Hz, **b** 0.167 Hz and **c** 0.500 Hz

5.5 Concluding Remarks

This chapter addresses an attempt to narrow the gap between theory and experiments in the field of simulation of human brain EEG signal dynamics. After careful revision of simulation and prognostication methods, we have elaborated on oscillation-based modeling and prognostication of short (several ms) and long (several seconds) time series of the real EEG signals at the localized brain areas (chosen EEG channels).

For this task we created and employed an iterative COEEM model, which is somewhat similar to the Kuramoto model but essentially differs in that the Kuramoto model oscillator amplitude is constant, while in the COEEM model it is dependent on the phase of the oscillator. In this regard, the coupled oscillator PLM has been proposed. The complex amplitudes, which represent EEG data, exhibit phase locking. We argue that such a coupled oscillation-based simulation approach reveals phase locking and energy exchange features which are characteristic of human brain EEG signal dynamics. In this way, the chosen approach gets much closer to the most difficult remaining problems in the field of neuroscience, i.e., the explanation of the dynamic relationships between different brain regions and the explanation of the partial coherence of EEG signals.

Hence, in this study we searched for the optimal *modus operandi* of the proposed COEEM approach. First, we calibrated the system of external oscillators, and then optimized them for the best prognostication performance. Afterwards, we followed the simulation study of the chosen EEG channels for stable mind states (A), (B), (C), and (D).

Prognostication was performed in two major stages. First, we employed an iterative COEEM scheme for short time EEG signal prognostication (till 92 ms) using a nonfiltered spectral range of 1–512 Hz (including power line noises) and fourth-order Runge–Kuta algorithms (RK4). In the second stage, for the long time prognostication we employed an iterative COEEM scheme using (i) filtered EEG data (without power noise) in the range of 1–38 Hz and (ii) superposition of prognostication results for very narrow spectral bandwidths (0.01–0.1 Hz).

In the first experimental research stage, short time EEG signal prognostications (up to 92 ms) of nonfiltered EEG signals provided correlations to the order of 0.76–0.8. Not quite satisfied with such prognostication results, we elaborated on the improved COEEM approach, which is presented in the second research stage below.

Hence, in the second research stage, we used an improved COEEM scheme, which is designed for the accurate prognostication of long periods. According to the new scheme, we (i) filtered out noise (leaving 1–38 Hz spectral range), (ii) applied the iterative COEEM scheme for the prognostication of narrow spectral bandwidths (0.01–0.1 Hz), (iii) performed superposition of the obtained (prognosticated) narrow bandwidth spectra, (iv) used an inverse Fourier transformation to obtain prognostication of the temporal EEG signal in the spectral range (1–38 Hz).

We applied this improved COEEM scheme for prognostication of filtered (1–38 Hz) 3 s time periods. In short, quite high values of the correlation coefficients (0.92–0.99) at different ranges of frequencies were obtained for all three investigated EEG signals (A), (B), and (C).

For the effective implementation of the improved COEEM scheme, we also investigated how prognostication results depend on the chosen narrow spectral bandwidth. We found that for better prognostication results we have to choose as narrow a bandwidth as possible, i.e., in our case in the range of 0.05–0.07 Hz.

We also applied the improved COEEM scheme for the prognostication of filtered (1–38 Hz) very long, i.e., 28 s time periods. In sum, the prognostication correlation coefficient remained very high (0.96–0.98) even for quite long prognostication periods. Regarding these results, it is evident that the proposed approach for longer prognostication periods is able to smooth out prognostication errors.

In another set of experiments, we revealed the relation between the size of chosen narrow spectral bandwidths and the superpositioned final prognostications results. Our findings show a clear tendency: the smaller the bandwidth the better the superpositioned final prognostication results.

In conclusion, based on the promising prognostication results for the real EEG signals, we infer that oscillatory model presented here well simulates phase locking and energy exchange features characteristic to localized human brain activations dynamics.

Additional research needs to be undertaken to examine, in detail, the applicability of the COEEM prognostication approach not only to the chosen (stable) mind states, but also to epileptic seizures and other mental disorders.

In short, investigation and further refinements of this biologically inspired experimental coupled oscillators energy exchange model (COEEM) helped to investigate real mind-field features of our artificially constructed OAM in the OSIMAS paradigm.

Chapter 6
Mind-Field States and Transitions: Quantum Approach

This chapter describes a conceptually novel modeling approach, based on quantum theory, to basic human mind states as systems of coherent oscillation. The aim is to bridge the gap between fundamental theory, experimental observation, and the simulation of agents' mind states. The proposed approach, i.e., an oscillating agent model (OAM), reveals possibilities of employing wave functions and quantum operators for a stylized description of basic mind states (BMS) and the transitions between them. In the OAM the BMS are defined using experimentally observed EEG spectra, i.e., brainwaves (delta, theta, alpha, beta, and gamma), which reveal an oscillatory nature of agents' mind states. Such an approach provides an opportunity to model the dynamics of BMS by employing stylized oscillation-based representations of the characteristic EEG power spectral density (PSD) distributions of brainwaves observed in experiments. In other words, the proposed OAM describes a probabilistic mechanism for transitions between BMS characterized by unique sets of brainwaves. The instantiated theoretical framework is pertinent not only for the simulation of the individual cognitive and behavioral patterns observed in experiments, but also for the prospective development of OAM-based multi-agent systems.

Admittedly, in neurodynamics there are two classes of effects: dynamic and structural. The duration and form of dynamic effects depends on the dynamical stability of the system in relation to perturbations of its states (i.e. how the system's trajectories change with the state). Structural effects depend on structural stability (i.e., how the system's trajectories change with the parameters). Systematic changes in the parameters can produce systematic changes in the response, even in the absence of input (David 2007).

© Springer International Publishing Switzerland 2016
D. Plikynas, *Introducing the Oscillations Based Paradigm*,
DOI 10.1007/978-3-319-39040-6_6

The proposed OAM approach has the potential to address both classes of neurodynamic effects, i.e., dynamic and structural, because it uses the quantum mechanical representation of self-organized mind states.[1] The OAM simulates the dynamics and structural effects of agents' BMS by employing stylized oscillations-based representations of experimentally observed characteristic EEG power spectral density (PSD) distributions of brainwaves (Buzsaki 2011; Nunez and Srinivasan 2005; Plikynas et al. 2014) in delta, theta, alpha, beta, and gamma spectral ranges.[2] Of course, we infer that PSD patterns can objectively identify BMS (Benca et al. 1999; Lutz et al. 2004; Plikynas et al. 2015).

We can elaborate further that the wave-like nature of coherent human mind-field states (BMS) can be approximated using the wave mechanics approach (Conte et al. 2009), i.e., wave function (also known as state function) and linear operators (Haven and Khrennikov 2013).[3] This assumption is based not only on recent theories such as Pribram and Bohm's holonomic brain theory (Pribram 1991, 1999) and Vitello's dissipative quantum model of the brain (Pessa and Vitiello 2004; Vitiello 2001), but also on recent evidence that shows a "warm quantum coherence" in plant photosynthesis, bird-brain navigation, our sense of smell, and the activities in microtubules of brain neurons (Engel et al. 2007; O'Reilly and Olaya-Castro 2014). This evidence corroborates well with the twenty-year-old 'Orch OR' (orchestrated objective reduction) theory of consciousness proposed by Stuart Hameroff and Sir Roger Penrose (Hameroff and Penrose 2014).

Following our earlier proposed OSIMAS (oscillations-based multi-agent system) paradigm (Plikynas et al. 2014, 2015) and other related research (Adamski 2013; Haven and Khrennikov 2013; Newandee and Reisman 1996; Orme-Johnson and Oates 2009; Pessa and Vitiello 2004), we designed the quantum based approach as a foundation for the OAM model setup.

---

[1]Emmanuel Haven and Andrei Khrennikov's recent monograph "Quantum Social Science" forms one of the very first contributions to a very novel area of related research, where information processing in social systems is formalized by the mathematical apparatus of quantum mechanics (Haven and Khrennikov 2013).

[2]We chose the following basic mind states (BMS)—sleeping, wakefulness, thinking, and resting. We make use of the fact that each BMS has a characteristic brainwave pattern, which can be identified using power spectral density (PSD) distribution analyses (Müller et al. 2008; Plikynas et al. 2014).

[3]In quantum mechanics, the wave function describes the quantum state of a particle and how it behaves in terms of its wave-like nature. Although the human mind consists of billions of particles, we assume that in BMS some sort of coherence mechanism prevails, binding these particles and causing them to oscillate coherently (in unison) as one. Therefore, in simplified terms we assume that in principle wave function can be used as an approximation of the representation of BMS.

6.1   Stylized Quantum States Model

Let us recall that the proposed individual EEG baseline tests were designed to test the basic conceptual OSIMAS assumption that different states of social agents can be represented in terms of coherent oscillations. In fact, EEG experimental results revealed how the theoretically deduced idea of agent as system of coherent oscillations can find solid empirical backup in terms of coherent delta, theta, alpha, beta, and gamma brainwave oscillations, which uniquely describe agents' mind states.

PSD patterns can objectively identify BMS (Wackermann et al. 2003). They provide universality and objectivity for the construction of an OAM. In this regard, it is important to emphasize that in the OAM basic mind states play a key role, as they are objectively measurable and universal in nature.

Hence, any human mind gets into one or another BMS depending on the inner (psychic) and outer (local and nonlocal environment) conditions, denoted in the OAM as C_{in} and C_{out} accordingly

$$\text{BMS}_n = f_n(C_{\text{in}}, C_{\text{out}}), \tag{6.1}$$

where

$$C_{\text{in}} = \{P_i\}, \quad C_{\text{out}} = \{P_j\},$$

here n is the number of the basic mind state $n = 1, 2, 3,$ and 4 (sleeping, wakefulness, thinking, and resting accordingly), $\{P_i\}$ is set i of inner parameters, and $\{P_j\}$ is set j of outer parameters. In the sense of parameters, BMS can be interpreted as the attractor basin (see Fig. 6.1).[4]

As we already mentioned in the introduction, we elaborate further that the wave-like nature of coherent human mind-field states (instantiated as BMS) can be approximated using the wave mechanics approach, i.e., wave function (also known as state function) and linear operators (Benenson et al. 2006; Buzsaki 2011; Conte et al. 2009). As was mentioned above this assumption is based on recent evidence that demonstrates "warm quantum coherence" in plant photosynthesis, bird-brain navigation, our sense of smell, and the microtubules of brain neurons (Engel et al. 2007; O'Reilly and Olaya-Castro 2014).[5] Following the 'Orch OR' (orchestrated objective reduction) theory of consciousness proposed by Stuart Hameroff and Sir Roger Penrose (Hameroff and Penrose 2014) and recent experimental evidence,

---

[4]In dynamic systems, an attractor is a set of properties toward which a system tends to evolve, regardless of the starting conditions of the system (Osipov et al. 2007). Property values that get close enough to the attractor values remain close even if slightly disturbed.

[5]In quantum mechanics, the wave function describes the quantum state of a particle and how it behaves in terms of its wave-like nature. Although the human mind consists of billions of particles, we assume that in BMS some sort of coherence mechanism prevails that binds these particles and causes them to oscillate coherently (in unison) as one. Therefore, in simple terms wave function in principle can be used as an approximation for the representation of BMS.

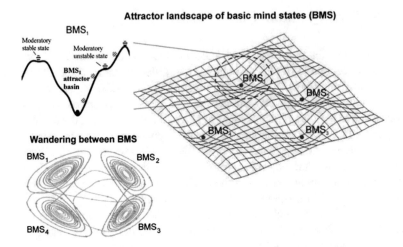

Fig. 6.1 Basic mind states: stylized illustration of attractor landscape, attractor basin and wandering between basic states

EEG rhythms (brain waves) derive from deeper level vibrations of microtubules (protein polymers inside brain neurons) in the megahertz frequency range,[6] which govern neuronal and synaptic function and connect brain processes to self-organizing processes on a fine scale (Georgiev and Glazebrook 2006). Hence, according to Orch OR theory, consciousness depends on biologically 'orchestrated' coherent quantum processes in collections of microtubules within brain neurons. These quantum processes correlate with, and regulate, neuronal synaptic and membrane activity. The continuous Schrödinger evolution of each such process terminates in accordance with the specific Diósi–Penrose (DP) scheme of 'objective reduction' (OR) of the quantum state (Hameroff and Penrose 2014). This orchestrated OR activity (Orch OR) is taken to result in moments of conscious awareness and choice. The DP form of OR is related to the fundamentals of quantum mechanics.

Consequently, in our conceptual model, following Orch OR theory, we assume that a Schrödinger type partial differential equation can be fitted for the description of the temporal evolution of mind-field states (BMS in our model case)

$$i\ell \frac{\partial}{\partial t}\psi = \hat{H}\psi, \tag{6.2}$$

---

[6]Despite a century of clinical use, the underlying origins of EEG rhythms remain a mystery. However, microtubule quantum vibrations (e.g., in the megahertz frequency range) appear to interfere and produce much slower EEG "beat frequencies." Clinical trials of brief brain stimulation—aimed at microtubule resonances with megahertz mechanical vibrations using transcranial ultrasound—have shown reported improvements in mood (Hameroff and Penrose 2014b).

where i is the imaginary unit, ℓ is constant, ψ is the wave function and $\hat{H}$ is the Hamiltonian operator, which is used to describe the total energy of the wave (state) function. Hence, we use the principle of the conservation of total energy E, which is equal to system's kinetic energy E_k plus potential energy E_p

$$H = E_k({}^{\backslash}m, {}^{\backslash}v) + E_p(r, t) = E, \qquad (6.3)$$

where ${}^{\backslash}m$ and ${}^{\backslash}v$ represent stylized mass and velocity in our model and r represents the virtual space coordinates. The interpretation of E_k and E_p in the framework of the proposed model is discussed later.

For the sake of clarity, the conceptual model rests upon the simplest expression for the wave function as the superposition of plane monochromatic waves with the discrete wave vector k

$$\psi(r, t) = \sum_{n=1}^{\infty} A_n(t) e^{i(k_n \cdot r - w_n \cdot t)}, \qquad (6.4)$$

where A_n denotes real amplitudes and w denotes the angular frequency of the plain wave.[7] It is important to note that in the OAM the wave function ψ can be derived via the superposition of brainwaves. Hence, each BMS has a mind-field described by the composition of characteristic brainwaves (delta, theta, alpha, beta, and gamma), which via superposition produce the brainwave function denoted as ψ_{BW}. Hence, the filtered EEG spectrum produces our wave function. In this way, the experiments meet the theory. We use the state (wave) function to represent the probability amplitude of finding the system in some particular BMS.

In contrast, an operator is a function acting on the BMS. Because of its action on a particular BMS, another BMS is obtained. Hence, we model the mind-field in the form of wave function ψ and use $\hat{Z}$ as linear operator for some observable Z,[8] which in our case represents one or another experimentally measured BMS

$$\hat{Z}\psi = z\psi, \qquad (6.5)$$

where z is the eigenvalue of the operator $\hat{Z}$, i.e., observable Z has a measured value of z (such as energy, entropy, etc.). If this relation holds, ψ works as the eigenfunction of $\hat{Z}$. Conversely, if observable does not have a single definite value, measurements of the observable Z will yield a certain probability for each eigenvalue. It can also be written using eigenvector representation

---

[7] The extension to three dimensions is straightforward, all position operators are replaced by their three-dimensional expressions and the partial derivative with respect to space is replaced by the gradient operator.

[8] The complex wave function contains information, but only its relative phase and magnitude can be measured. An operator $\hat{Z}$ extracts this information by acting on the wave function ψ.

$$\hat{Z}\psi = \hat{Z}\psi(r) = \hat{Z}\langle r|\psi\rangle = \langle r|\hat{Z}|\psi\rangle,$$
$$z\psi = z\psi(r) = z\langle r|\psi\rangle = \langle r|z|\psi\rangle, \tag{6.6}$$

where $|\psi\rangle$ is an eigenvector for our wave function ψ. Due to linearity, the above expression provides vectors that can be defined in any number of dimensions, where each component of the vector acts on the function separately (Benenson et al. 2006). We exploit this property, assuming two virtual dimensions in our model, i.e., inner (psychic) and outer (local and nonlocal environment). Consequently, for our two-dimensional virtual space we get

$$e\hat{Z} = \sum_{j=1}^{2} e_j\hat{Z}_j = e_1\hat{Z}_1 + e_2\hat{Z}_2, \tag{6.7}$$

where e denotes a chosen unit vector, and e_1 and e_2 are basis vectors. In our model case, an operator $\hat{Z}_j$ is the function acting in the virtual space of mind-field states. Because of its action on one mind-field state, another mind-field state is obtained. Hence, operator $\hat{Z}_j$ transforms mind states.

In our model, mind states are conditioned by the inner (psychic) and outer (local and nonlocal environment) factors. In this regard, mind states altering the inner (psychic) C_{in} and outer (local and nonlocal environment) C_{out} conditions are attributed to the $\hat{Z}_1$ and $\hat{Z}_2$ operators accordingly. Hence, we can rewrite Eq. 6.7 as follows:

$$\hat{Z} = e_{in}\hat{C}_{in} + e_{out}\hat{C}_{out}, \tag{6.8}$$

where e_{in} and e_{out} are basis vectors corresponding to each component operator $\hat{C}_{in}$ and $\hat{C}_{out}$ of inner and outer virtual spaces.[9] Now Eq. 6.5 can be rewritten:

$$\left(e_{in}\hat{C}_{in} + e_{out}\hat{C}_{out}\right)\psi = (e_{in}c_{in} + e_{out}c_{out})\psi,$$
or
$$\langle r|e_{in}\hat{C}_{in} + e_{out}\hat{C}_{out}|\psi\rangle = \langle r|e_{in}c_{in} + e_{out}c_{out}|\psi\rangle, \tag{6.9}$$

where c_{in} and c_{out} correspond to the eigenvalues (measurable values) of the two-dimensional operator $\hat{Z}$.

The expectation (mean) value of a particular BMS depends on the operator $\hat{Z}$. It is the average measurement of an observable for the particular BMS in the

---

[9]In the OAM model, inner and outer conditions are located in the virtual space, where the inner dimension is orthogonal to the outer. Hence, both dimensions are located not in the physical space but in the mental, i.e., mind-field space.

Common mind-field space in spherical coordinates (r,θ,φ)

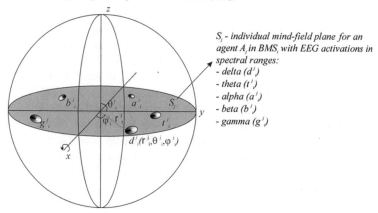

Fig. 6.2 Illustration of the common mind-field space in the adapted spherical coordinates (r, θ, φ) and depiction of the individual mind-field plane S_j for agent A_j, characterized by the EEG-recorded brainwave amplitude A (depicted as r in spherical coordinates) and frequency v (depicted as φ in spherical coordinates) for delta, theta, alpha, beta, and gamma spectral ranges

mind-field region R. The expectation value $\langle \hat{Z} \rangle$ of the operator $\hat{Z}$ is calculated using the equation

$$\langle \hat{Z} \rangle = \int_R \psi^*(r)\hat{Z}\psi(r)\mathrm{d}^3 r = \langle \psi | \hat{Z} | \psi \rangle. \tag{6.10}$$

This can be generalized to any function F of an operator $\hat{Z}$:

$$\langle F(\hat{Z}) \rangle = \int_R \psi^*(r)[F(\hat{Z})\psi(r)]\mathrm{d}^3 r = \langle \psi | F(\hat{Z}) | \psi \rangle.$$

Leaving mathematical notation aside, Fig. 6.2 helps to visualize an agent's A_j individual BMS in the common (collective) mind-field space. This illustration is based on Eq. 6.9 than local and nonlocal environment factors (in other words, neighboring and distant agents accordingly) do not influence a mind state of the chosen agent A_j, i.e., when observable $\hat{C}_{\text{out}} = 0$, which provides the equation

$$e_{\text{in}}\hat{C}_{\text{in}}\psi = e_{\text{in}}c_{\text{in}}\psi,$$

or

$$\langle r | e_{\text{in}}\hat{C}_{\text{in}} | \psi \rangle = \langle r | e_{\text{in}}c_{\text{in}} | \psi \rangle. \tag{6.11}$$

Hence, in this simplified case we ignore the influence of neighboring and distant agents,[10] leaving only inner mind-state factors characterized by the measurable c_{in}, which we associate with the experimentally (EEG) measured individual brain-waves, i.e., their amplitude and frequency (A, v) for the appropriate spectral ranges.

Hence, the visualization proposed in Fig. 6.2 not only shows the momentary individual mind-field activations of the EEG-recorded brainwave patterns (see plane S_j) but is also capable of depicting momentary collective mind-field activations for a group of people. That is, the intrinsic dynamics of the mind-field states of agent A_j are depicted on the individual plane S_j only. Whereas, each agent's individual plane S_j is angled by the θ_j that separates these planes. In this way, individual planes rotate around the x axis (see Fig. 6.2). For instance, for neighboring agents the alteration $\Delta\theta_j$ is small and for distant agents it is large, depending on the chosen "distance" measures.[11]

Theoretically, the proposed visualization can be used to depict the state of collective consciousness only if simultaneous EEG measurements for a large group of people is possible (Plikynas et al. 2014; Standish et al. 2004). Moreover, such an approach could be employed not only for a momentary visualization of mind-field states but also for the dynamic view and even for real-time monitoring. Besides, by analyzing inner spheres shell by shell and by angular regions one could observe and compare not only collective mind states (CMS) but also the tendencies of coherent activations, transitions, and synchronicities in the mind-field of the collective consciousness during, e.g., team work, linguistic communication, behavioral interactions, mass-media broadcasting, and many other cases.

In summary, we base our OAM on the EEG-recorded brainwave dynamics, assuming that there is a direct correspondence between brainwaves and mind states (Fingelkurts and Fingelkurts 2001).[12] In this regard, we base our assumptions not only on our earlier research results but also on the recent advances in brain-imaging techniques, which help to discriminate mind states, e.g., by using EEG spatial and power spectral density (PSD) distributions (Cantero et al. 2004; Müller et al. 2008; Travis and Arenander 2006; Travis and Orme-Johnson 1989).

Hence, the very foundations of OAM construction are in the experimental findings, which show that mind states are dynamic and wander between BMS (for the illustration, see the bottom left-hand corner of Fig. 6.1).

In the wider context of multi-agent system (MAS) development, the proposed OAM modeling is an intermediate step. An OAM-based MAS aims to simulate

---

[10]We reduce the autopoietic properties (Maturana 1980) of the social agent, leaving only the inner network of mind-state transformation processes, which through their interactions continuously regenerate and realize the network of processes (relations) that produced them.

[11]In fact, "distance" can be used in the mental as well as spatial sense, e.g., family ties, psychographic features, demographic, firmographic variables, etc.

[12]At this stage, however, we do not extrapolate on yet another layer of sophistication, which links mind states with observed behavioral patterns. We assume that for the BMS there is a direct correspondence between brainwaves, mind states and behavioral patterns.

CMS that emerge as a result of individual BMS. Naturally, we also take into account that via feedback, CMS influence individual states too.

In reality, a state of mind is a very complex multi-dimensional matrix of many parameters (Buzsaki 2011; Kurooka et al. 2000; Nunez and Srinivasan 2005). In this regard, for the sake of simplicity we propose to employ just few of the most fundamental factors in our model, i.e., energy and information. That is, kinetic/potential energy and entropy/negentropy. We will elaborate on these fundamental factors in more detail in the next section.

6.2 Constraints and Conditions

For the sake of clarity, let us have a stylized example. We will formalize an agent's A_i transition between the first and second BMS,[13] e.g., $BMS_1 \rightarrow BMS_2$. Hence, first we admit that any transition between corresponding BMS can take place only if certain marginal (state-related) conditions are reached. In our case, we denote $M(1, 2)$ as marginal conditions, which initialize the transition between states

$$BMS_{n=1} \xrightarrow{M(1,2)} BMS_{n=2}, \tag{6.12}$$

where $n = 1, 2, 3, 4$ indicates BMS, i.e., sleeping, wakefulness, thinking, and resting, accordingly.

Next, constructing an OAM requires an answer to the following fundamental question—what is the essential nature of these BMS? As mentioned earlier, nontangible mind states cannot be described in terms of particle physics, as they are not of a material but rather a wave-like nature (Hameroff and Penrose 2014; McFadden 2002; Nunez and Srinivasan 2005). In this sense, for the BMS definition we proposed the concept of mind-field, which takes an observable form of coherently radiated energy in delta, theta, alpha, beta, and gamma spectral ranges (see Eq. 6.11). In other words, each BMS can be recognized by its characteristic PSD (power spectral density) distribution.

Hence, this clue gives rise to the idea that *each BMS is a different mental (cognitive) process P,[14] which consequently generates an experimentally measurable characteristic PSD distribution pattern* (2008; Plikynas et al. 2014). Hence, Eq. 6.12 becomes

$$(P_1 \Rightarrow PSD_1)_{BMS_1} \xrightarrow{M(1,2)} (P_2 \Rightarrow PSD_2)_{BMS_2}. \tag{6.13}$$

---

[13]In order to simplify the modeling, at this stage we will investigate just one case of transition from one state to another, without paying attention to other possible transitions or loops.

[14]In general, inner mental processes include perception, introspection, memory, creativity, imagination, ideas, beliefs, reasoning, volition, and emotion. However, in our case we use the most basic states of the mind in the sense of human cognitive behavior.

In essence, any process is related to the canonical transformation of inputs to outputs. In the case of cognitive processes, two fundamental transformations take place. One is related to the transformation T_E between kinetic E_k and potential E_p energies and the other T_N is related to the conversion between entropy S and negentropy N

$$(P_{[T_E,T_N]_1} \Rightarrow \mathrm{PSD}_1)_{\mathrm{BMS}_1} \xrightarrow{M(1,2)} (P_{[T_E,T_N]_2} \Rightarrow \mathrm{PSD}_2)_{\mathrm{BMS}_2},$$

where transformations are conditioned by (see Eq. 6.3)

$$\begin{aligned}
T_E(E_k, E_p) &: E_k + E_p = U, \\
T_N(S, N) &: S_{\max} - S = N,
\end{aligned} \tag{6.14}$$

where U is the total internal energy of the system, which for closed systems is constant, $S_{\max}$ denotes the maximum entropy and N denotes negentropy.[15] Following the principles of thermodynamics,[16] there is a close connection between free energy A and negentropy N (Benenson et al. 2006)

$$\begin{aligned}
A &\equiv U - TS, \\
A &\equiv U - T(S_{\max} - |N|),
\end{aligned} \tag{6.15}$$

where T is the temperature and U denotes internal energy (the total energy contained in a thermodynamic system), which is the sum of E_k and E_p

$$A \equiv E_k + E_p - T(S_{\max} - |N|). \tag{6.16}$$

The thermodynamic temperature

$$T = \left(\frac{\partial U}{\partial S}\right)_{V,N}, \tag{6.17}$$

which provides for Eq. 2.14

$$A_U \equiv U - \left(\frac{\partial U}{\partial S}\right)_{V,N} S,$$

---

[15] According to Willard Gibbs, negentropy is the amount of entropy that may be increased without changing an internal energy or increasing its volume. In other words, it is the difference between the maximum possible entropy, under assumed conditions, and its actual entropy. Usually negentropy is denoted as negative; therefore, we use the modulus value.

[16] We have adapted one of the major laws of thermodynamics because of its universal applicability to interacting systems of particles, which resemble the brain's NNs on a large scale.

or in the expressed form

$$A_U \equiv E_k + E_p - \left(\frac{\partial(E_k + E_p)}{\partial S}\right)_{V,N} (S_{\max} - |N|). \tag{6.18}$$

This expression is fundamental to the OAM. It provides us with all the fundamental variables needed and establishes relations between them. Let us examine this in more detail. Hence, free energy A is the source of useful work, which expresses itself in the expendable energy forms E_k and E_p. However, the nature of the source of free energy is beyond the current scope of our research. In this regard, we will only mention that all processes in the brain are fed by the bloodstream, which supplies the brain's neural networks (NNs) with the necessary biochemical raw energy sources, i.e., flow of nutrients Ω. The stream of nutrients provides the inflow of free energy $\dot{A}_\Omega$

$$\begin{aligned} A_\Omega &= f(\Omega), \\ \dot{A}_\Omega &= f'(\Omega). \end{aligned} \tag{6.19}$$

Hence, the inflow of free energy $\dot{A}_\Omega$ balances the expendable outflow $\dot{A}_U$ producing

$$\dot{A} = \dot{A}_\Omega - \dot{A}_U, \tag{6.20}$$

which, if $\dot{A}_\Omega > \dot{A}_U$ results in $\dot{A} > 0$, i.e. an accumulation of free energy A in the dynamic system, and if $\dot{A}_\Omega < \dot{A}_U$ results in $\dot{A} < 0$, i.e. a drain of free energy A in the dynamic system.

Now let us examine the expendable side A_U in the free energy expression (see Eq. 6.18), which is used for E_k and E_p. Both energy forms compose internal energy U and can be described by the Hamiltonian operator H (see Eqs. 6.2 and 6.3). In the proposed OAM, the kinetic energy E_k of the brain's NN system expresses the sum of the kinetic motions of all the particles of the system; whereas the potential energy E_p expresses the potential expressed via the electromagnetic field, which is measured by EEG.[17] As noted earlier, the filtered EEG spectrum produces our wave function. Hence, we can say that E_p denotes mind-field energy and E_k corpuscular kinetic energy.

It is important to observe in Eq. 6.18 that the right-hand part of the equation, i.e. $\left(\frac{\partial(E_k + E_p)}{\partial S}\right)_{V,N} (S_{\max} - |N|)$ is not expendable (cannot be exploited for useful work)

---

[17]EEG registers the synchronized activity of membrane potentials or rhythmic patterns of action potentials for large numbers of neurons (neural ensembles) close to the surface of the scalp (Buzsaki 2011b). However, in some our later research we assume that electromagnetic fields are also a form of kinetic expendable energy, whereas, potential energy is not expressed; it is stored and ready to be used in the kinetic form.

in any way. It follows from the universal entropy maximization principle for closed systems, which is practically manifested in the above-mentioned principle of minimization of free energy A. In this way, during the process of transformation part of the free energy becomes heat that is of no use.

In Eq. 6.18 the partial derivatives $\left(\frac{\partial E_k}{\partial S}\right)$ and $\left(\frac{\partial E_p}{\partial S}\right)$ represent the relationship between changes in kinetic and potential energy with respect to the change in entropy (in thermodynamics these ratios have a meaning of temperature; see Eq. 6.18). Multiplied by the entropy, they provide heat energy. In both cases, these relationships are nonlinear and interrelated. Concrete analytical forms, however, depend on the modeling setup.

The OAM aims to model the dynamics of transitions between BMS. In this regard, temporal dynamics of the OAM can be formally described by

$$A(t) \equiv E_k(t) + E_p(t) - \left(\frac{\partial(E_k(t) + \partial E_p(t))}{\partial S(t)}\right)_{V,N} S(t),$$

$$A(t) \equiv E_k(t) + E_p(t) - \left(\frac{\partial(E_k(t) + \partial E_p(t))}{\partial(S_{max} - |N(t)|)}\right)_{V,N} (S_{max} - |N(t)|).$$

$$(6.21)$$

However, this expression has to be further clarified in order to take closed-form analytical expression in the concrete applied case. It requires additional explanations regarding how these fundamental variables can be interpreted for the BMS_n and transitions between them. Hence, Table 6.1 provides a few stylized considerations in this regard.

A concrete OAM simulation model can be constructed based on the interrelationships provided in Table 6.1. For the effective implementation the above-mentioned parameters have to be adjusted so that marginal conditions and universal laws of entropy and energy minimization apply. Let us discuss few implications of the proposed model below.

The OAM model is designed to simulate mind transitions between BMS, which act like attractors in the state space; see Fig. 6.1 (bottom left). The crux of the proposed dynamic simulation approach rests in the observation that the mind never returns to exactly the same BMS_n, that is, $BMS_n(t_i) \neq BMS_n(t_j)$. It means that each basic mind state has a set of characteristics (a unique composition of E_k, E_p, T and S), which make even the same BMS_n each time different.

Let us recall that transitions between BMS_n are initialized after reaching marginal conditions (see Table 6.1). However, the act of transition from one state to another is realized via the previously mentioned set of operators $\hat{Z}_{i-j}$ (see Fig. 6.3).

The observable operators $\hat{Z}$ can be "back engineered" from the measured values z, which are eigenvalues of the operator $\hat{Z}$ (see Eq. 6.5). In practical terms, we can employ such a scheme

Table 6.1 BMS$_n$ processes and conditions for transitions between states

BMS$_n$	A(t)	$E_k(t)$ and $E_p(t)$	N(t)	T^*	Marginal conditions and transitions**
BMS$_1$ (deep sleep)	Positive growth: $\dot{A} > 0$, Strong increase: $A \equiv \int \dot{A} dt \to A_{max}$	E_p active: $E_k(t) \to (E_k)_{min}$ $E_p(t) \to (E_p)_{max}$	Positive growth: $|\dot{N}| > 0$ Strong increase: $|N| \equiv \int |\dot{N}| dt \to |N_{max}|$ $|N_{max}| < S_{max}$	$\lim_{N \to N_{max}} T \to T' \geq 1$	Marginal conditions: $A(t) \xrightarrow{T' \cdot S \to min} A_{max}$ or N_{max} Transitions: $A_{max} : BMS_1 \to BMS_2$ $N_{max} : BMS_1 \to BMS_3$
BMS$_2$ (physically active wakefulness)	Negative growth: $\dot{A} < 0$ Strong decrease: $A \equiv \int \dot{A} dt \to A_{min}$ $A \geq 0$	E_k active: $E_k(t) \to (E_k)_{max}$ $E_p(t) \to (E_p)_{min}$	Negative growth: $|\dot{N}| < 0$ Strong decrease: $|N| \equiv \int |\dot{N}| dt \to |N_{min}|$ $|N_{min}| > 0$	$\lim_{N \to N_{min}} T \to T''$ $T'' \cdot S \to max$	Marginal conditions: $A(t) \xrightarrow{T'' \cdot S \to max} A_{min}$ or N_{min} Transitions: $A_{min} : BMS_2 \to BMS_1$ $N_{min} : BMS_2 \to BMS_4$ $A_{min}, N > \bar{N} : BMS_2 \to BMS_3$
BMS$_3$ (thinking)	Negative growth: $\dot{A} < 0$ Strong decrease: $A \equiv \int \dot{A} dt \to A_{min}$ $A \geq 0$	E_k and E_p active: $E_k(t) \to (E_k)_{max}$ $E_p(t) \to (E_p)_{max}$	Negative growth: $|\dot{N}| < 0$ Strong decrease: $|N| \equiv \int |\dot{N}| dt \to |N_{min}|$ $|N_{min}| > 0$	$\lim_{N \to N_{min}} T \to T'''$ $T''' \cdot S \to max$	Marginal conditions: $A(t) \xrightarrow{T''' \cdot S \to max} A_{min}$ or N_{min} Transitions: $A_{min} : BMS_3 \to BMS_1$ $N_{min} : BMS_3 \to BMS_4$ $N_{min}, A > \bar{A} : BMS_3 \to BMS_2$
BMS$_4$ (resting)	Positive growth: $\dot{A} > 0$, $A \equiv \int \dot{A} dt \to \bar{A}$	$E_k(t) \to \bar{E}_k$ $E_p(t) \to \bar{E}_p$	Positive growth: $|\dot{N}| > 0$ $|N| \equiv \int |\dot{N}| dt \to \bar{N}$	$\lim_{N \to \bar{N}} T \to {}'T \geq 1$ ${}'T \cdot S \to {}'\bar{T} \cdot \bar{S}$	Marginal conditions: $A(t) \xrightarrow{{}'T \cdot S \to {}'\bar{T} \cdot \bar{S}} \bar{A}$ or $\bar{N}$ Transitions: $\bar{A}, \bar{E}_k : BMS_4 \to BMS_2$ $\bar{A}, \bar{E}_p : BMS_4 \to BMS_3$ $\bar{N}, A < \bar{A} : BMS_4 \to BMS_1$

$^*T$ denotes $\left(\frac{\partial(E_k(t) + E_p(t))}{\partial(S_{max} - N(t))}\right)$; see Eq. 6.21

**The probabilities of transitions are shown in decreasing order; in each transition the expected probabilities also depend on $E_p(t)$, $E_k(t)$ and $T \cdot S$

Fig. 6.3 Illustration of
transitions between BMS_n,
where corresponding
transitional operators are
denoted as $\hat{Z}_{i-j}$

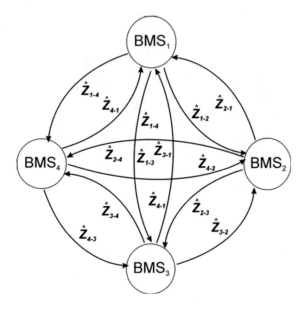

$$f(PSD_n) \rightarrow z_n \rightarrow \hat{Z}_{i-j}, \qquad (6.22)$$

where PSD_n denotes the experimentally measured (EEG) power spectral density.
After doing so, we obtain

$$\psi \xrightarrow{\hat{Z}_{i-j}} BMS_n. \qquad (6.23)$$

It is quite natural that transitions of the mind-field in the state space begin from
the chosen initial BMS and proceed to the other BMS, depending on the OAM
setup (see Table 6.1). The probabilistic nature of the parameters (in Eq. 6.21)
makes the state dynamics look like a complex branching tree; see Fig. 6.4.

Figure 6.4 gives an idea of how an agent's mind travels from state-to-state. The
transitions between states are initiated whenever marginal conditions are met. In
such a model the cyclical nature of state dynamics occurs naturally. In this sense, it
resembles agents' real cycles, e.g., after sleeping people are in a waking or thinking
state, after a period of time they have to rest or get some sleep, and afterwards they
are ready to work again, etc.

We should also note that time-by-time bifurcations occur. This happens when
one state can lead to two other states (see Table 6.1) such as those depicted in
Fig. 6.4.

$$BMS_2(t)|_{A_{\min}} \begin{array}{l} \rightarrow P_1(BMS_1) \\ \rightarrow P_2(BMS_3) \end{array}, \qquad (6.24)$$

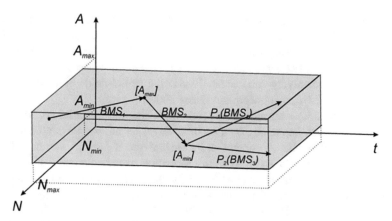

Fig. 6.4 Illustration of the state dynamics in the dimensions of free energy A, negentropy N and time t. States change whenever marginal conditions, i.e. A(min, max) or N(min, max) are reached (see the *shaded rectangular* "tunnel")

where P_1 and P_2 denote the respective probabilities of transition from the first to the second and third states. Indeed, after several transitions, bifurcations make the probabilities of finding a mind-field in a particular state too small. This can be explained in terms of the complexity of the state space (despite the fact that we used only a few BMS).

In the OAM transitions from one state to another depend only on the current state and not on the sequence of events that has preceded it. The proposed scheme of state dynamics conforms to the Markov chain in mathematics. In fact, directional graphs with vertices (BMS as processes) and nodes (transition points between different BMS) can be employed to visualize state dynamics as well.

Mathematical notation aside, the motivation behind the involvement of intangible negentropy can also be explained. We refer to our previous research (see Chap. 2), where negentropy is indirectly exhibited in the mind-field context as ordered oscillations (PSD distributions), which essentially differ from the even white noise spectra (Nunez and Srinivasan 2005). In a general sense, we can interpret negentropy as a source of self-organized coherent brainwaves, or mind-field in short.

Following the OSIMAS paradigm and other related research, we do not refute the possibility of some sort of entanglement between individual mind-fields. This could lead to the explanation of some not only paired but also collective coherent states. This assumption, however, requires further empirical research so that a conclusive and experimentally proven theory can evolve.

We also argue that the universal law of increasing entropy (in closed systems) matches its opposite law of increasing negentropy for self-organizing living (open) systems. It is truism that negentropy can be preserved in living systems because of the constant influx of free energy. Some metabolic biological processes perform useful work in order to use this free energy to maintain negentropy, i.e., to export entropy away [see "heating pump" (see Chap. 2)]. This is especially relevant to the highly organized and coherent central nervous system, in which complicated brain processes are effectively orchestrated via the mind-field biofeedback loop (McFadden 2002; Orme-Johnson and Oates 2009).

Based on the above considerations, we imply that the unconscious "orchestration" process is also happening on a social scale. After all, people devote their free energy to maintain one or another order on the social scale. Consequently, the process of negentropy is involved in the formation of the coherence of individual mind-fields on the social level (Paolo and Jaegher 2012; Meijer and Jakob 2013). Coherence of the individual mind-field states in social mediums is imperative to maintain social order. This leads to the distributed cognition approach, which can be exploited in nonlocal multi-agent systems (Chatel-Goldman et al. 2013). Simulations under correct assumptions can help us understand how distributed cognitive processes synchronize and actually work in coherence.

In sum, we assumed that the wave-like nature of coherent human mind-field states can be approximated by taking the quantum mechanics approach. That is, in the proposed model the wave-like nature of human mind-field states (instantiated as basic mind states—BMS) is approximated using a quantum wave function (also known as state function) and linear operators.

We envisage the stylized wave function as a superposition of experimentally observable EEG brainwaves. Such an approach somewhat bridges the gap between EEG spectral representations and quantum theory. It does so by reducing complex mind-field states into the observable form of BMS, which are commonly characterized by the well-known EEG brainwaves (delta, theta, alpha, beta, and gamma). Superposition of these brainwaves produces a stylized state (wave) function. Hence, each BMS is a specific mental (cognitive) process, which as a state function generates experimentally measurable characteristic power spectral density (PSD) distribution patterns.

In other words, we propose to use state (wave) function to represent the probability amplitude of finding the system in particular BMS. In addition, operators are acting on the BMS to produce the dynamics of these states. In the OAM for the sake of simplicity we propose to employ just a few constraining factors, i.e., energy (kinetic versus potential energy) and information (entropy versus negentropy).

The proposed visualization technique shows not only momentary individual mind-field activations of the brainwave patterns but also momentary collective mind-field activations for a group of people. Hence, the instantiated theoretical framework is pertinent not only to the simulation of the experimentally observed individual cognitive and behavioral patterns, but also to the prospective development of OAM-based multi-agent systems, in which the coherence of the individual mind-field states in social mediums is imperative to maintain social order. This approach leads to distributed cognition in nonlocal multi-agent systems.

Simulations under the correct assumptions can help us understand how distributed cognitive processes synchronize and actually work in coherence. However, further empirical research is required so that OAM-based conclusive and experimentally proven social simulation theory is able to evolve.

In the next two chapters, we target our efforts toward the expansion of the quantum-based approach to the OAM and OAM-based MAS modeling.

Chapter 7
OAM Simulation Setup and Results Using Entropy and Free Energy Terms

Hereafter, a description and some explanatory sources are provided pertaining to the oscillating agent model (OAM), which is an intrinsic part of the oscillations-based OSIMAS paradigm. Supported by empirical neuroscience observations, the OAM strives to model a social agent in terms of a set of basic mind-brain states and the dynamics between them. The proposed multidisciplinary approach is based on an experimental electroencephalography (EEG) spectral data analysis of brainwaves, which paves the way for a putative link between the conceptual oscillations-based OSIMAS framework and the empirically observed coherent bioelectromagnetic oscillations of ensembles of neurons in the brain (EEG spectra).

7.1 OAM in Terms of Free Energy and Negentropy

Hence, in referring to our earlier quantum mechanics-based OAM approach (Plikynas 2015), we further elucidate the free energy term A, but this time from a classical physics perspective (i.e., thermodynamics and statistical mechanics). According to common understanding, free energy is a universal energy source used for every kind of cellular and consequently neural metabolic activity. We admit that each agent's mind-brain states are essentially dependent on a means of using available free energy. In this way, coordinated neural processes in the mind-brain states can be interpreted in terms of their specific use of free energy.

© Springer International Publishing Switzerland 2016
D. Plikynas, *Introducing the Oscillations Based Paradigm*,
DOI 10.1007/978-3-319-39040-6_7

Consequently, for the representation of the state function, we can apply the basic principles of thermodynamics by employing the free energy term

$$A = U - TS, \tag{7.1}$$

where the free energy term A denotes the amount of work that a thermodynamic system (state function) can perform;[1] U is the total internal energy of the mind-brain system $U > A$, which for closed systems is constant; T is an intensive measure in thermodynamics called temperature, which in a more general sense, is equal to the partial derivative of the internal energy U with respect to the entropy S

$$T = \left(\frac{\partial U}{\partial S}\right)_{V,N}. \tag{7.2}$$

Instead of entropy for the biological systems, in following Schrödinger, 1955, we can also use the term of negentropy, N

$$\begin{cases} |N| = S_{\max} - S \\ S_{\max} = \text{const.} \end{cases} \tag{7.3}$$

Here, $S_{\max}$ denotes the maximum possible entropy in the mind-brain system and N denotes negentropy.[2] For the sake of simplicity, we can interpret negentropy as order and entropy as disorder (see Chap. 2). As the entropy $S \rightarrow S_{\max}$, the order (negentropy) in the system disappears. Whereas, in the case of $S \rightarrow 0$, the negentropy $|N| \rightarrow S_{\max}$.

Admittedly, following the principles of thermodynamics, there is a close connection between free energy A and negentropy N (Benenson 2006)

$$A = U - \left(\frac{\partial U}{\partial S}\right)_{V,N} \cdot S = U - \left(\frac{\partial U}{\partial (S_{\max} - |N|)}\right)_{V,N} \cdot (S_{\max} - |N|). \tag{7.4}$$

In essence, this expression provides an important insight about ambivalent nature of the use of free energy in the mind-brain system. That is, it shows that free energy A (left side of Eq. 7.4) can be exploited for a useful task as a form of the total available internal energy U (right side of Eq. 7.4). However, U is affected by

---

[1]Like the total internal energy, the free energy is a thermodynamic state function. Free energy refers to the energy contained within the system, while excluding the kinetic energy of the motion of the system as a whole and the potential energy of the system as a whole due to external force fields.

[2]According to Willard Gibbs, negentropy is the amount of entropy that may be increased without changing the internal energy or increasing its volume. In other words, it is the difference between the maximum possible entropy, under assumed conditions, and its actual entropy. Usually, negentropy is denoted as negative; therefore, we use the modulus.

the negative second term $\left(\frac{\partial U}{\partial S}\right)_{V,N} \cdot S$, which diminishes total internal energy U. This term, in a marginal case, yields

$$
\begin{cases}
A = U & \text{than} \quad \left(\dfrac{\partial U}{\partial S}\right)_{V,N} \cdot S \Rightarrow 0 \\[3mm]
A = 0 & \text{than} \quad \left(\dfrac{\partial U}{\partial S}\right)_{V,N} \cdot S \Rightarrow U
\end{cases}
\tag{7.5}
$$

Hence, the product of $\left(\frac{\partial U}{\partial S}\right)_{V,N} \cdot S$ plays an important role. The term $\left(\frac{\partial U}{\partial S}\right)_{V,N}$ denotes the sensitivity of the change in total internal energy ∂U with respect to the change of entropy ∂S, see Eq. 7.2. Let us denote this sensitivity of the mind-brain's internal energy to the change in entropy by η. This sensitivity parameter is individual and can vary for each agent, depending on the level of entropy $\eta = f(S)$.

In our model, we also propose two fundamental ingredients for the total internal energy U, i.e., pseudo kinetic E_k and pseudo potential E_p energy

$$
U = E_k + E_p,
\tag{7.6}
$$

where E_k denotes all experimentally measurable corpuscular types of energy at the level of molecular movements. Whereas, E_p denotes a form of the field-like "potential" energy, e.g., electromagnetic field energy emitted in the case of the EEG measurements,[3] i.e., the spectral power of the measured fields of brainwaves (delta, theta, alpha, beta and gamma spectral bands).

Hence, free energy A is the source of useful energy (work), which expresses itself in the expendable energy forms of E_k and E_p. However, the nature of the source of free energy itself is beyond the current scope of our research.[4]

Following the above considerations, we yield a more concise free energy expression

$$
A = U - TS = (E_k + E_p) - \left(\frac{\partial U}{\partial(S_{max} - |N|)}\right)_{V,N} (S_{max} - |N|).
\tag{7.7}
$$

Obviously, this expression exhibits the fact that the mind-brain is not a closed system. Admittedly, such inference stems from the effect of the negative second term TS [see also Exp. (4) and (5)], which pertains to the expenditure of free energy during the mind-brain functioning process and is called heat waste in thermodynamics. This is an inevitable waste of free energy, emitted into the environment in

---

[3] Of course, an EEG only measures a tiny part of the total emitted bioelectromagnetic energy in the entire mind-brain system. However, following recent neuroscience developments, we assume that analyses of this tiny EEG registered part of the entire bioelectromagnetic field present in the brain is capable of differentiating basic mind-brain states (Buzsaki 2011a; Thatcher 2010).

[4] In short, all processes in the brain are fed by the bloodstream, which supplies the brain's neural networks with the necessary biochemical raw energy sources, i.e. flow of nutrients.

the form of bioelectromagnetic fields. In fact, such electromagnetic radiation naturally occurs and it enables, for instance, EEG-based measurements of the bioelectromagnetic fields emitted from the human cranium (Buzsaki 2011).

Hence, mind-brain produced bioelectromagnetic fields are emitted into the environment. However, this is not a one-way road, so to speak. The environment is full of different kinds of electromagnetic fields, including the bioelectromagnetic fields emitted from other humans. Admittedly, these other electromagnetic fields from the environment can affect our mind-brain systems too (Radin 2004; Standish et al. 2004; Travis and Arenander 2006).

Presumably, mind states, experimentally measured as EEG bioelectromagnetic field correlates of consciousness, can be directly (via external electromagnetic field effects on the mind-brain system) affected by the outside natural, social, or technological environment. Let us note the close connection of this effect to the heat waste process, which is about the irreversible loss of free energy radiated outside from the mind-brain in a field-like fashion. A reverse process is similarly at work, i.e., mind-brain absorption of field-like radiation stemming from the external environment. We can denote it by simply adding a heat-like term $\tau \cdot \Delta N$ (see Eq. 7.7)

$$A = U - TS + \tau \Delta N = U - T(S_{\max} - N) + \tau \Delta N, \qquad (7.8)$$

here, ΔN signifies the change in negentropy in the mind-brain system after the absorption of external fields.[5] In the case of $\Delta N > 0$, the inner entropy $S = S_{\max} - |N| - \Delta N$ decreases, and vice versa in the case of $\Delta N < 0$.

The relationship between $S_{\max}, |N|, S$ can be interpreted by using a normalized EEG power spectral density (PSD) plot, where $S_{\max}$ can be represented by the total area of the EEG plot; the summed area of the registered spectral bands represents $|N|$, i.e., the coherent (ordered) oscillations of corresponding ensembles of neurons; and the total EEG plot area (represented by $S_{\max}$) minus the summed area of the registered spectral bands (represented by $|N|$) indicates entropy S, see Fig. 7.1.

Following Fig. 7.1, The PSD distribution indicates coherent oscillations, i.e., negentropy of the mind-brain system in a chosen location (EEG channel) (Plikynas et al. 2014). However, integration of the PSD curve only provides an approximate empirical estimate of the negentropy of the local mind-field. This is mainly due to (i) the normalization problem and (ii), the EEG methodological constraints.[6] In the

---

[5]It is quite natural, that even very weak but coherent external fields can trigger large qualitative changes in the negentropy of the mind-brain system (Radin 2004; Standish et al. 2004a; Travis and Orme-Johnson 1989).

[6]A standard EEG setup measures bioelectric fields on the surface of the cranium. It does not measure electric fields in the deeper layers of the brain. Therefore, we can't say that the obtained PSD represents the entire mind-brain spectral energy distribution. It only shows a tiny part of it. However, most probably due to the holographic nature of the mind-brain system (Pribram 1999), even this small portion of the registered energy distribution is informative enough for the recognition of mind-brain states. Hence, in some way, it reflects the entire mind-brain picture.

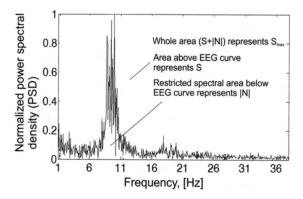

Fig. 7.1 EEG spectrum: an example of normalized power spectral density (*PSD*) distribution. In an empirical sense, negentropy can be understood as order in a form of coherent oscillations of ensembles of neurons. In some sense, the PSD indicates the spectral distribution of negentropy $|N(\omega)|$. Integration of the PSD curve gives the total value of negentropy. The area above the PSD curve indicates entropy S, whereas, the entire area of the EEG plot *gives* $|N| + S = S_{max}$

former case, for the normalization, we used the maximum PSD value that corresponds to the max$|N(\omega)|$, assuming that the entropy at this point approaches $S \rightarrow 0$. However, in reality it approaches $S \rightarrow S_{min}$, where S_{min} is not necessarily close to zero.

In other words, the maximum PSD value for the given mind-brain state indicates the highest possible level of synchronized neural oscillations, i.e., the maximum order or negentropy. In other spectral regions, however, neural oscillations do not disappear. They have similar intensity, but more random phases and, therefore, do not interfere so strongly. These disordered (incoherent) neural oscillations represent entropy S. Hence, the proposed empirical method for the estimation of negentropy/entropy for the mind-brain system is based on the direct estimates of the coherence level from the ordered neural oscillations registered via the EEG spectra. However, due to the above-mentioned drawbacks, such an estimation method can only be considered as a first-order approximation.

Let us remember that the OAM approach is based on the OSIMAS paradigm, where order pertains to the coherent oscillations in the mind-brain system (Fingelkurts et al. 2006; Plikynas et al. 2014; Travis and Arenander 2006; Travis and Orme-Johnson 1989). Hence, a higher order represents a greater synchronization of oscillations of ensembles of neurons, which correspondingly shows up in the EEG brainwave PSD distributions and the summed spectral power estimates (Plikynas et al. 2014; Šušmáková and Krakovská 2008).

Coming back to Eq. 7.8, we added the environmental influence effect (heat gain term $T \cdot \Delta N$) on the individual mind-field.[7] We inferred that the external (physical and social) environment emits fields, which interact with the individual mind-field.

---

[7]The term "mind-field", following McFadden's CEMI (conscious electromagnetic information) field theory, is understood as a field-like correlate of consciousness (McFadden 2002).

This interaction takes place at the level of neural oscillations. Positive interference renders $\Delta N > 0$, then the internal negentropy (ordered neural oscillations) increases and vice versa for $\Delta N < 0$. In short, we infer that fields stemming from the external environment at some sensitive conditions are capable of increasing or decreasing the total internal energy U of the individual mind-field, see Eq. 7.8.

Such an approach provides good correspondence with the numerous, although controversial, experimental findings of, e.g., coherent mind-brain states for distant persons (Haven and Khrennikov 2013; Newandee and Reisman 1996; Nummenmaa et al. 2012; Radin 2004; Standish et al. 2004; Travis and Orme-Johnson 1989). These and many other studies indicate the existence of the putative field-like nonlocal connection mechanism between individual mind-fields.

At this point, we would like to add a few more considerations pertaining to the putative wave-like interaction mechanism (WIM) or entanglement between distant mind-brain systems. Admittedly, the external environment can influence the inner processes of the mind-brain system in a direct field-to-field, i.e., nonsensory way. Following the OSIMAS holistic line of interpretation, very weak external fields are capable of changing, although very slightly, the internal order/disorder (entropy/negentropy) balance in the mind-field. Consequently, via this very sensitive and close psychosomatic relationship between the mind-field and brain, correspondingly gross changes take place at the neuronal activation level, which shows up in the experimental findings too. Hence, the mind is a subtle field-like instrument attuned to influences on the brain. And in the reverse aspect, both the mind and brain are complimentary parts of one holistic mind-brain system (Plikynas et al. 2014; Radin 2004; Standish et al. 2004; Travis and Orme-Johnson 1989).

However, there are many controversial findings in this field of research. Some studies point toward nonlocal interaction effects beyond the classical interpretation of bioelectromagnetics (Chatel-Goldman et al. 2013; Engel et al. 2007; Fingelkurts and Fingelkurts 2001; Haas and Langer 2014). It therefore seems plausible that the mind-field is not only able to interact with the very close, but also with the distant environment. In any case, this challenging field of research requires thorough further investigation.

Hence, the above-mentioned experimental and theoretical mind-field underpinning supports the implementation of the proposed OAM approach. It also supports the previously mentioned conceptual ideas concerning a WIM and pervasive information field (PIF). The WIM was introduced in the Chaps. 2 and 3. In short, it recognises a wave-like interaction between agents, whereas the PIF serves as a virtual field-like information medium, where all information is stored. In this way, contextual information is distributed in fields, and these fields—although expressing some global information—can be locally perceived by agents.

Following the perspective of classical physics, synchronization effects occurring inside an individual mind-field and between such individual mind-fields are neither well investigated nor well understood. However, in one of our earlier research works, we created a coupled oscillators energy exchange model (COEEM), which simulates neural synchronization and energy exchange effects by using a coupled oscillators energy exchange approach (see Chap. 4). By and large, the COEEM simulates experimentally observed human brain EEG signal dynamics. This model is in some ways similar to the Kuramoto model (Pessa and Vitiello 2004), but essentially differs in that with the Kuramoto model, oscillator amplitude is constant, while with the COEEM model, oscillator amplitude is dependent on the phase of the oscillators. The reasoning behind the COEEM model construction is based on an energy exchange and synchronization simulation in a localized brain area using (i) the coupled oscillators approach and (ii) experimental (non-filtered and filtered) EEG observations. In conclusion, based on the promising prognostication results obtained with real EEG signals, we inferred that the oscillatory model well simulates the energy exchange features characteristic to localized human brain activation dynamics.

Enabled by recent neuroscience and especially electroencephalogram (EEG) studies, in one of our other research directions we designed a conceptually novel modeling approach, based on quantum theory, to basic human mind states as systems of coherent oscillation. The aim was to bridge the gap between fundamental theory, experimental observation and the simulation of agents' mind states. The proposed conceptually novel OAM approach deals with the possibilities of employing wave functions and quantum operators for a stylized description of basic mind states (BMS) and transitions between them.

In the OAM, the BMS are defined using experimentally observed EEG spectra, i.e., brainwaves (delta, theta, alpha, beta, and gamma), which reveal the oscillatory nature of agent mind states. Such an approach provides an opportunity to model the dynamics of BMS by employing stylized oscillation-based representations of the characteristic EEG PSD distributions of brainwaves observed in experiments. In other words, the proposed OAM describes a probabilistic mechanism for transitions between BMS characterized by unique sets of brainwaves. The instantiated theoretical framework is pertinent not only for the simulation of the individual cognitive and behavioral patterns observed in experiments, but also for the prospective development of OAM-based multi-agent systems.

Following this line of research, we also foresee an employment of statistical mechanics, where the modeling of complex coordinated dynamics is usually based on mean field theory (MFT) approaches, which consider the interaction of a large number of small individual components (or in our case neurons or their ensembles). The effect of individual neurons on all the other neurons is approximated by a single averaged effect, thus reducing a many-body problem to a one-body problem (Kadanoff 2009).[8] Then it naturally follows that if a neuron's bioelectromagnetic field exhibits many interactions with its neighboring neurons, the MFT will be more accurate for such a system.

This is true in cases of high dimensionality, or when the Hamiltonian includes long-range forces. The free energy of a system with a Hamiltonian of $H = H_0 + \Delta H$ has the following upper bound $F \leq F_0 \overset{\text{def}}{=} \langle H \rangle_0 - TS_0$, where S_0 is

---

[8]The n-body system is replaced by a 1-body problem with a chosen good external field. The external field replaces the interaction of all the other neurons with an arbitrary neuron (Kadanoff 2009).

the entropy and where the average is taken over the equilibrium ensemble of the reference system with a Hamiltonian of H_0. In the special case where the reference Hamiltonian is that of a noninteracting system, it can thus be written as $H_0 = \sum_{i=1}^{N} h_i(\xi_i)$, where ξ_i is shorthand for the degrees of freedom of the individual components of our statistical system (neurons or their ensembles).

A review of the literature has revealed some other related, although, different approaches. For instance, the adaptive resonance theory (ART) (Grossberg 2013) provides functional and mechanistic links between the mental processes of consciousness, learning, expectation, attention, resonance, and synchrony during both unsupervised and supervised learning. A concise review of modeling mental states and a concept of a cyclic mental workplace is given in the Meijer work (Meijer and Jakob 2013). Some other researchers provide preliminary evidence of quantum like behavior in the measurement of mental states (Conte et al. 2009; Conte et al. 2008). Another line of investigation and modeling—studies of the brain as a conscious system (Marcer and Schempp 1998). Whereas, on the social distributed cognitive systems scale, there are also some new research frontiers, e.g., the interactive interpersonal brain (Paolo and Jaegher 2012), nonlocal mind from the perspective of social cognition (Chatel-Goldman et al. 2013), emotional promotion of social interaction by synchronizing brain activity across individuals (Nummenmaa et al. 2012) and many other related multidisciplinary studies.

7.2 Bridging the Gap: Quantum Beats Approach

The mind-brain-body issue has a long history of multidisciplinary investigation. Recently, many research frameworks have been proposed to reconcile the mental and the quantum (physical) states. For an introductory purpose, just a few general concepts about brain functioning based on recent neurobiological investigations and quantum mechanical (QM) approaches are briefly mentioned below. However, quantum mechanics and its formalisms should be viewed hereafter more like metaphors, in that the particular features of quantum theoretical formalism are realized in a nonphysical context (Atmanspacher 2011).

A recent review of quantum modeling of the mental states was made by Meijer and Korf (2013). They noted that some QM theories have been justified by their basic elements of: First, uncertainty, which means that mental phenomena are not governed by classical physical laws of determinism and causation, so that there is room for intuition and free will; Second, a universal consciousness with individual consciousness acting as a participating agent; Third, consciousness, even seen as a nonphysical entity, finally acts in the physical domain and may exert causal power; Fourth, explaining mental transitions from apparently nonconscious thoughts and processes to conscious ones and vice versa; Fifth, a potential difference between internal time perception and external physical time.

As we can notice, QM theories can potentially open a vast new field of research providing a totally novel perspective toward mental phenomena and agency as well.

However, this short current review is less concerned about the quantum theories used for probability-based brain processing (applied for cognitive processes such as decision making). We are interested about quantum theories that see the mind defined by energy/fields to describe and understand the mental aspects of reality (Atmanspacher 2011; Hu and Wu 2010; Tarlacı 2010; Vannini and Orpo 2008).

It is also important to note that the neurobiological and the QM concepts may function in a complementary fashion. The classical neurological interpretation of the brain can therefore be seen as a flow of information through molecular to neuronal networks, whereas quantum approaches are based on more fundamental atomic and subatomic coherent and entangled states implying a nonlocal and unbroken reality. A multilayered organization of the brain is most likely, where the personal mind is mostly envisioned as a local expression of a cosmic (universal) mind through interfacing individual and universal consciousness through the quantum information field (Meijer and Jakob 2013). In fact, it closely resembles the OSIMAS approach, which in that regard, uses a PIF concept (see Chaps. 2 and 3).

A number of theoretical and empirical studies have suggested the appropriateness of the QM approach (Meijer and Jakob 2013). For instance, psychological observations and recent experiments have led to nondeterministic models of brain function based on quantum mechanics and chaos theory (Freeman and Vitiello 2006). Whereas according to Stapp, consciousness can also be usefully construed as the collapse of a superposition of brain states. He provides a number of highly technical examples of exactly how this process might work neurophysiologically (Stapp 2009). Other recent research has been focused on a modeling of decision making (Busemeyer and Bruza 2014), which pertains to QM, because they use terms such as superposition, entanglement, and collapse as they are used in quantum physics to denote an undetermined state (a wave function), which after it collapses, behaves as a particle. The classical experiments of Libet and later by others (Haggard 2005; Libet 2006) might be regarded as decision-making experiments, and their readiness potential can be seen in reflecting the superposition state of the brain. Some have emphasized the continuity of mind, rather than sequential states (Spivey 2008). Their modeling is based on laboratory experiments, which are compatible with the idea that the central nervous system (the subject) develops a kind of superposition state before making a choice (the "collapse"). There are also some other quantum interpretations like quantum holographic criminology, which explains the key concepts of quantum and holography theories, and then shows their relevancy for criminology (Milovanovic 2014). In fact, it is impossible to cover the full range of other related work here, but some of the examples above provide a taste of emergent trends in this highly prospective multidisciplinary research field.

In the light of the other QM concepts, theories and applications, a conceptually novel mind state modeling approach is presented below. It aims to bridge recent neuroscience discoveries with classical physics and QM theory. In this sense, an OAM construction is extended using (i) OSIMAS premises (see Chap. 3) and (ii) the 'Orch OR' (orchestrated objective reduction) theory of consciousness proposed by Stuart Hameroff and Sir Roger Penrose (Hameroff and Penrose 2014).

Following Orch OR, conscious states have been assumed to correlate with physiological EEG oscillations, which might come about, namely as beat frequencies, arising when OR is applied to superposition of the quantum states of slightly different energies.[9]

If there is a state Ψ which is a superposition of two slightly different states Ψ_1 and Ψ_2, each of which would be stationary on its own, but with very slightly different respective energies E_1 and E_2, then the superposition would not be quite stationary. Its basic frequency would be the average $(E_1 + E_2)/2\hbar$ of the two, corresponding to the average energy $1/2(E_1 + E_2)$, but this would be modulated by a much lower classical frequency ('beats') which is the difference between the two, namely $|E_1 - E_2|/h$, as follows, very roughly, from the following mathematical identity

$$e^{ia} + e^{ib} = 2e^{i(a+b)/2}\cos(a-b)/2, \qquad (7.9)$$

where we may take $a = -E_1 t/\hbar$ and $b = -E_2 t/\hbar$ to represent the quantum wave functions of the two energies. The superposition of the complex quantum oscillations has a frequency which is the average of the two, but this is modulated by a classical oscillation as given by the cosine term with a much lower frequency determined by the difference between the QM frequencies E_1 and E_2 of the two individual states Ψ_1 and Ψ_2 (Hameroff and Penrose 2014).

Thus, according to Orch OR, we may consider conscious moments as occurring with the beat frequencies $|E_1 - E_2|/h$, rather than the primary frequencies E_1/h and E_2/h.[10] The primary frequencies may be around 10 MHz, with time periods of $\sim 10^{-8}$ s. Decoherence might need to be avoided for a mere ten-millionth of a second with consciousness occurring at far slower beat frequencies.[11] For example, if E_1/h and E_2/h were 10.000000 and 10.000040 MHz, respectively, a beat frequency of 40 Hz (by $|E_1 - E_2|/h$) could correlate with discrete conscious moments.

---

[9]The specific Diósi–Penrose scheme of 'objective reduction' ('OR') of the quantum state (Hameroff and Penrose 2014a).

[10]Conscious moments are associated with beat frequencies, which indicate the synchronized and modulated firing rate of ensembles of neurons, when the quantum states in the dendrites and soma of a particular neuron could entangle with microtubules in the dendritic tree of that neuron, and also in the neighboring neurons via dendritic–dendritic (or dendritic–interneuron–dendritic) gap junctions, enabling quantum entanglement of the superposed microtubule tubulins among many neurons. This allows (via the synchronized and modulated firing rate of ensembles of neurons) unity and binding of conscious content moment to moment.

[11]For Orch OR, the environmental decoherence, i.e. the interaction of a quantum superposition with its environment 'erodes' the quantum states, so that instead of a single wave function being used to describe the state, a more complicated entity is used, referred to as a density matrix. Indeed, it is essential for Orch OR that some degrees of freedom in the system are kept isolated from environmental decoherence, so that OR can be made use of by the system in a controlled way. If, however, a quantum superposition is (1) 'orchestrated', i.e. adequately organized, imbued with cognitive information, and capable of integration and computation, and (2) isolated from a non-orchestrated, random environment long enough for the quantum superposition to evolve, then Orch OR will occur and this, according to the scheme, will result in a moment of consciousness (Hameroff and Penrose 2014a).

Indeed, a relatively high rate of discrete conscious moments at about 40 Hz is located in the EEG gamma band which is mostly associated with the brain binding into a coherent system for the purpose of complex cognitive or motor functions. Intensive states of conscious alertness need higher rates of conscious moments, i.e., higher beat frequencies. However, in the restful or sleeping states, our brains do not need high conscious alertness. In these states, conscious moments produced by the beat frequencies become less frequent, moving toward the alpha and theta spectral ranges. In the case of deep sleep, conscious moments of alert awareness move into the very infrequent delta spectral range, i.e., almost cease at all.

We construct the OAM model based on the above-mentioned conceptual ideas, which are aimed at the modeling of basic mind states (BMS, e.g., the wakeful, dreaming and deep sleep states) and the transitions between them in terms of the EEG power spectral redistribution. In the OAM, the BMS are interpreted as the above-mentioned beat frequency correlates of consciousness.

Hence, the OAM postulates the following (Plikynas et al. 2014):

- Each BMS is a twofold dynamic process, when (i) following a specific pattern, power spectral energy is gradually redistributed between the spectral ranges, and (ii) coherent information either increases or decreases in the mind.
- Change in the dynamic process (transitions between the states) depends on reaching the marginal constraints of (a) potential energy $\lim E_p = [\max, \min]$, (b) kinetic energy $\lim E_K = [\max, \min]$, and (c) negentropy $N[\max, \min]$.
- The marginal constraints for $\lim E_p = [\max, \min]$ operate in the lower delta, theta and alpha1 spectral ranges [1–10 Hz] and for $\lim E_K = [\max, \min]$, operate in the higher alpha2, beta and gamma ranges [10–40 Hz].
- *The BMS can be characterized as:*

  (a) a deep sleep state, when E_p is exponentially accumulated at the expense of E_k and negentropy N exponentially increases,[12]
  (b) a wakeful state, when E_k is exponentially accumulated at the expense of E_p and negentropy N exponentially decreases,
  (c) a dreaming state, when E_k is linearly accumulated at the expense of E_p and negentropy N linearly increases.

The proposed model extends Orch OR theory by proposing a mechanism for the emergence of two slightly different states Ψ_1 and Ψ_2 with respective energies E_1 and E_2. The corresponding primary frequencies are E_1/h and E_2/h. According to Orch OR theory, the difference between these frequencies, depending on the state, leads to the much lower beat frequencies observed in the delta, theta, alpha, beta, and gamma EEG spectral ranges.

Whereas according to proposed interpretation, there is only one stationary primary state Ψ_p with the respective energy E_p and corresponding primary frequency

---

[12]Negentropy N is a measure opposite to entropy, and can be interpreted as the spectral power dissipation level between the spectral ranges. A low dissipation level indicates high negentropy and vice versa (Plikynas et al. 2014).

$\omega_p = E_p/h$. The other slightly different reflected state is denoted as Ψ_r with the respective energy E_r and corresponding primary frequency $\omega_r = E_r/h$. The reflected state is obtained as the primary state's reflection from environment. Hence, Ψ_r depends directly on environmental factors. The biological mechanism for the evolvement of primary state Ψ_p and its energy and frequency correlates of consciousness has been explicitly described in the Stuart Hameroff and Sir Roger Penrose Orch OR theory[13] (Hameroff and Penrose 2014). However, the Orch OR theory has little to say about the nonstationary reflected state which we denoted as Ψ_r. It assumes the existence of some endogenous factors at the microtubules level which produce two slightly different quantum states, but it does not elaborate as to why such a difference occurs.

We admit that Ψ_r emerges as a consequence of some complex processes taking place in the microtubules of neural cells, when primary inner cellular state Ψ_p interacts with environment ε and produces its own reflected quantum state Ψ_r. Following this line of thought, reflected state Ψ_r can be interpreted as the high frequency (in the megahertz range) correlate of conscious awareness of the environment at the cellular level. Consequently, primary stationary state Ψ_p interacts with its own reflected and slightly perturbed (by the environment) state Ψ_r. In this way, superposed state Ψ emerges, which is the superposition of both primary and reflected quantum states. Like in the case of Orch OR, superposed state Ψ has a frequency that is determined by the average of the two (ω_p and ω_r), but this is modulated by a cos term with a much lower beat frequency determined by the difference between ω_p and ω_r, see Eq. 7.9.

Hence, superposed state Ψ has two levels of correlates of conscious awareness, i.e., at the cellular level in the megahertz range and at the multicellular organizational level of the synchronized and modulated firing (beat frequency) rate [1–100 Hz] of ensembles of neurons (they are called conscious moments in the Orch OR theory). The latter correlate of conscious awareness is achieved by the unity and binding of conscious content with the beat frequency rate, which is recorded in the EEG, fMRI, and MEG signals.

---

[13]They originate from the primary frequencies of deeper level microtubule (protein polymers inside brain neurons) vibrations in the megahertz frequency range, which govern neuronal and synaptic function, and also connect the brain processes to self-organizing processes in the fine scale (Georgiev and Glazebrook 2006b). Hence, according to the 'Orch OR' theory, consciousness depends on biologically 'orchestrated' coherent quantum processes in collections of microtubules within the brain neurons. These quantum processes correlate with, and regulate, neuronal synaptic and membrane activity.

The recent discovery of warm-temperature quantum vibrations in microtubules inside brain neurons by the research group led by Anirban Bandyopadhyay, PhD, at the National Institute of Material Sciences in Tsukuba, Japan (and now at MIT), corroborates the pair's theory and suggests that EEG rhythms also derive from deeper level microtubule vibrations. In addition, work from the laboratory of Roderick G. Eckenhoff, MD, at the University of Pennsylvania, suggests that anesthesia, which selectively erases consciousness while sparing nonconscious brain activities, acts via microtubules in brain neurons.

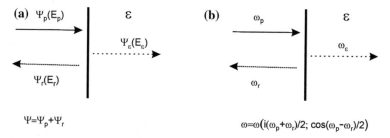

Fig. 7.2 In the figure, area **a** shows the quantum superposed state Ψ of the stationary primary state Ψ_p with the respective energy E_p, and the nonstationary (reflected from environment) state Ψ_r with the respective energy E_r; while area **b** shows the resulting frequencies. The resulting state of lost energy from interacting with environment ε is represented by $\Psi_\varepsilon(E_\varepsilon)$ and the corresponding frequency ω_ε

According to the Orch OR theory, discrepancies between both quantum states take place during a quantum decoherence period starting at around ten-millionth of a second (Hameroff and Penrose 2014). These discrepancies naturally tend to accumulate for longer decoherence periods, producing larger mismatches between both quantum states. The process of quantum decoherence is faster in rapidly changing environments,[14] which initiate complex motor and cognitive activities. In order to maintain increased cognitive activities, the binding of conscious content across the whole brain takes place at a higher rate of beat frequencies, i.e., in the beta and gamma brainwave ranges. However, the process of quantum decoherence decreases considerably during the deep sleep state, which is not perturbed by fast sensory and thinking processes in the brain. During the deep sleep state, the beat frequencies are very low, and standing (resonating) waves occur with a relatively high spectral power in the low frequency delta range.

It is important to note yet another implication. According to the OAM, non-stationary environment ε is a major driving factor, which influences nonstationary state Ψ_r and consequently, quantum superposed state Ψ, see Fig. 7.2.

In principle, the superposition of stationary primary state Ψ_p and nonstationary reflected state Ψ_r proposed here can also be interpreted in holographic terms. However, in this case, the interference of the reference and reflected (from the object) beams is not projected as usual in the 2D plane, but instead takes place in the 3D energy space of the mind-field. Moreover, instead of the beams' interference, we are dealing with the superposition of the quantum functions or so-called quantum states, see Fig. 7.3.

According to Orch OR theory, after the superposition of two similar, but not identical quantum states—in our case it applies to Ψ_p and Ψ_r—low frequency beats are observed (Hameroff and Penrose 2014). We infer that empirically, this can be

---

[14] In defining environmental change, we not only recognize changes in the external, but also in the internal environments, i.e. the body and even the brain itself. For instance, during the dreaming state, we experience wakeful-like states in terms of the observed EEG spectra even though there are no any real physical actions involved, because body consciousness is not active.

Fig. 7.3 Holographic principle in action: 3D hologram formation from the superposed quantum states in the mind-field

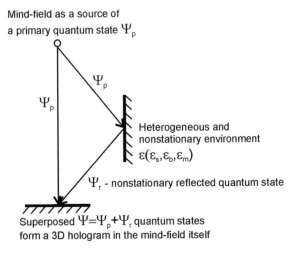

Mind-field as a source of a primary quantum state Ψ_p

Ψ_p

Ψ_p

Heterogeneous and nonstationary environment
$\varepsilon(\varepsilon_s, \varepsilon_b, \varepsilon_m)$

Ψ_r - nonstationary reflected quantum state

Superposed $\Psi = \Psi_p + \Psi_r$ quantum states form a 3D hologram in the mind-field itself

observed in the composition of the low frequency EEG spectra, see Eq. 7.9. In fact, these oscillations are produced by ensembles of neurons in the brain. It is imperative to also notice that due to the superposition of such quantum states, the mind-field undergoes adaptive changes that help to reflect the nonstationary environment in order to construct and store an internal image of external reality. Therefore, the novel interpretation of the Orch OR theory proposed here sheds new light on the longstanding mind-body problem debates.[15]

We maintain that the mind-field state in some way reflects the state of the environment. Moreover, the very closely concerted mind-brain-body system, through genetically inherited and learned psychosomatic connections, reacts to the different states of the mind-field (McFadden 2002). The opposite is also true, states of the body influence the states of the brain and consequently, the mind-field states too. In some sense, the mind-field reflects the environment, whereas, the body through psychosomatic links learns how to react to the different states of the mind-field. Admittedly, this mind and body relation is intermediated by the brain, peripheral nervous system, hormone glands, etc.

Hence, like in the holographic case, a superposed analogy of a quantum state Ψ is obtained when the primary and reflected quantum states are superposed. That is, in the following analogy, quantum state Ψ can be interpreted as a hologram, whereas primary state Ψ_p as a reference beam and state Ψ_r as an illumination beam from the object (the environment). Consequently, the physiological correlates of

---

[15]The mind-body problem in philosophy examines the relationship between the mind (consciousness) and the brain. Today's cognitive science has also become increasingly interested in the embodiment of human perception, thinking, and action. Interest has shifted to interactions between the material human body and its surroundings, and to the way in which such interactions shape the mind or vice versa. Proponents of this approach have expressed the hope that it will ultimately dissolve the Cartesian divide between the immaterial mind and the material existence of human beings (Gallagher 2006).

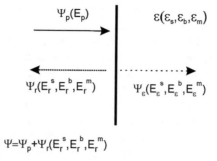

$$\Psi = \Psi_p + \Psi_r(E_r^s, E_r^b, E_r^m)$$

Fig. 7.4 Superposition of the primal Ψ_p and reflected state Ψ_r in the heterogeneous nonstationary environment $\varepsilon(\varepsilon_s,\ \varepsilon_b,\ \varepsilon_m)$ produces complex energy patterns $\left(E_r^s, E_r^b, E_r^m\right)$

consciousness recorded with an EEG and other brain imaging techniques (like fMRI, MEG, etc.) empirically reveal superposed state Ψ as a hologram of the mind-field's reaction to environmental stimuli.

We maintain that environment ε is heterogeneous, i.e., it embraces (1) external sensory environment ε_s, (2) the body as internal environment ε_b, and (3) the mind as the deepest internal environment, ε_m, of stored emotions, feelings, memories, thinking patterns, etc., see Fig. 7.4. All these environments $\varepsilon = \varepsilon(\varepsilon_s,\ \varepsilon_b,\ \varepsilon_m)$ simultaneously affect the primary quantum state Ψ_p, producing reflected state Ψ_r with complex EEG spectral energy patterns $\left(E_r^{\varepsilon_s}, E_r^{\varepsilon_b}, E_r^{\varepsilon_m}\right)$ in the delta, theta, alpha, beta, and gamma ranges.

In Fig. 7.4, the quantum state Ψ_ε is also included, which indicates that some part of primary quantum state Ψ_p is broadcasted into the environment. For simplicity, let assume now that the energy loss $\left(E_\varepsilon^s, E_\varepsilon^b, E_\varepsilon^m\right)$ to the environment is negligible compared with the energy reflected back $\left(E_r^s, E_r^b, E_r^m\right)$.

Keeping in mind the above-mentioned considerations, the dynamics of the mind-field states can be generally described using an iterative mental cycle:

1. For the chosen time moment, the mind-field initiates its primary (referential) and stationary quantum state $\Psi_p(t, \varphi)$; here, φ denotes the characterizing parameter of the current mind-field state.
2. Part of $\Psi_p(t, \varphi)$ interacts with the heterogeneous environment ε (this includes external stimuli, the internal senses of the body and even the mind state as the deepest internal inner environment), which produces the reflected nonstationary quantum state $\Psi_r(t)$.
3. Superposition of the primary Ψ_p and reflected $\Psi_r(t)$ quantum states produces superposed quantum state $\Psi(t) = \Psi_p(t, \varphi) + \Psi_r(t)$; it forms a spatial 3D hologram in the mind-field.[16]

---

[16]In one of our earlier research publications, the initial experimental research results were presented, which testify to the specific redistribution of the EEG power spectral density patterns that takes place during transitions between mind states (Plikynas et al. 2014).

4. The spatial 3D hologram of the mind-field affects the brain's complex neural network, leaving an energy imprint.[17]
5. The brain-body consciously or unconsciously reacts to the induced energy imprint, which can involve activation of the body through the peripheral nervous system using genetically inherited and learned psychosomatic connections.
6. After the induction of the energy imprint and the chain of events described above (primal quantum state → reflected quantum state → superposed quantum state → spatial 3D hologram in the mind-field → energy imprint on the brain's neural networks → body response), one or another response follows from the environment. It can trigger a feedback cycle: body → brain → mind-field.
6. The mental cycle repeats for the next time moment $t + 1$.

Let us see how the proposed scheme may work in the case of remembering (working memory). Undoubtedly, the holographic imprints of the past states are stored in the brain's neural networks. Following the scheme presented above, to activate (recall) a specific memory, i.e., the energy imprinted on the neural networks, it is first required to consciously or unconsciously initiate its primary quantum state $\Psi_p(t, \varphi)$. The presence of the recalled specific Ψ_p automatically affects the spatial 3D hologram in the mind-field. It can be done by recollecting two main parameters—the time moment t and characteristic mind-field state parameter φ. With these two parameters, the primary (referential) quantum state $\Psi_p(t, \varphi)$ can be invoked in the mind-field. Next, $\Psi_p(t, \varphi)$ is superposed on to the existing spatial 3D hologram, which activates specific neural networks and it brings into the foreground the corresponding past mental imprints for the observing subject. In this way, the remembrance of the past experience takes place.

According to the proposed approach, the process of remembrance invokes a past experience in all its levels of self-organization including the physical one (the level of neural network activation). It is a truism in this research field that during the remembrance process, the subject experiences the past mental experience again as if re-living it, although usually with less intensity (Sternberg 2011). Admittedly, similar mental processes take place during REM sleep, dreaming, recognition, etc. However, it is likely that more complicated mental processes take place during perception, thinking, awareness, volition, and emotional arousal. It most probably involves a more complex scheme than the one described above.[18]

---

[17]It works in a similar fashion as the currently available technology of holographic information storage. It records information in a parallel fashion throughout the whole volume of the medium and is capable of recording multiple images in the same area utilizing light at different angles. The so-called 'dynamic range' determines how many holograms may be multiplexed in a single volume of data (currently a few thousand). The stored data is read through the reproduction of the same reference beam used to create the hologram (Glass and Cardillo 2012).

[18]One possible candidate to explain and model more sophisticated mental processes is the GWT— Global Workspace Theory, see Appendix B. The GWT resembles the concept of Working Memory, qualitatively accounting for a large set of matched pairs of conscious and unconscious processes. The GWT lends itself well to computational modeling. It can also successfully model a number of characteristics of consciousness and trigger a vast range of unconscious brain processes.

It is worth noting that the key ideas contained in the approach presented here are also very close to the well-known autopoietic theory, which deals with the dynamics of the transformation processes of mind states, which through their interactions, continuously regenerate and reorganize the network of processes (relations) that produced them (Maturana 1980; Plikynas 2015). In this regard, it is quite natural to conclude that idea of the cyclical nature of the self-referential mind-field presented above is very close to the autopoietic theory. In this approach, it is also stated that the transformation process of the mind states continuously regenerates itself too.

Hence, the OAM concept is not only very similar in many ways to some of the other cutting edge theories, but it also provides new important implications, which are backed up by the experimental observations. For instance, in neuroscience EEG research, it is a well known fact that that the EEG spectrum has a much higher spectral content located in the lower spectral regions. However, there are no conclusive explanations as to why this is so, whereas the OAM based on the Orch OR theory of quantum beats naturally explains this, see Fig. 7.5.

In short, the superposition differences between the self-centered primary quantum state Ψ_p and its reflections from the multiple environment (Ψ_r^m, Ψ_r^b, Ψ_r^s) produce the beat frequencies ω^m, ω^b and ω^s, respectively. In essence, these coherent beat frequencies represent the mind-field's response to the decoherence rate, which naturally tends to be faster in the outer layers of the mind-field, where changes take place. While in the more centered layers of the mind-field, the need for the tuning process diminishes, e.g., in the state of the deep sleep, the mind-field is totally withdrawn from the senses (i.e., from the outer environment) and is more self-centered, so that the primary quantum state Ψ_p is not disturbed. In the deep sleep case, the superposition of Ψ_p with its very close reflection Ψ_r produces a very low rate of beat frequencies. As they closer approach a perfect match, the rate of the beat frequencies approaches zero. Admittedly, in the deep sleep state, so-called delta brainwaves are observed (0–4 Hz), i.e., beat frequency rates, whereas, in the rest states, the spectral power of the brainwaves slightly shifts towards the theta range. Hence, the hypothesis is that an EEG registers the tuning frequencies of the mind-field state according to the decoherence rate invoked by the changing environment.

Such a hypothesis is able to explain some phenomena, i.e., high activation in the β and γ spectral ranges during active thinking, focus, high-alertness, and complex cognitive functions, see Fig. 7.5. In these mental states, the mind-field has to focus and interconnect multiple complex external mental objects into a meaningful and comprehensive system. That is, during this process, a mainly self-centered mind-field, represented by the primary quantum state Ψ_p, expands outwards into the external layers of the environment and is reflected back in the form of the quantum state $\Psi_r(\Psi_r^m, \Psi_r^b, \Psi_r^s)$. The reflected state Ψ_r undergoes comparatively

---

(Footnote 18 continued)

The GWT also specifies "behind the scenes" contextual systems, which shape conscious content autonomically, without ever becoming conscious, such as the dorsal cortical stream of the visual system (Robinson 2009).

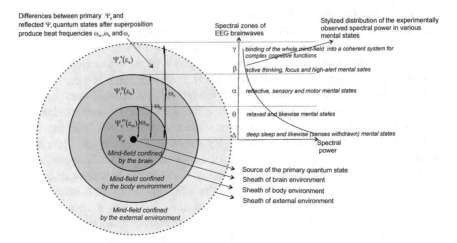

Fig. 7.5 Following the OAM and Orch OR conceptual ideas, an agent is represented as a distributed mind-field based on the cognitive process of primary and reflected quantum state superposition, which produces the low frequency beats empirically registered as brainwaves

larger changes, because it reaches the outer external layers, which can be considerably and continuously affected by the volatile external environment.

Since the decoherence period between Ψ_p and Ψ_r is relatively small, the mind-field tends to increase the intensity and frequency of the external environment scanning process, which is reflected in the relatively higher activation of the β and γ brainwave spectral ranges. However, according to the experimental data, the activated spectral power in these spectral ranges remains much smaller compared to the lower brainwave ranges. This indicates the self-centered nature of the personal mind-field.

It therefore seems plausible that the Orch OR-based quantum beats approach can be related with the experimentally observed EEG spectra for various mental states. However, from the modeling point of view, the question is how to construct such a model. The authors discuss this issue further on in this chapter.

The idea is centered on the creation of a model for relating the phase shift $\Delta\varphi(\omega)$ between the primary and reflected quantum states Ψ_p and Ψ_r, depending on the mind-field state.

7.3   OAM Construction Framework

According to the above argumentation, this section examines the OAM's implementation issues. Before delving deeper into the subject, let us remember that in the general scheme, the OAM is an integral if not central part of the OSIMAS [oscillations-based multi-agent system (MAS)] paradigm. In this sense, the implementation of the OAM is crucial for the effective development of the novel MAS approach and its applications in the social domain.[19]

The main general scheme of the OSIMAS-based OAM implementation framework is provided in Fig. 7.6. It brings into attention three major stages: initiation, mental state process, and outcomes (see on the left side of Fig. 7.6). In short, the initiation stage initializes variables, individual parameters, marginal conditions, and other mind state-related representations. The mental state process stage not only describes the major relationships between the variables, parameters, and marginal conditions, but also provides algorithms for temporal dynamics, transitions, etc. In the outcomes stage, modeling and simulation results are provided there for temporal plots of 2D/3D $BMS(t)$ dynamics, evolution of transitions, and tables of the relationship between simulated bands of oscillations with the Orch OR beat frequencies (Hameroff and Penrose 2014); SPD (spectral power distribution) diagrams for the various BMS (Plikynas et al. 2014), etc.

For the effective implementation of the OAM simulation framework, two major processes are distinguished:

1. The BMS as a process of the mind-field's gradual change in aggregated energy and negentropy over time (see the first column in Fig. 7.6).
2. Transitions between the BMS as a process of abrupt energy and negentropy redistribution when certain marginal conditions are reached (see the second column in Fig. 7.6).

The modeling of the BMS and the transitions between them compose the core of the OAM simulation framework.[20] However, the general scheme presented is

---

[19]Modern ICT allows the almost instantaneous broadcasting of multichannel information. The mass media facilitates excitement through waves of news, commercials, novelties, emotions, fashion, propaganda, etc. that resonate within populations, where the agents' act as resonating mind-fields throughout the worldwide social networks. In this regard, the multidisciplinary OSIMAS approach aims to not only provide conceptual, but also practical insight on how to model these individual and social oscillations as well. It facilitates the development of multiagent systems (MAS) using the many-to-one, one-to-many and many-to-many interaction approach (Kezys and Plikynas 2014; Plikynas et al. 2014; Raudys et al. 2014).

[20]According to the idealistic understanding (in part represented by early platonism, Advaita Vedanta philosophy and the more recent research of W. James, C. Jung, I. Kant, G.W.F. Hegel, J. G. Fichte, Spinoza, F. Wilhelm, J. Schelling, A. Schopenhauer, G. Leibnitz, E. Schrodinger, D. Bohm, etc.) and the OSIMAS paradigm (Plikynas et al. 2014), an individual mind-brain based system is limited spatially (via the brain) and temporally (via an individual lifespan). However, as a tool, the mind-brain based system serves in terms of a local workspace for a universal consciousness, which operates beyond the individual constraints of time and space. In fact, people always experience a very similar set of basic conscious experiences (sensitive sentience, awareness, emotions, wakefulness, having a sense of selfhood, etc.). Animals also have similar experiences. Hence, it can be assumed that consciousness in terms of states does not belong to individuals, although individuals can subjectively experience their consciousness their mind-brain system states. Therefore, we could even infer that consciousness is intrinsic and universal in nature, which is locally expressed and experienced through the individual mind-brains. Our final goal is to construct a simulation platform where individual consciousness works out in terms of the mind-brain states "moving" throughout the manifold universal consciousness.

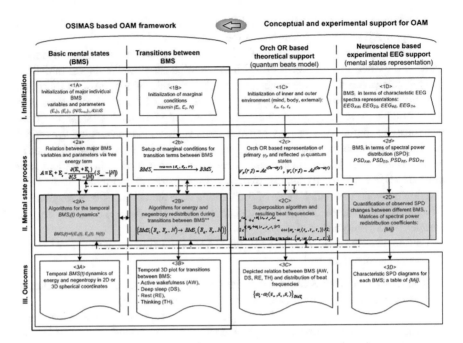

Fig. 7.6 General scheme of the OSIMAS-based OAM implementation framework together with the conceptual (Orch OR based) and experimental (EEG based) supporting groundwork. The proposed scheme pertains to the modeling of agent A_i as an oscillating dynamic process of changing basic mental states (*BMS*)

constructed in such a way as to also depict important conceptual and experimental support for the OAM framework, which provides a strong basis for the approach. In this regard, the third column (in Fig. 7.6) indicates the relation of the beat frequencies, stemming from the superposed quantum states of the Orch OR theory (Hameroff and Penrose 2014), with the OAM approach. Moreover, the fourth column provides a solid experimental foundation employing the neuroscience-based EEG research of BMS represented in terms of the spectral power distributions of the brainwaves (SPD) (Benca et al. 1999; Cantero et al. 2004; Fingelkurts et al. 2006; Newandee and Reisman 1996; Nummenmaa et al. 2012; Plikynas et al. 2014; Travis and Orme-Johnson 1989). In the general scheme, the dotted lines indicate the stages where the conceptual (Orch OR based) and experimental (EEG based) approaches contribute to the OAM development. More details about the scheme and the associated algorithms are described step-by-step below.

Hence, following above-described theoretical framework, let us name the main preconditions:

1. OAM is based on the OSIMAS mind-field concept.
2. It is aimed at modeling the BMS and the time dependent transitions between them.

3. It simulates the basic mind-brain states as time t dependent dynamic processes.
4. The model employs two major independent variables, i.e., entropy S (or negentropy N) and time t. Other variables are dependent on them, e.g., the total internal energy $U(S)$, and the correspondingly kinetic energy $E_k(S)$ with the potential energy $E_p(S)$, and partial derivatives $\delta E_k/\delta S$ and $\delta E_p/\delta S$.
5. The model assumes that an agent's mind-field is an open system $U \neq$ const, but under the condition that free energy $A =$ const (see Eq. 7.7).[21]
6. Following the description of the conceptual model, the total internal energy has two major forms, i.e., potential and kinetic. Potential energy E_p is related to the superposition of primary and reflected quantum states Ψ_p and Ψ_r, which yields beat frequencies according to the Orch OR theory (Hameroff and Penrose 2014). In essence, E_p can be interpreted as a field-like type of implicit energy operating in the mind-brain system.
7. The source of kinetic energy E_k is related with all other types of energy (biochemical, diffusion of electric charges, neuropeptides, etc.). In essence, E_k can be interpreted as a corpuscular type of explicit energy operating in the mind-brain system.
8. Following Schrödinger's (1955) interpretation, the employed negentropy term can be understood as the order of coherent oscillations opposite to entropy (see Chaps. 2 and 3). Negentropy has a key role in the proposed approach, because it represents information practically measured in terms of the magnitude of the ordered oscillations in the neuronal ensembles, which, according to the Orch OR theory (Hameroff and Penrose 2014), are caused by the superposition of the primary Ψ_p and reflected Ψ_r quantum states.
9. Transitions between the BMS are probabilistic in nature.
10. The above preconditions apply for all BMS:

   (a) deep sleep BMS_{DS},
   (b) physically active wakefulness BMS_{AW},
   (c) thinking (contemplating) BMS_{TH},
   (d) resting BMS_{RE}.

According to the general scheme (see the first and second columns in Fig. 7.6), the modeling begins with the setup of initial conditions and major relationships between entropy and energy. Consequently, it proceeds with a description of the marginal conditions. In the next section, functions are provided for the temporal BMS process dynamics, marginal conditions for transitions and their relation to the experimental EEG observations. All together, the above steps generate the simulated behavior for an agent A_i in terms of the BMS and dynamic transitions between them. The probabilistic nature of the transitions between the mind-brain states ensures the unique patterns of the states' dynamics. Such an approach enables the probabilistic modeling of circadian rhythms (Dijk 1999).

---

[21]The source of free energy, i.e. the inflow of nutrients via the bloodstream is kept steady in the brain at all times (Plikynas 2015).

The reiteration of the above-mentioned steps n times, using different driving parameters and probabilities, produces a unique set of agents $\{A_i\}$ with different behavior patterns in terms of mind state dynamics. In this way, a BMS-based agent approach is constructed that can be applied for the consequent development of the MAS simulation platform. Such a MAS is oriented towards simulating group-wide mind-brain state synergetic interaction, synchronicities, clustering, excitation (e.g., novelties, news) propagation patterns, etc.

In returning to the OAM setup, we start below by naming the main conditions and inequalities encountered in following the free energy term $A = U - TS$

$$ U \gg TS, \quad A < U, \quad T > 0, \quad S > 0, $$

where entropy is related to the negentropy via the relation $S = S_{max} - N$.[22] Besides, $TS > 0$ as mind processes are associated with heat loss. Consequently, $\delta U/\delta S \cdot S \neq 0$ and $\delta U/\delta S \neq 0$. The latter inequality provides a very important consequence as $U(t) \neq$ const, i.e., the total internal energy varies in time.

Following Eqs. 7.2 and 7.4, it also provides another important observation for the derivatives

$$ A = U - TS = E_k + E_p - \left(\frac{\partial(E_k + E_p)}{\partial S}\right)S = E_k + E_p - \left(\frac{\partial E_k}{\partial S} + \frac{\partial E_p}{\partial S}\right)S, \quad (7.10) $$

that is, $\delta E_k/\delta S \neq \delta E_p/\delta S$; $\delta E_k/\delta S > 0$ and $\delta E_p/\delta S < 0$, as the increase in entropy (decoherence of oscillations) increases the kinetic energy, but decreases the potential energy. In order to meet the above-mentioned $T > 0$ condition, the rate of change for E_k and E_p with respect to the change in S has to be different, i.e.,

$$ |\delta E_k/\delta S| - |\delta E_p/\delta S| > 0. \quad (7.11) $$

Such an asymmetric response to the change in S pertains to the model's condition $U(t) \neq$ const, i.e., $U(S(t))$. The rationale for such reasoning relates to the nature of entropy, which is related to the level of decoherence of the neural oscillators, i.e., to the less coherent mind field, which is produced by the superposed fields of individual neuronal ensembles. In the OSIMAS paradigm, the potential energy refers to the mind-field type of energy, which only accounts for a minor portion of the all internal energy U. In this regard, higher entropy pertains to the slight reduction in absolute terms of the potential mind-field energy. However, the model assumes that it renders an adverse effect on the kinetic energy, which is positively more sensitive to the change of entropy because (a) it composes the major part of the internal energy and (b) Eq. 7.11 has to be obeyed in order to comply with the condition $T > 0$.

---

[22] In the OSIMAS paradigm, N is interpreted as the level of the phase coherence of neuronal oscillators.

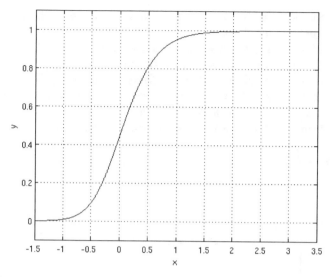

Fig. 7.7 Generalized logistic function, which is used for a depiction of the relationship between $E_p(S)$ and $E_k(S)$

While looking for basic stylized mathematical functions suitable to represent these energy relationships, with respect to the entropy, we employed nonlinear logistic functions, which are observed naturally in various biological systems.[23] For instance, the generalized logistic function (S-curve or Richards' curve), which was originally developed for the growth modeling inspired by nature, is able to depict the various nonlinear energy dependencies on S (see Fig. 7.7)

$$E(S) = L + \frac{P - L}{\left(1 + \Omega e^{-k(S-M)}\right)^{1/\nu}}, \qquad (7.12)$$

where L denotes the lower asymptote of the S-curve; P denotes the upper asymptote; k denotes the growth rate, Ω depends on the value of $E(S_0)$; ν denotes the zone near which asymptote maximum growth occurs and denotes the time moment of the maximum growth.

The generalized logistic function has plenty of parameters that allow its flexibility and ability for adaptation in many applied cases. However, in the simplified OAM version, we apply the classical logistic function

$$E(S) = \frac{L}{\left(1 + e^{-k(S-S_0)}\right)}, \qquad (7.13)$$

---

[23]This function finds many applications throughout a vast range of fields, including biology, neural networks, ecology, biomathematics, chemistry, economics, geosciences, sociology, political sciences, etc.

where S_0 denotes the moment of maximum growth, L denotes the curve's maximum value and k denotes the steepness of the curve.

The classical logistic function has sufficient shape forming parameters, which can be arbitrarily chosen for each agent A_i. Hence, the logistic function is employed for the formation of the main $E_k(S)$ and $E_p(S)$ dependencies, see Fig. 7.8. All other dependencies (mainly $U(S)$, $A(S)$, $-TS$) are obtained from these two, see Fig. 7.8. As we can see, the $E_k(S)$ and $E_p(S)$ dependencies are chosen in such a way as to obtain the major condition for free energy term $A = \text{const} = 1$.

In order to generate the E_k and E_p curves described in Fig. 7.8, we used the parameters of $S_{max} = 1$, $S_{min} = 0$, $dS = 0.01$, $K_s = 15$, $E_{kmin} = 0.5$, $E_{kmax} = 0.9$,

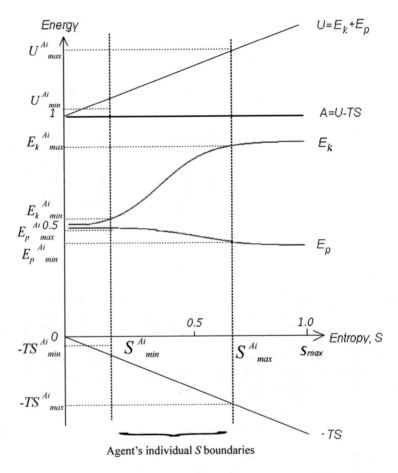

Fig. 7.8 Principal scheme of basic relationships between the fundamental independent factor S and the dependant forms of energy $E_k(S)$, $E_p(S)$, $A(S)$, $U(S)$, and $-T(S) \cdot S$. Arbitrarily chosen entropy values $\left(S_{min}^{A_i}, S_{max}^{A_i}\right)$ are assigned for each agent A_i (see *vertical lines*); these values determine an agent's marginal conditions in terms of energy

$S_{0E_k} = 0.45$, $K_k = 15$, $E_{p\ min} = 0.4$, $E_{p\ max} = 0.5$, $S_{0\,E_p} = 0.55$, and $K_p = 14$. The other curves $U(S)$ and $A(S)$ were obtained using the term provided below.

In short, the other three (derivative) dependencies are obtained in the following way (see Eqs. 7.1 and 7.6)

$$\begin{cases} U(S) = E_k(S) + E_p(S), \\ T(S) \cdot S = U(S) - A(S) = U(S) - 1, \\ A(S) = U(S) - (U(S) - 1) = 1. \end{cases} \qquad (7.14)$$

In this way, we do not explicitly calculate the $T(S) \cdot S$ term, which is, however, very important for a deeper understanding of the underlying processes in terms of the $\delta E_k/\delta S$ and $\delta E_p/\delta S$ dynamics. A more detailed derivation is provided a few paragraphs below.

At this point we have to stress some important considerations concerning the application of the free energy term in the mind-brain system. First, we are not modeling a closed physical thermodynamic system, but the open biological system of a mind-brain, where the assumption $U(S) \neq$ const and $A(S) =$ const holds. Hence, it does not preclude changes in the total internal energy $U(S)$, see Fig. 7.8. The second assumption is due to the previously mentioned observation, that the inflow of nutrients via the bloodstream and the blood pressure itself is kept almost steady in the brain at all times in order to provide free energy ready to be exploited as useful work in any immediate circumstances. That is, we assume that there is a negentropy-based, autopoietic, self-organizing mechanism to keep $A(S) =$ const despite the momentary differences caused by neuronal activations during the various mind-brain states or during daily circadian cycles. In different mental states, free energy is simply transformed into the reversible form of different spectral representations. However, this biological free energy balancing mechanism is beyond the framework of the current model, although it plays an essential role for the functioning of the mind-brain system (Maturana 1980).

In contrast, if the strictly closed physical system assumption is held, then the free energy term (see Eq. 7.10) gives $\delta U/\delta S = 0$, therefore $A = U$ and no entropy-related dynamics occur. Besides, the term $\delta U/\delta S = \delta E_k/\delta S + \delta E_p/\delta S = 0$ holds only if $\delta E_k/\delta S = -\delta E_p/\delta S$, i.e., the moduli of the partial derivatives are equal. This is in contrast to the basic OAM assumptions, where we hold that field-based energy part $\delta E_p/\delta S$ is a considerably smaller part of the total internal energy, although, it is essential for binding the corpuscular-based neuronal activities ($\delta E_k/\delta S$ part of energy derivative) scattered between the different regions of the neural networks (McFadden 2002).

Still, in physical systems, A is optimized to its minimum value, i.e., A varies in time and $A(S) \neq$ const. However, the OAM model adapts the free energy term from the major thermodynamic equilibrium usually employed for modeling physical systems. The biological nature of the self-organizing (negentropic) OAM is not bounded by some major assumptions inherent for nonliving matter. Hence, we employ Eq. 7.14 and a more detailed derivation as proof below in adapting the free energy term for the functioning of the hypothetical mind-brain system. In this way, based on the main thermodynamic equation, and using the dependencies between the major energy forms and entropy, we mathematically support the foundations of the OAM modeling.

As was mentioned a few paragraphs above, we are now going to provide more detailed reasoning and solutions concerning Eq. 7.14 and the corresponding Fig. 7.8. Hence, for the free energy term A = const assumption $\partial A/\partial S = 0$, which renders, according to Eq. 7.10

$$\frac{\partial^2 U}{\partial S^2} = \frac{\partial}{\partial S}\left(\frac{\partial}{\partial S}(E_k + E_p)\right) = \frac{\partial^2 E_k}{\partial S^2} + \frac{\partial^2 E_p}{\partial S^2} = 0. \tag{7.15}$$

Following the standard approach toward the solution of partial equations, we obtain

$$\frac{\partial}{\partial S}(E_k + E_p) = C_1,$$

and

$$E_k + E_p = C_1 \cdot S + C_2, \tag{7.16}$$

where C_1 and C_2 are constants. It renders a solution in a form of

$$E_k = C_1 \cdot S + C_2 - E_p. \tag{7.17}$$

Using Eq. 7.17 for the $A = U - TS$ and $T = \partial U/\partial S$ terms gives the following

$$\begin{aligned} A &= C_1 \cdot S + C_2 - E_p + E_p - \frac{\delta}{\delta S}(C_1 \cdot S + C_2) \cdot S \\ &= C_1 \cdot S + C_2 - C_1 \cdot S = C_2. \end{aligned} \tag{7.18}$$

Let us remember that, following Eq. 7.13, we have chosen a logistic curve to represent $E_p(S)$

$$E_p(S) = L_1 - \frac{L_2}{1 + e^{-k(S-S_0)}}, \tag{7.19}$$

where the coefficients $L_1 = 0.5$ (denotes E_p initial value), $L_2 = 0.1$ (denotes E_p maximum value), $k = -14$ and $S_0 = 0.55$. Using Eq. 7.19, revision of the first initial condition when $S\,(t = 0) = 0$ provides a value for the coefficient C_2

$$E_{k0} = E_{p0} = L_1 - \frac{L_2}{\left(1 + e^{-k(S-S_0)}\right)},$$

$$L_1 - \frac{L_2}{\left(1 + e^{-k(S-S_0)}\right)} = C_1 \cdot 0 + C_2 - \left(L_1 - \frac{L_2}{\left(1 + e^{-k(S-S_0)}\right)}\right), \qquad (7.20)$$

$$C_2 = 2 \cdot \left(L_1 - \frac{L_2}{\left(1 + e^{-k(S-S_0)}\right)}\right).$$

Inspection for the second condition renders the following considerations

$$S = S_{\min} + \Delta S = \Delta S; \quad S_{\min} = 0,$$
$$E_{k1} = (1 + \alpha) \cdot E_{k0}; \quad \alpha \ll 1, \qquad (7.21)$$

where α is freely chosen. Next, following Eq. 7.17 and the initial condition $E_{k0} = E_{p0} = E_p(S_{\min})$, then $S_{\min} = 0$ for $t = 0$ we obtain

$$C_1 = \left(E_{k1} - C_2 + E_{p1}\right)/(S_{\min} + \Delta S), \qquad (7.22)$$

where $E_{p1} = E_p(S_{\min} + \Delta S)$.

The latter expression provides a whole family of possible values for the coefficient C_1 as $\Delta S \to 0$ and therefore, C_1 can be assigned with a set of small values. In this way, following Eq. 7.17 and the initial analytical form for E_p (see Eq. 7.13) we can draw a family of possible curves for E_k, see Fig. 7.9. In fact, exactly the same solution can be obtained not only in the analytical, but also in the numerical form too.[24]

Of course, we have to admit that the true form of the $E_k(S)$, $E_p(S)$ dependencies could hardly be ascertained from the experimental data. It can be only inferred from the theoretical reasoning, as we did above in assuming the logistic $E_p(S)$ term, which provided a unique expression for the term $\partial U/\partial S$ which is commonly used in the thermodynamics of nonliving matter.[25] The proposed approach leads to the realization of the OAM model below.

In returning to the OAM in Fig. 7.8, the tendencies depicted remain the same for all agents. However, each agent is allowed to have small individual perturbations in terms of their logistic curve steepness and asymptotes.

Now, let us have an example of numerical estimates for the construction of the deep sleep process, BMS_{DS}. It requires a description of the main stages $\langle 1A\rangle$, $\langle 2a\rangle$, $\langle 2A\rangle$, and $\langle 3A\rangle$, see Fig. 7.6. As mentioned above, the major relationships between the variables are determined by Fig. 7.8. The implication of the partial derivatives is commented in Eqs. 7.5, 7.10, and 7.11. The definition of the initial values and

---

[24]The numerical simulation generates an identical family of curves for $U = C_1 S + C_2$, see Fig. 7.9.
[25]The major difference of the proposed approach compared with most of the nonliving matter physical models is related to the interpretation of the $U = E_k + E_p$ term, where $E_k + E_p = \text{const}$ and $|\partial E_k/\partial S| = |\partial E_p/\partial S|$.

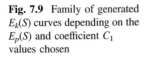

Fig. 7.9 Family of generated $E_k(S)$ curves depending on the $E_p(S)$ and coefficient C_1 values chosen

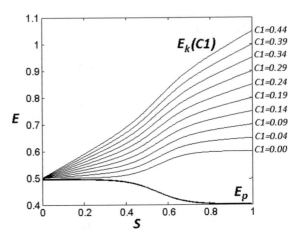

constraints for the variables and parameters, following Eq. 7.7, is summed up below:

1. According to neuroscience research, it is assumed that the free energy of the mind-brain system is at equilibrium and retains a constant value $A = 1$.[26] Experimental observations show that the internal metabolism of the mind-brain system keeps free energy A constant at all times. This is due to the survival need to have a stable source of free energy, which can almost be instantaneously employed for useful mental work if needed.[27]
2. The parameter of maximum entropy $S_{max} = 1$, see Eq. 7.3.
3. The kinetic energy $E_k(t)$ obtains an interval of values

$$E_k(t) = [\min(E_k); \max(E_k)] = [0.5; 0.9].$$

---

[26]In neuroscience research, it is experimentally estimated that the local task-related increase in energy consumption relative to the baseline (resting state) is less than 5 % (Congedo et al. 2010). Hence, neuroscience does not preclude the assumption that in general, energy consumption throughout the entire brain remains more or less stable in different mental states. It does not preclude complex internal transformations between the different types of internal energy, e.g. E_k and E_p.

[27]The second law of thermodynamics is in place for closed systems (with constant external parameters and entropy) that follow the principle of minimum internal energy at the equilibrium state. However, external parameters and entropy are not constant in the mind-brain system, because the constant supply of nutrients (and correspondingly energy) via a stable (in the sense of flow and internal pressure) supply of blood and nerve signals coming through the spinal cord counterbalance the loss in energy and entropy in the mind-brain system. In other words, the whole organism is at work to keep the total internal and free energy of the brain as stable as possible at all times even in the different mental states (Plikynas 2015). Admittedly, the mind-brain system itself has internal mechanisms for making free energy stable. The stability of A is associated with energy redistribution during different mental states.

The parameters $\min(E_k)$ and $\max(E_k)$ probabilistically vary ± 0.05 for each agent.

4. The potential energy $E_p(t)$ obtains an interval of values[28]

$$E_p(t_0) = \left[\min(E_p); \max(E_p)\right] = [0.4; 0.5].$$

The parameters $\min(E_p)$ and $\max(E_p)$ probabilistically vary ± 0.05 for each agent.

5. In following the above considerations, $E_p + E_k \neq$ const, however, $E_p + E_k - TS =$ const, where $TS = \left[\min(E_k) + \max(E_p) - 1; \max(E_k) + \min(E_p) - 1\right] = [0; 0.4]$.

The term TS has much more meaning than just heat waste as the T term links energy with the internal informational processes, see Eq. 7.10. Accordingly, in Eq. 7.1 and the above-mentioned inequalities, the TS product is negative and denotes the loss of energy. It plays an important role in the MAS model, where the communication effects between agents take place mainly via the coherence and decoherence principles using this sensitivity term, see Eq. 7.8.

It is imperative to note that in the proposed model, all energy forms depend on the mind-brain system's entropy S, which varies over time $S(t)$ following the specific temporal process of the invoked mind state.

This is due to the previously explained reasoning that each mind state is a characteristic dynamic process of entropy/negentropy change which renders the transformational processes in terms of energy too. For instance, in the case of the deep sleep state BMS_{DS}, we assume such a characteristic process

$$\begin{cases} S(t) \to S_{\min} \text{ or } N(t) \to N_{\max}, \\ E_k(t) \to E_{k\ \min}, \\ E_p(t) \to E_{p\ \max}. \end{cases} \quad (7.23)$$

Following the general scheme (see the second column in Fig. 7.6), these marginal conditions and the skew of the curves $E_k(S)$ and $E_p(S)$ can be probabilistically determined at chosen intervals for each agent A_i. In general, however, the marginal conditions for each agent are determined in the following way

1. Arbitrarily chosen entropy values $\left(S_{\min}^{A_i}, S_{\max}^{A_i}\right)$ are assigned for each agent A_i; these values determine each agent's marginal conditions in terms of allowable entropy limits; note, that $0 \leq S_{\min}^{A_i} - S_{\min} \leq \varepsilon_{\min}$,

---

[28]The potential energy is less than the kinetic due to the assumption that greatest part of the total internal energy U is stored in the corpuscular energy form of molecular movements. Besides, the interval of allowed values for the potential energy is four times smaller than the kinetic due to the paramount role of the stability of the field-like coordination and binding mechanisms across the whole brain during all mental states at all times.

$$0 \leq S_{\max} - S_{\max}^{A_i} \leq \varepsilon_{\max},$$

where $\varepsilon_{\min}$ and $\varepsilon_{\max}$ are arbitrarily chosen (by default $\varepsilon_{\min} = 0.3$ and $\varepsilon_{\max} = 0.3$),

2. Following the dependencies in Fig. 7.8, the intersections of vertical lines $\left(S_{\min}^{A_i}, S_{\max}^{A_i} \right)$ with the E_k, E_p, $-TS$ and U curves show each agent's individual marginal conditions in terms of energy, i.e.

$$\begin{cases} \left[E_{k\,\min}^{A_i}, \; E_{k\,\max}^{A_i} \right], \\ \left[E_{p\,\min}^{A_i}, \; E_{p\,\max}^{A_i} \right], \\ \left[-TS_{\min}^{A_i}, \; -TS_{\max}^{A_i} \right], \\ \left[U_{\min}^{A_i}, \; U_{\max}^{A_i} \right]. \end{cases} \tag{7.24}$$

3. These agent-specific marginal conditions are used as benchmarks for transition moments between the BMS, see the next section.

In the next section, according to the general scheme (see Fig. 7.6), functions for the temporal BMS process dynamics are provided, including the marginal conditions for transitions between the BMS and their relation to the EEG experimental observations.

7.4   BMS as Dynamic Processes

This section links the OAM with the experimental EEG observations of PSD redistribution across brainwaves in different mind states. It also provides the temporal dynamic functions for the BMS processes and describes the marginal conditions for the transitions between them.

In our previous experimental EEG research, we found that qualitatively different mind states are similar in terms of the spatial distribution of the activated brain zones, but the spectral power activations are located across different spectral ranges. Although each individual has a unique distribution of spatial-temporal activations during different mind states, the spectral mind-field fingerprints follow the same relative distribution of the spectral activation patterns found in the Δ, Θ, α, and β brainwave frequency ranges. These and other findings confirm basic OSIMAS assumptions concerning the oscillatory nature of the social agents' states (see Chap. 3).

People have specific and unique spatial distributions of activated electric field potentials for each mind state, but independently of the person, each mind state has common features that can be recognized in terms of their dominant frequency ranges. For instance, the transition from meditation to the thinking state is characterized by the transition of the dominant brainwaves from Δ to α frequencies (i.e., from lower to higher frequencies).

These observations can be inferred from the OSIMAS paradigm, where different mental states are perceived as part of one and the same multivariate mental field composed of a set of dominant frequencies which are called natural resonant frequencies. According to the OSIMAS paradigm, the brain is able to store free energy in these superimposed spectral bands. In this regard, an agent's states can be associated with the dominant frequencies of the mind-field, i.e., with the experimentally observed Δ, θ, α, and β brainwaves.

From the OSIMAS paradigm, it follows that various combinations of activated dominant frequencies in the common mind-field yield unique mental states. Our initial experimental data for some basic states validate this assumption. Our results and those of others observed spectral energy redistribution over frequency ranges for different mind states (Conte et al. 2008; Travis and Arenander 2006; Travis and Orme-Johnson 1989).

Based on the above-mentioned experimental EEG research results, we infer that the BMS can be expressed in terms of stylized PSD distributions using major variables from the OAM model setup, i.e., entropy/negentropy and free energy. Of course, such an approach pertains to a simplification of the interpretation of real data.[29] However, in principle, it provides a unique chance to relate the OAM approach with the experimental neuroscience results.

For this reason, we looked for a statistical distribution function with two controlling parameters, which could be associated with the entropy S and total internal energy U variables from the OAM model in such a way, that these parameters could change shape and shift the distribution curve along the frequency axis following the observed experimental PSD distributions for different mind states (see Chap. 3). Such a statistical distribution function, depending on the values of controlling parameters (S, U), should approximate the observed PSD distributions for all BMS.[30]

---

[29]Stylized PSD distributions only fit for the BMS characterization. In order to recognize more subtle mental states, e.g. emotions or even thoughts, more detailed and advanced EEG PSD analyses are needed (Murugappan et al. 2008). Of course, this would produce more complex distribution patterns that can be approximated, for instance, using the first and second kind of Bessel functions, which are especially appealing because of their adaptability for a wide range complex distribution patterns of wave propagation and static potentials.

[30]It seems plausible that statistical distribution functions could be employed to approximate the observed experimental PSD distributions as they are, according to the theoretical Orch OR assumptions adapted here (Hameroff and Penrose 2014) and generated probabilistically as beat frequencies, which are obtained after the superposition of the primal and reflected quantum functions (see earlier sections). The probabilistic nature of quantum beats naturally produces the observed distribution patterns.

We found that the gamma function $\Gamma(k,\ \mu)$ with a shape parameter k and scale parameter μ fits well for such a purpose.[31] Both $\Gamma(k,\ \mu)$ parameters are real positive numbers. Shape parameter k and scale parameter μ can be naturally associated with the relative levels of entropy S and potential field type energy E_p correspondingly

$$\Gamma(k,\mu) \rightarrow \Gamma(s,e),$$
$$s = \kappa S, \tag{7.25}$$
$$e = \rho E_p$$

here, κ and ρ denote the scaling coefficients $\kappa = 15$ and $\rho = 4$, which are calculated by setting the $S_{\max} = 1$ and $E_{p\ \max} = 0.5$ values.

Here is an example for interpreting the s and e controlling parameters. Higher levels of entropy S in the mind-brain system are usually associated with more alert states, e.g., thinking and physical activity when α and β brainwaves dominate correspondingly and the field type of coherent E_p energy diminishes in the process. Lower levels of entropy are usually associated with rest and deep sleep when θ and Δ brainwaves dominate correspondingly and the field type of coherent E_p energy (see Fig. 7.8) increases in the process.

For the sake of clarity, all four characteristic gamma distributions $\Gamma(s,\ e)$ have been generated for each BMS (deep sleep, active wakefulness, thinking, and resting), see Fig. 7.10. The gamma distributions $\Gamma(s,\ e)$ depict the averaged PSD distributions experimentally observed during EEG tests for various mental states.

It is worth noting, that the experimentally verified $\Gamma(s,\ e)$ distributions employ controlling parameters, which are directly related to the OAM variables S and E_p. It indicates that the proposed conceptually novel BMS simulation approach is based on a solid neuroscience experimental basis and lends itself well to computational modeling.

After all, the OAM subject is about the simulation of BMS as time dependant processes $BMS(t)$. For that reason, we have to provide a $S(t)$ for each BMS. Having such $S(t)$ dependencies allows us to obtain all the rest of energy-related dependencies (see Fig. 7.8), i.e.

$$\begin{cases} E_k(S(t)), \\ E_p(S(t)), \\ U(t) = A - TS(t) = 1 - (\mathrm{d}E_k(S(t))/\mathrm{d}S(t) - \mathrm{d}E_p(S(t))/\mathrm{d}S(t)) \cdot S(t). \end{cases} \tag{7.26}$$

In fact, the first two terms are fundamental and the third term in Eq. 7.26 is derivative, see Fig. 7.8. It is evident that $S(t)$ is a key element in all the above dependencies. It describes the essential entropy-related dynamic process that takes

---

[31]The gamma distribution is often used in econometrics, Bayesian statistics and neuroscience (to describe the distribution of inter-spike intervals). It describes maximum entropy probability distribution for a random variable. Its cumulative distribution function matches the logistic curve.

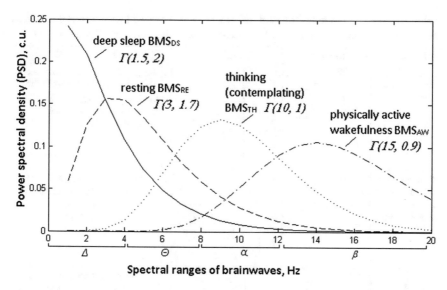

Fig. 7.10 Statistical gamma distributions $\Gamma(s, e)$ adapted for depiction of the averaged EEG power spectral density distributions across the spectral ranges in different BMS

place in the BMS. We infer that each BMS has a characteristic time-dependent process $S(t)$, i.e., $S_{BMS_{DS}}(t)$, $S_{BMS_{AW}}(t)$, $S_{BMS_{TH}}(t)$, $S_{BMS_{RE}}(t)$.

In Table 7.1, the characteristic tendencies are depicted in terms of the S (t) functions for each BMS. Corresponding time dependant $S(t)$ processes are distinguished as continuous power functions,

$$S(t) = t^c, \tag{7.27}$$

where S and c vary in the interval [0; 1] and [0; 10] accordingly, see the characteristic graphs in Table 7.1. Meanwhile, t values vary in the interval [0; 1], however, for the real-life simulations, we adapt another time scale T

$$T = \tau \cdot t, \tag{7.28}$$

where t is multiplied by the factor of $\tau = 1440$, that is, the total number of iterations in one simulation round, which is equal to the number of minutes in a day, 1440 min = 24 h * 60 min/h.

In the OAM, the power functions were chosen to reflect the nonlinear nature of the BMS processes. Each mind state's $S(t)$ dynamics are characterized by the specific power function pertinent to the empirically observed tendencies. Hence, for the mutually related time dependent mental processes BMS_{DS} and BMS_{RE}, the same power function was employed

Table 7.1 The BMS as time dependant $S(t)$ processes are bounded by the marginal individual conditions ($S_{max}^{A_i}$ and $S_{min}^{A_i}$), which invoke probabilistic transitions to other mind states

Mind state	Characteristic process	Characteristic $S(t)$ function	Transitions to other BMS[a]
BMS$_{DS}$ (deep sleep)	$\begin{cases} S(t) \to S_{min}^{A_i}, \\ E_k(t) \to E_{k\,min}^{A_i}, \\ E_p(t) \to E_{p\,max}^{A_i}. \end{cases}$	$S_{DS}(t)=1-t^c,\ 0 \leq c \leq 1$ *Marginal points of transition* *Points of transition in-between*	Transition takes place when the marginal value $S_{min}^{A_i}$ is reached, or takes place in-between the probabilistic transition
BMS$_{AW}$ (physically active wakefulness)	$\begin{cases} S(t) \to S_{max}^{A_i}, \\ E_k(t) \to E_{k\,max}^{A_i}, \\ E_p(t) \to E_{p\,min}^{A_i}. \end{cases}$	$S_{AW}(t)=t^c,\ 1 \leq c \leq 10$ *Marginal points of transition* *Points of transition in-between*	Transition takes place when the marginal value $S_{max}^{A_i}$ is reached, or takes place in-between the probabilistic transition
BMS$_{TH}$ (thinking)	$\begin{cases} S(t) \to S_{max}^{A_i}, \\ E_k(t) \to E_{k\,max}^{A_i}, \\ E_p(t) \to E_{p\,min}^{A_i}. \end{cases}$	$S_{th}(t)=t^c,\ 0 \leq c \leq 1$ *Points of transition in-between* *Marginal points of transition*	Transition takes place when the marginal value $S_{max}^{A_i}$ is reached, or takes place in-between the probabilistic transition
BMS$_{RE}$ (resting)	$\begin{cases} S(t) \to S_{min}^{A_i}, \\ E_k(t) \to E_{k\,min}^{A_i}, \\ E_p(t) \to E_{p\,max}^{A_i}. \end{cases}$	$S_{RE}(t)=1-t^c,\ 1 \leq c \leq 10$ *Marginal points of transition* *Points of transition in-between*	Transition takes place when the marginal value $S_{min}^{A_i}$ is reached, or takes place in-between the probabilistic transition.

[a]The detailed conditions and probabilities of the transitions to other states are described in the next section

$$S_{DS}(t) = 1 - t^c, \quad 0 \leq c \leq 1,$$
$$S_{RE}(t) = 1 - t^c, \quad 1 \leq c \leq 10. \tag{7.29}$$

Whereas, for the mutually related time dependent mental processes BMS$_{AW}$ and BMS$_{TH}$, the following power function was employed

$$S_{AW}(t) = t^c, \quad 1 \leq c \leq 10,$$
$$S_{TH}(t) = t^c, \quad 0 \leq c \leq 1. \tag{7.30}$$

In fact, Eqs. 7.29 and 7.30 are opposite (reflections) of each other. It corresponds well with the opposite nature of both pairs of states.

In Table 7.1, the curves of the time dependent mental processes are depicted as well as (i) basic marginal conditions for transitions between states and (ii) small probabilistic transitions, "in-between" marginal conditions, which depend on the tangent of the $S(t)$ function.

For instance, let us examine a few cases. At the beginning of the deep sleep state BMS_{DS}, a sharp reduction in mental activity takes place, which is reflected in a sharp reduction in the spectral activity seen in the EEG spectrum ranges (except the delta range) (Plikynas et al. 2014; Werth et al. 1997). In the corresponding $S(t)$ function, the rate of change $dS/dt \ll 0$, but in time it gradually changes until $dS/dt \rightarrow 0$ and the lower limit S_{min} is reached (maximum negentropy or order is restored in the brain).

In the BMS_{AW} state, various mental sensory activities disperse the PSD over a wide spectral range and, therefore, disorder (denoted as S) gradually increases, see Table 7.1. It slowly starts from the $dS/dt \rightarrow 0$ rate, and gradually accelerates until reaching a tangent and, following the power law, becomes $dS/dt \gg 0$, i.e., the exhaustion limit S_{max} is reached and the probabilistic transition to another mind state takes place.

In the thinking state BMS_{TH}, we assume that a concentrated mental effort takes place during which the mental disorder with the power law increases, gradually saturates and reaches its maximum. According to the proposed model, the thinking (contemplating) process BMS_{TH} in the sense of its intensity and character of entropy dynamics $S(t)$ is opposite to the deep sleep state BMS_{DS}, see Table 7.1. Similarly, physically active wakefulness BMS_{AW} is opposite to the resting state BMS_{RE}. Besides the similarity in their nature, pairs of states like (BMS_{DS}, BMS_{RE}) and (BMS_{AW}-BMS_{TH}) mainly differ by the different rate of change of the $S(t)$ curve.

Additionally, the OAM model envisages small in-between probabilities, where the expression "in-between" means the probability of transition while the mind-brain system is still in a state between marginal entropy limits $\left(S_{min}^{A_i}, S_{max}^{A_i}\right)$, see Table 7.1. It is important to note that these probabilistic transitions are intrinsic for all four BMS. The probabilities of the in-between transitions depend on the state and time of the day, see circadian rhythms in the next section.

According to the proposed model, each agent has characteristic patterns of transitions, which could differ from the statistical averages. In the next section, the practical simulation results of the proposed model are provided. However, before discussing the OAM simulation results, first, we will briefly review some of the other most widely used alternative models in this context.

According to the OAM assumptions, the BMS as time dependant $S(t)$ processes are bounded by the marginal conditions, which invoke probabilistic transitions from one mind state to another, see the fourth column in Table 7.1. As previously mentioned, arbitrarily chosen marginal entropy limits $\left(S_{min}^{A_i}, S_{max}^{A_i}\right)$ are assigned for each agent A_i,[32] see graphs in the third column in Table 7.1. That is, the

---

[32]They are like the upper and lower limits of the mental capacity to retain entropy or in other words—ordered information (negentropy).

probabilistically determined transition to other states takes place then the mind state process reaches one of the marginal entropy limits.

Historically, the research of mind state dynamics was basically related with impaired alertness and cognition, which is mathematically associated with the fundamental sleep–wake cycle of restful and active mind states.[33] The mathematical two-process model was introduced by the well-known seminal work of Borbély, Daan and Beersma (Daan et al. 1984) and extended by Borbély and Acherman (1999).

As indicated by its name, the two-process model proposes that the sleep–wake cycle can be understood in terms of two processes, a homeostatic process and a circadian process.[34] The homeostatic process takes the form of a relaxation oscillator that results in a monotonically increasing 'sleep pressure' during the time awake that is dissipated during sleep. Switching from wake to sleep and from sleep to wake occurs at the upper and lower threshold values of the sleep pressure respectively, when the thresholds are modulated by an approximately sinusoidal circadian oscillator.

This model has proved compelling for both its physiological grounding and graphical simplicity and has been used extensively (Skeldon et al. 2014). As a matter of fact, the two-process model considers a homeostatic pressure $H(t)$ that decreases exponentially during sleep

$$H(t) = H_0 \, e^{(t_0 - t)/\chi_s}, \qquad (7.31)$$

and increases when awake

$$H(t) = \mu + (H_0 - \mu)e^{(t_0 - t)/\chi_w}. \qquad (7.32)$$

The parameter μ is known as the 'upper asymptote' (Skeldon et al. 2014), and this is the value that the homeostatic pressure H would reach if no switch to sleep occurred. Similarly there is a 'lower asymptote' of zero. Switching between wake and sleep occurs when the homeostatic pressure $H(t)$ reaches an upper threshold, $H^+(t)$, that consists of the mean value of H_0^+ modulated by a circadian process $C(t)$

---

[33]Disrupted or irregular sleep-wake cycles have been correlated with increases in a diverse range of medical problems including all-cause mortality, cardio-vascular disease, diabetes and also metabolic, inflammatory, immune and stress related disorders (Möller-Levet et al. 2013; Nielsen et al. 2011).

[34]A circadian rhythm is any biological process that displays an endogenous, entrainable oscillation of about 24 h. In mammals, endogenous rhythms are generated by the suprachiasmatic nuclei (SCN) of the anterior hypothalamus. Entrainment, within the study of chronobiology, occurs when rhythmic physiological or behavioral events match their period and phase to that of an environmental oscillation. A common example is the entrainment of circadian rhythms to the daily light-dark cycle. Of the several possible cues, which can contribute to entrainment, daily bright light is by far the most effective. The activity/rest (sleep) cycle in animals is only one set of circadian rhythms that normally are entrained by environmental cues (Roenneberg et al. 2003).

Fig. 7.11 Sleep–wake cycles generated by the two-process model. The times when sleep occurs (*H* decreasing) are shaded (Daan et al. 1984)

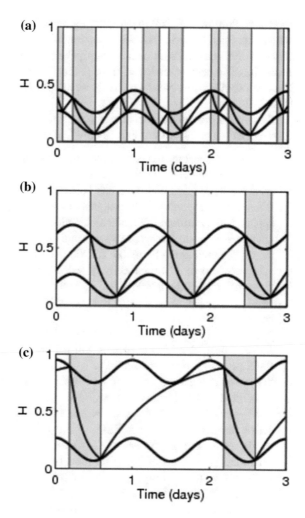

$$H^+(t) = \mu + H_0^+ + aC(t). \tag{7.33}$$

The switch between sleep and wake occurs when $H(t)$ reaches a lower threshold $H^-(t)$

$$H^-(t) = \mu + H_0^- + aC(t), \tag{7.34}$$

where $C(t)$ is a periodic (circadian) function with a period of 24 h. The typical results of this model illustrating its rich dynamics are shown in Fig. 7.11.

Advances in neurophysiology have led to a proliferation of models that aim to extend the two-process model to a more physiological setting. A recent review is given in (Booth and Diniz Behn 2014). The most extensively tested of these is the model of Phillips and Robinson (Phillips and Robinson 2007) (the PR model), which has been used to explain sleep fragmentation experiments (Fulcher et al. 2008), differences in mammalian sleep patterns (Phillips et al. 2010) and the PR model has also been extended to allow for feedback of the sleep–wake cycle on the circadian oscillator in order to explain spontaneous internal desynchrony (Phillips et al. 2011).

The major pros and cons in comparing the OAM and other BMS dynamic models mentioned above are listed below:

1. The prevailing other models employ time dependent, exponential, two-process growth, and decline functions (see Eqs. 3.13–3.16 and Fig. 3.7), which are bounded by the external circadian harmonic function. Hence, transitions occur when the exponents of the vaguely explained term "homeostatic pressure" approach the harmonic (circadian) function. In essence, it means that the marginal values for the two-process states are determined externally and vary during the day following the harmonic function. This is in sharp contrast to the OAM approach, where marginal entropy values (used instead of the "homeostatic pressure" term) are admitted as intrinsic and fundamental properties of the mind-brain system and are not so strongly dependent on the external environment.

2. The circadian cycles in the OAM approach occur naturally as a consequence of (a) state dependent $S(t)$ dynamics and (b) probabilistic transitions between the BMS processes (see next section).

3. The other models only employ two basic sleep–wake mind states (processes), whereas the OAM employs four BMS processes, namely (a) the deep sleep BMS_{DS} and resting BMS_{RE} states instead of a single sleep state, and (b) the active physical wakefulness BMS_{AW} and thinking BMS_{TH} states instead of a single wake state. In this sense, the OAM extends the earlier approaches.

4. The conceptual basis of the other models is poorly supported by fundamental insights into the underlying nature of the mind-state processes. For instance, they assume that "The homeostatic process takes the form of a relaxation oscillator that results in a monotonically increasing 'sleep pressure' during wake that is dissipated during sleep."(Skeldon et al. 2014). That is, these mathematical models do not explain the terms of "homeostatic process," "relaxation oscillator," or "sleep pressure" in biophysical terms, whereas the OAM is backed by some fundamental insights stemming from:

    (a) the OSIMAS premises (see Chaps. 1–3);
    (b) the Orch OR (orchestrated objective reduction) theory of consciousness proposed by Stuart Hameroff and Sir Roger Penrose (Hameroff and Penrose 2014), where the conscious states have been assumed to correlate with the physiological EEG oscillations, which might come about, namely as beat frequencies, arising when the OR (the specific Diósi–Penrose scheme of 'objective reduction' of the quantum state) is applied to the superposition of quantum states of slightly different energies, see Eq. 7.9 and Figs. 7.1, 7.2, 7.3 and 7.4;

(c) the modeling of the open mind-state process, where the OAM employs the basic principles of thermodynamics by using a universal free energy term, where entropy (or negentropy) plays the major role, see Fig. 7.8 and Eqs. 7.1–7.8 and 7.10–7.15 In essence, the OAM is an entropy-based modeling of the BMS dynamics. All the other variables (e.g., kinetic corpuscular and potential field-like energies) are derivatives of the most fundamental entropy term. Consequently, the OAM is able to explain the homeostatic term "sleep pressure" used in the two-process models as an entropy term according to the OAM approach.

5. In other conceptual mathematical models, vague neurophysiological considerations are used about the two groups of neurons (promoting arousal or sleep accordingly), for instance, in the PR model, whereas the OAM is based on experimental EEG observations of PSD redistributions, see Fig. 7.10.

6. Most importantly, the other models are limited in their perspective of only using a two-process mind-state simulation for a single agent's basic dynamics. Whereas, according to the OSIMAS paradigm, the OAM is constructed for the simulation of the so-called distributed cognitive systems or social cognition in a multi-agent system simulation setup. Hence, OSIMAS offers a much wider perspective for applications in the areas of social neuroscience, team neurodynamics, and neuroeconomics research in terms of collective consciousness modeling using the OAM approach and MAS setup.

For the sake of clarity, we also have to note some major drawbacks of the proposed OAM approach:

(d) the absence of an explanatory link between Orch OR beat frequencies and the OAM EEG-based PSD observations in the different BSM

(e) not all of the explanations are sufficient for the proposed probabilistic estimates of transitions, which should stem from the proposed free energy model and Orch OR quantum beat frequency model.

In the next section, an OAM setup is presented, aimed at simulation of the circadian daily rhythms, which occur endogenously as a result of the application of a novel type of homeostatic BMS-related processes.

7.5 Simulation Setup: Circadian Cycles Case

The main goal of this section is to adapt the previously described BMS state (process) modeling setup for simulation of the mind-state dynamics involved in the circadian (day-night) rhythms. In essence, the OAM provides a novel method for simulating the daily rhythms of mind states. The proposed model extends the two-process sleep–wake daily transitions described in the earlier models (Booth and Diniz Behn 2014; Skeldon et al. 2014) with the four-process daily transitions between the BMS.

The proposed model setup uses circadian daily cycles, which affect (a) the entropy $S(t)$ curve for each BMS, and (b) the probabilistic in-between transitions between the BMS (see graphs in Table 7.1).[35] Hence, the proposed model distinguishes between the two characteristic daily periods:

(a) *the daytime active period*, when the BMS_{AW} and BMS_{TH} states are alternately dominating with the BMS_{RE} restoring state occasionally occurring in between,[36] and
(b) *the nighttime restoring period*, when the BMS_{DS} and BMS_{RE} states are alternately dominating with some BMS_{AW} occasionally occurring in between.[37]

During a day, a simulated agent passes through multiple entropy-related processes (states). According to the proposed model setup, the time spent in one state can vary in the interval [0; 100] min. In this way, the minimum number of transitions during a day is 14. However, it can also reach 50 or more if in-between probabilistic transitions are involved too. According to the model setup, in-between transitions can occur probabilistically (the user can choose the probability level). By default, for every time interval of 10 min, these small probabilities are applied and if the probability is 10 % (0.1) then after 100 min, an in-between transition will statistically occur.

This flexible setup for the duration of mind processes and the transitions between them is constructed for the simulation of various experimentally observed physiological mind-state behavioral patterns, e.g., (i) repeating REM (BMS_{RE}), NREM (BMS_{DS}) and awaking (BMS_{AW}) rhythms during the nighttime, and (ii) repeating rhythms between BMS_{AW}, $BMS_{TH,}$ and occasionally, BMS_{RE} during the daytime.

So first, let us discuss changes in the $S(t)$ dynamics depending on the time of day. Concerning the dependence of the $S(t)$ curves on the daily circadian rhythm, the OAM uses the external circadian 24 h harmonic function, but in contrast to the earlier models, this function does not modulate the upper and lower margins of the "homeostatic pressure" (i.e., the entropy according to the OAM setup). It simply determines the daytime and nighttime periods.

However, the OAM applies a mind state specific analog of the homeostatic function defined as $c(t)$, which regulates the shape of the $S(t)$ function for each BMS

---

[35]Let us remember that during the day, the mind is involved in the BMS processes and the transitions between them. So occasionally, the mind transitions multiple times between one or another BMS state during the daily circadian cycle. The different periods of the day presumably have a characteristic influence on the states' $S(t)$ dependencies. This forms families of entropy curves for each BMS, each fitted for the defined period of the day. .

[36]Admittedly, during the daytime, the mind state jumps many times between the BMS_{AW} and BMS_{TH} states, with some restoring breaks spent in the BMS_{RE} state.

[37]It is a truism in mind state research (Fulcher et al. 2008; Phillips et al. 2010) that during the nighttime, the human (and mammals in general) mind alternates between NREM and REM sleep (REM is usually interrupted a few times by short wakeful states), i.e. between the BMS_{DS} and BMS_{RE} states, with short intervals in the BMS_{TH} state.

depending on the period of the day, see Eq. 7.27. There are many possible ways to mathematically describe such functions, following the corresponding BMS specific $S(t)$ processes, see Table 7.1. For example, we can use harmonic functions like

$$
\begin{aligned}
c_{DS}(t) &= 1 - \sin \pi v t = 1 - \sin \pi t, & 0 \leq c_{DS}(t) \leq 1, \\
c_{AW}(t) &= 1 + 9|\sin \pi v t| = 1 + 9|\sin \pi t|, & 1 \leq c_{AW}(t) \leq 10, \\
c_{TH}(t) &= |\sin \pi v t| = |\sin \pi t|, & 0 \leq c_{TH}(t) \leq 1, \\
c_{RE}(t) &= 1 + 9|\sin \pi v t| = 1 + 9|\sin \pi t|, & 1 \leq c_{RE}(t) \leq 10,
\end{aligned}
\tag{7.35}
$$

where parameter $v = 1$ denotes the frequency of the homeostatic function, which is equal to one because we only simulate a one day period.[38] Time t varies during the day between [0; 1]. We picked 10 equidistant t values in the intervals [1, 10] and [0, 1] in order to obtain corresponding sets of $c(t)$ curves,[39] which are used in Eqs. 7.29 and 7.30. The obtained families of the $S(t; c(t))$ curves for all BMS are depicted in Fig. 7.12.

In Fig. 7.12, a normalized time scale is applied, i.e., the usual time of the state's duration is reduced to the mathematical interval [0; 1].[40] Hence, the normalized time axes in Fig. 7.12 do not denote a 24 h day period, but a normalized period of a state. Thus, the denoted state time period $t = [0; 1]$ corresponds to a 100 min time interval, which is the maximum possible time period for a BMS-related process duration before it reaches the marginal entropy limit and the transition to another state occurs. That is, depending on the day time, the BMS related process duration varies, but it is limited by the maximum 100 min time interval. Actually, this period is close to the well-known brain hemispheres' period of alternate activations, which activate the different mind states (Stančák and Kuna 1994).

It is important to note that the proposed model can regulate the average duration of the states via the chosen level of the two main parameters S_{max} and S_{min}, see Fig. 7.12. These marginal entropy values set the primary constraints on the dynamics of $S(t)$. In this way, the OAM not only has the possibility to model the duration of staying in each BMS, depending on the time of the day, but it also differentiates the states in terms of how the marginal constraints are reached. Such properties of the proposed model ensure multiple probabilistically repeating occurrences of the same BMS during the day. It corresponds well with behavioral observations, e.g., multiple repeating cycles of altering duration for the REM (BMS$_{RE}$ as dreaming), NREM (BMS$_{DS}$), and wakeful (BMS$_{AW}$) states during the nighttime (Dijk 1999; Möller-Levet et al. 2013; Nielsen, Danielsen and Sørensen

---

[38]The frequency term is included in order to have a nondimensional argument in the trigonometric function, i.e. $v \cdot t = 1/T \cdot t$, where T = day, [day^{-1} * day = 1].

[39]In fact, the examples of the $S(t)$ curves given for various states in the third column of Table 7.1 are just single cases for one value of the function $c(t)$.

[40]In the proposed OAM setup, the maximum duration of the mind state can reach up to 100 min. Hence, the time interval [0, 100] min is reduced to the normalized duration of the state related time interval [0, 1].

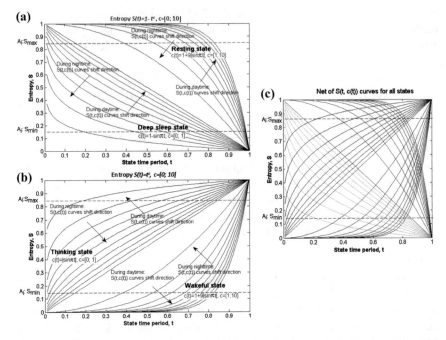

Fig. 7.12 $S(t, c(t))$ curves obtained for all BMS using the homeostatic function $c(t)$, which determines the shape of the $S(t)$ curves during the daytime and nighttime periods. In graph A, the families of ten $S(t, c(t))$ curves are depicted for the similar in kind BMS_{DS} and BMS_{RE} states. In graph B, the families of $S(t, c(t))$ curves are depicted for the similar in kind BMS_{AW} and BMS_{TH} states. In graph C, all the $S(t, c(t))$ curves are depicted, which form the mind states' entropy "space" map, where the $S(t)$ processes of all the mind states can be visualized. The time scale does not indicate the whole day period. Instead, it indicates the maximum normalized period of staying in one mind state

2011) or similarly, during the daytime—multiple repeating cycles of altering duration for the BMS_{AW}, BMS_{TH} and BMS_{RE} states.

Thus, the proposed state entropy space simulation approach has some clear advantages over the earlier models. For instance,

(a) the $S(t, c(t))$ map covers the BMS-related entropy space, which can be employed to reveal the agent's daily dynamics of states and the multiple transitions between them.
(b) the daytime dependent shift of the $S(t, c(t))$ curves for each BMS renders different moments of intersection with the marginal values S_{max} and S_{min} (see Fig. 7.12), which regulates the different time periods of staying in a particular state, depending on the time of the day, e.g.,
(f) for the BMS_{DS}: during the nighttime, the occasionally repeating deep sleep state periods gradually become shorter ($c(t)$: $1 \rightarrow 0$); at the end of the daytime, the occasionally repeating deep sleep periods become longer ($c(t)$: $0 \rightarrow 1$) as the mind gets tired and occasionally goes into deep sleep; such properties of the model correspond well with the observed physiological data,

(g) for the $\mathrm{BMS_{AW}}$: during the nighttime, the active wakeful periods become longer ($c(t)$: $1 \to 10$) and during the daytime, the active wakeful periods become shorter ($c(t)$: $10 \to 1$) as mind gets tired and tends to be in other states); such properties also correspond well with the empirical data,

(h) for the $\mathrm{BMS_{TH}}$: during the nighttime, the thinking periods occasionally become longer $c(t)$: $0 \to 1$ (this is confirmed by the sleep studies) and during the daytime, the thinking periods become shorter $c(t)$: $1 \to 0$ (the mind gets tired from the concentrated mental efforts),

(i) for the $\mathrm{BMS_{RE}}$: during the nighttime, the dreaming periods become longer ($c(t)$: $10 \to 1$) and during the daytime, the dreaming periods become shorter ($c(t)$: $1 \to 10$); this also agrees with the physiological observations.

According to the literature cited above, under normal circumstances, during the nighttime there are (1) 3–4 deep sleep cycles ($\mathrm{BMS_{DS}}$) with diminishing duration, (2) 4–5 cycles of the REM dreaming state ($\mathrm{BMS_{RE}}$) with increasing duration, and (3) around 4 cycles of wakeful state ($\mathrm{BMS_{AW}}$) with increasing duration. The two-process models described earlier are, in principle, not able to simulate such rhythms. Whereas, the OAM approach provides a method for the realization and investigation of such rhythms under various BMS conditions and probabilistic transition terms.

Moreover, the model uses an additional powerful tool for modeling the transitions. These are the previously mentioned so-called in-between probabilistic transitions that occasionally happen in all BMS processes before they reach the marginal entropy conditions (S_{min} and S_{max}). In each BMS, they occur probabilistically and accumulate during the state's time period. Besides this, according to the model, their probabilities not only depend on the state, but also on the period of the day (daytime $t = [0; 0.5]$, and the nighttime $t = [0.5; 1]$ periods have distinct functions for the in-between probabilities).[41]

The proposed method for the in-between probabilistic transitions is purely probabilistic. It introduces some chaos, stochasticity, and order (synchronization) into the BMS dynamics.[42] Thus, it has an important role in the foreseen multi-agent simulation setting, where the group-wide dynamics of the agents' states become synchronized via the coherence mechanism of the purely probabilistic in-between transitions.

In the proposed model, the general expression for the in-between probabilistic transitions, depending on daytime or nighttime, can be expressed using a simple harmonic function

$$p'(t) = p_{\mathrm{BMS}}'^{\max} \cdot |\sin(2\pi v't + \varphi)| = p_{\mathrm{BMS}}'^{\max} \cdot |\sin(\pi t + \varphi_{\mathrm{BMS}})|, \qquad (7.36)$$

---

[41] According to the simulation setup, the daytime period lasts 12 h [9 AM–9 PM], and correspondingly, the nighttime period is [9 PM–9AM].

[42] Let us remember that the probabilities of the in-between transitions are periodically applied (by default, every 10 min.) when the mind state has not reached its marginal entropy conditions.

Table 7.2 Phase shifts for the sinusoidal function (see Eq. 7.36)

	Night owls				Early birds			
	BMS_{DS}	BMS_{AW}	BMS_{TH}	BMS_{RE}	BMS_{DS}	BMS_{AW}	BMS_{TH}	BMS_{RE}
$\varphi_{DS \to AW}$	$\pi/2$				0			
$\varphi_{DS \to RE}$	0				0			
$\varphi_{DS \to TH}$	0				$\pi/2$			
$\varphi_{AW \to TH}$		0				$\pi/2$		
$\varphi_{AW \to RE}$		$\pi/2$				0		
$\varphi_{AW \to DS}$		0				0		
$\varphi_{TH \to AW}$			0				$\pi/2$	
$\varphi_{TH \to RE}$			$\pi/2$				0	
$\varphi_{TH \to DS}$			0				0	
$\varphi_{RE \to AW}$				$\pi/2$				0
$\varphi_{RE \to TH}$				0				$\pi/2$
$\varphi_{RE \to DS}$				0				0

where $p_{BMS}^{\prime max}$ denotes the arbitrarily chosen maximum probability value for each BMS (by default, it equals 0.05); $2\pi v'$ denotes angular frequency ω, where the frequency term $v' = 1/2$ splits the day period ($v = 1$) into two parts, i.e., daytime and nighttime periods; t denotes the time of day; φ_{BMS} denotes the phase shift of the sinusoidal function, see Table 7.2.

We apply the sinusoidal function (Eq. 7.36) with the corresponding phases (as depicted in Table 4.1) and obtain the time varying, circadian, in-between probabilistic transitions $p'(t)$, see Fig. 7.13. The chosen sinusoidal form of the $p'(t)$ curves can be explained in the following way. First of all, the in-between transitions are dedicated to expand the OAM circadian rhythms beyond the limitations of the marginal transitions. For instance, strict marginal transitions do not envisage transitions between similar states, e.g., $BMS_{DS} \leftrightarrow BMS_{RE}$ and $BMS_{AW} \leftrightarrow BMS_{TH}$, because technically speaking, such transitions do not make sense. That is, when a state process reaches the upper entropy limit, there is no reason that it should immediately transition to a state with the same upper entropy limit, because an opposite process has to take place, which leads to the lower entropy limit. Although it sounds correct technically, it omits a lot of the transitions that take place in reality. In this regard, the presented scheme for the probabilistic in-between transitions helps to sort things out according to the real-life observations, allowing for transitions between similar in kind states.

Another reason why the presented scheme of the in-between transitions makes sense is related with the fact that daytime and nighttime transitions have some specific patterns and, therefore, differ, see Fig. 7.13. The previously denoted $S(t, c(t))$ dependencies make changes to the period of staying in a state, depending on the period of the day (daytime or nighttime), see Eq. 7.35 and Fig. 7.12. However, that is not enough to model well-known circadian cycles like the repeating REM (BMS_{RE}), NREM (BMS_{DS}), and awaking (BMS_{AW}) rhythms during nighttime, and

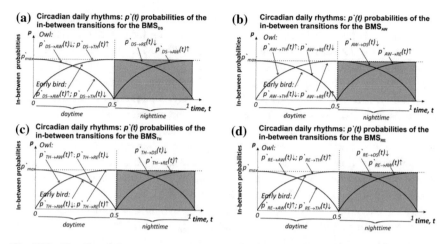

Fig. 7.13 Circadian daily rhythms: daytime and nighttime probabilities $p'(t)$ of the in-between transitions for all BMS. The two chronotypes of people—night owls and early birds—are distinguished by their different daytime set of probabilistic transitions

the repeating rhythms between BMS_{AW}, BMS_{TH} and occasionally, BMS_{RE} during daytime. An additional probabilistic scheme has to be employed to understand such transitions. That is why the proposed scheme of in-between transitions has been designed. Under appropriate conditions, it can simulate the complex dynamics of the experimentally observed transitions between the mind-brain states.

According to the literature review, the mainstream simulation of an agent's chronotype only encompasses the sleep–wake cycles (Booth and Diniz Behn 2014; Borb and Achermann 1999; Daan et al. 1984; Fulcher et al. 2008; Phillips et al. 2011) with little reference to the behavioral manifestation of the underlying circadian rhythms. For this reason, the OAM model takes a pioneering step into the simulation of the behavioral manifestation of the underlying circadian rhythms in terms of people's chronotypes (night owl or early bird), see Fig. 7.13.[43]

Hence, the presented model admits the existence of two different types of people, i.e., "night owls" and "early birds." Night owls tend to feel most energetic just before they go to sleep at night. Early birds (e.g., a lark) as opposed to night owls, feel more energetic early in the daytime and tend to feel sleepy at a time that is considered early. Researchers also use the terms "morningness" and "eveningness" for the two chronotypes (Horne and Ostberg 1976). In this sense, for each chronotype, two different sets of circadian $p'(t)$ graphs are depicted for each state, see Fig. 7.13. In this way, the proposed OAM model is extended to simulate the circadian rhythms of two different types of people chronotypes.

---

[43]The causes and regulation of chronotypes have yet to be determined. However, research is beginning to shed some light on these questions, such as the relationship between age and chronotype. There are candidate genes (called clock genes) that exist in most cells in the body and brain, referred to as the circadian system that regulates physiological phenomena (hormone levels, metabolic function, body temperature, cognitive faculties, and sleeping).

For the sake of clarity, the chosen tendencies of the circadian daily rhythms have to be explained too. For instance, in the BMS_{DS} during the daytime period, which starts at $t = 0$ (9 A.M.) and ends at $t = 0.5$ (9 P.M.), two time dependent, in-between transitions with probabilities $p'_{DS \to AW}$ and $p'_{DS \to TH}$, are initiated, see Fig. 7.13a. In the case of the "night owls" chronotype, the probabilities of these transitions decrease ($\downarrow$) and increase ($\uparrow$) accordingly. This is due to the observations that night owls tend to be least mentally alert in the morning and are most capable of concentrating (thinking) just before they go to sleep at night.[44] Whereas, for the early birds chronotype, we can apply an opposite pattern of behavior, i.e., their probabilities of transitions increase ($\uparrow$) and decrease ($\downarrow$) accordingly. Admittedly, both chronotypes behave in a similar way during the nighttime restoring period, which starts at $t = 0.5$ (9 P.M.) and ends at $t = 1$ (9 A.M.).

Let us examine the BMS_{AW}. In this state, two time dependent in-between transitions with probabilities $p'_{AW \to TH}$ and $p'_{AW \_ RE}$, are initiated, see Fig. 4.2b. In the case of the "night owls" chronotype, the probabilities of these transitions increase ($\uparrow$) and decrease ($\downarrow$) accordingly, as "night owls" tend to be more mentally active and capable of concentrating in the evening time. The opposite is true for the "early birds," as they tend to be more mentally active in the morning and get tired (looking for rest) in the evening time.

A similar rationale for the probabilistic in-between transitions is applied to the other two states. Namely, the probabilistic in-between transitions from the BMS_{TH} (see Fig. 4.2c) are similar in kind to the transitions in the BMS_{AW} (see Fig. 4.2b), and the transitions from BMS_{RE} (see Fig. 4.2d) are similar in kind to the transitions in the BMS_{DS} (see Fig. 4.2a).

Hence, the proposed model incorporates circadian rhythms based on the above considerations and statistical observations. In sum, the marginal and in-between probabilistic transitions for all states are presented in Table 7.3.

Hence, the dynamics of daily transitions between BMS are fully described in Table 7.3. The probabilistic nature of the marginal and in-between transitions is meant to reproduce the stochastic and chaotic nature of the observed main patterns of BMS dynamics. The presented simulation scheme is dedicated to reproducing the experimentally observed (a) circadian, (b) homeostatic, and (c) chronotype behavioral patterns for different types of agents. The proposed simulation scheme is flexible enough to generate a wide range of different agents and their behavioral patterns as well. Depending on the chosen parameters, the proposed OAM model is capable of generating different mind-brain state dynamics for the daytime and nighttime periods (see next section).

For the depiction of all possible transitions between the BMS we can employ the directed graphs approach, see Fig. 7.14. Transitions invoked by the marginal

---

[44]According to the OAM assumptions, thinking state BMS_{TH} is more intense compared to the active wakefulness state BMS_{AW}, which is more related with physical activities and sensory processing.

Table 7.3 Probabilities of marginal and circadian in-between transitions for all BMS

BMS_{DS}	BMS_{AW}
I. Transitions invoked after reaching the marginal condition $$S_{\min}^{A_i} : \begin{cases} \text{BMS}_{\text{DS}} \longrightarrow p_{\text{DS}\to\text{AW}}\text{BMS}_{\text{AW}} \\ \text{BMS}_{\text{DS}} \longrightarrow p_{\text{DS}\to\text{TH}}\text{BMS}_{\text{TH}} \end{cases},$$ $p_{\text{DS}\to\text{AW}} + p_{\text{DS}\to\text{TH}} = 1,$ $p_{\text{DS}\to AW} > p_{\text{DS}\to\text{TH}}.$	**I. Transitions invoked after reaching marginal condition** $$S_{\max}^{A_i} : \begin{cases} \text{BMS}_{\text{AW}} \longrightarrow p_{\text{AW}\to\text{RE}}\text{BMS}_{\text{RE}} \\ \text{BMS}_{\text{AW}} \longrightarrow p_{\text{AW}\to\text{DS}}\text{BMS}_{\text{DS}} \end{cases},$$ $p_{\text{AW}\to\text{RE}} + p_{\text{AW}\to\text{DS}} = 1,$ $p_{\text{AW}\to\text{RE}} > p_{\text{AW}\to\text{DS}}.$
II. In-between probabilistic transitions During daytime $t = [0; 0.5]$: $$\text{BMS}_{\text{DS}} \underset{\text{earlybird}: p'_{\text{DS}\to\text{AW}}\uparrow}{\overset{\text{owl}: p'_{\text{DS}\to\text{AW}}\downarrow}{\longrightarrow}} \text{BMS}_{\text{AW}},$$ where $p'_{\text{DS}\to\text{AW}} \downarrow = p_{\text{DS}}'^{\max} \cdot |\sin(\pi t + \pi/2)|,$ $p'_{\text{DS}\to\text{AW}} \uparrow = p_{\text{DS}}'^{\max} \cdot |\sin(\pi t)|.$ $$\text{BMS}_{\text{DS}} \underset{\text{earlybird}: p'_{\text{DS}\to\text{TH}}\downarrow}{\overset{\text{owl}: p'_{\text{DS}\to\text{TH}}\uparrow}{\longrightarrow}} \text{BMS}_{\text{TH}},$$ where $p'_{\text{DS}\to\text{TH}} \uparrow = p_{\text{DS}}'^{\max} \cdot |\sin(\pi t)|,$ $p'_{\text{DS}\to\text{TH}} \downarrow = p_{\text{DS}}'^{\max} \cdot |\sin(\pi t + \pi/2)|.$ During nighttime $t = [0.5; 1]$: $$\text{BMS}_{\text{DS}} \overset{p'_{\text{DS}\to\text{AW}}\downarrow}{\longrightarrow} \text{BMS}_{\text{AW}},$$ where $p'_{\text{DS}\to\text{AW}} \downarrow = p_{\text{DS}}'^{\max} \cdot |\sin(\pi t)|,$ $$\text{BMS}_{\text{DS}} \overset{p'_{\text{DS}\to\text{RE}}\uparrow}{\longrightarrow} \text{BMS}_{\text{RE}},$$ where $p'_{\text{DS}\to\text{RE}} \uparrow = p_{\text{DS}}'^{\max} \cdot |\sin(\pi t + \pi/2)|.$	**II. In-between probabilistic transitions** During daytime $t = [0; 0.5]$: $$\text{BMS}_{\text{AW}} \underset{\text{earlybird}: p'_{\text{AW}\to\text{TH}}\downarrow}{\overset{\text{owl}: p'_{\text{AW}\to\text{TH}}\uparrow}{\longrightarrow}} \text{BMS}_{\text{TH}},$$ where $p'_{\text{AW}\to\text{TH}} \uparrow = p_{\text{AW}}'^{\max} \cdot |\sin(\pi t)|,$ $p'_{\text{AW}\to\text{TH}} \downarrow = p_{\text{AW}}'^{\max} \cdot |\sin(\pi t + \pi/2)|,$ $$\text{BMS}_{\text{AW}} \underset{\text{earlybird}: p'_{\text{AW}\to\text{RE}}\uparrow}{\overset{\text{owl}: p'_{\text{AW}\to\text{RE}}\downarrow}{\longrightarrow}} \text{BMS}_{\text{RE}},$$ where $p'_{\text{AW}\to\text{RE}} \downarrow = p_{\text{AW}}'^{\max} \cdot |\sin(\pi t + \pi/2)|,$ $p'_{\text{AW}\to\text{RE}} \uparrow = p_{\text{AW}}'^{\max} \cdot |\sin(\pi t)|.$ During nighttime $t = [0.5; 1]$: $$\text{BMS}_{\text{AW}} \overset{p'_{\text{AW}\to\text{DS}}\downarrow}{\longrightarrow} \text{BMS}_{\text{DS}},$$ where $p'_{\text{AW}\to\text{DS}} \downarrow = p'^{\max}_{\text{AW}} \cdot |\sin(\pi t)|,$ $$\text{BMS}_{\text{AW}} \overset{p'_{\text{AW}\to\text{RE}}\uparrow}{\longrightarrow} \text{BMS}_{\text{RE}},$$ where $p'_{\text{AW}\to\text{RE}} \uparrow = p'^{\max}_{\text{AW}} \cdot |\sin(\pi t + \pi/2)|.$
BMS_{TH}	BMS_{RE}
I. Transitions invoked after reaching the marginal condition $$S_{\max}^{A_i} : \begin{cases} \text{BMS}_{\text{TH}} \overset{p_{\text{TH}\to\text{RE}}}{\longrightarrow} \text{BMS}_{\text{RE}} \\ \text{BMS}_{\text{TH}} \overset{p_{\text{TH}\to\text{DS}}}{\longrightarrow} \text{BMS}_{\text{DS}} \end{cases},$$ $p_{\text{TH}\to\text{RE}} + p_{\text{TH}\to\text{DS}} = 1,$ $p_{\text{TH}\to\text{RE}} > p_{\text{TH}\to\text{DS}}.$	**I. Transitions invoked after reaching the marginal condition** $$S_{\min}^{A_i} : \begin{cases} \text{BMS}_{\text{RE}} \overset{p_{\text{RE}\to\text{AW}}}{\longrightarrow} \text{BMS}_{\text{AW}} \\ \text{BMS}_{\text{RE}} \overset{p_{\text{RE}\to\text{TH}}}{\longrightarrow} \text{BMS}_{\text{TH}} \end{cases},$$ $p_{\text{RE}\to\text{AW}} + p_{\text{RE}\to\text{TH}} = 1,$ $p_{\text{RE}\to\text{AW}} > p_{\text{RE}\to\text{TH}}.$
II. In-between probabilistic transitions During daytime $t = [0; 0.5]$: $$\text{BMS}_{\text{TH}} \underset{\text{earlybird}: p'_{\text{TH}\to\text{AW}}\downarrow}{\overset{\text{owl}: p'_{\text{TH}\to\text{AW}}\uparrow}{\longrightarrow}} \text{BMS}_{\text{AW}},$$ where $p'_{\text{TH}\to\text{AW}} \uparrow = p_{\text{TH}}'^{\max} \cdot |\sin(\pi t)|,$ $p'_{\text{TH}\to\text{AW}} \downarrow = p_{\text{TH}}'^{\max} \cdot \sin(\pi t + \pi/2)|.$ $$\text{BMS}_{\text{TH}} \underset{\text{earlybird}: p'_{\text{TH}\to\text{RE}}\uparrow}{\overset{\text{owl}: p'_{\text{TH}\to\text{RE}}\downarrow}{\longrightarrow}} \text{BMS}_{\text{RE}},$$ where $p'_{\text{TH}\to\text{RE}} \downarrow = p_{\text{TH}}'^{\max} \cdot \sin(\pi t + \pi/2)|,$ $p'_{\text{TH}\to\text{RE}} \uparrow = p_{\text{TH}}'^{\max} \cdot |\sin(\pi t)|.$	**II. In-between probabilistic transitions** During daytime $t = [0; 0.5]$: $$\text{BMS}_{\text{RE}} \underset{\text{earlybird}: p'_{\text{RE}\to\text{AW}}\uparrow}{\overset{\text{owl}: p'_{\text{RE}\to\text{AW}}\downarrow}{\longrightarrow}} \text{BMS}_{\text{AW}},$$ where $p'_{\text{RE}\to\text{AW}} \downarrow = p_{\text{RE}}'^{\max} \cdot |\sin(\pi t + \pi/2)|,$ $p'_{\text{RE}\to\text{AW}} \uparrow = p_{\text{RE}}'^{\max} \cdot |\sin(\pi t)|.$ $$\text{BMS}_{\text{RE}} \underset{\text{earlybird}: p'_{\text{RE}\to\text{TH}}\downarrow}{\overset{\text{owl}: p'_{\text{RE}\to\text{TH}}\uparrow}{\longrightarrow}} \text{BMS}_{\text{TH}},$$ where $p'_{\text{RE}\to\text{TH}} \uparrow = p_{\text{RE}}'^{\max} \cdot |\sin(\pi t)|,$ $p'_{\text{RE}\to\text{TH}} \downarrow = p_{\text{RE}}'^{\max} \cdot |\sin(\pi t + \pi/2)|.$

(continued)

Table 7.3 (continued)

BMS_{DS}	BMS_{AW}
During nighttime $t = [0.5; 1]$:	During nighttime $t = [0.5; 1]$:
$BMS_{TH} \xrightarrow{p'_{TH \to DS}\downarrow} BMS_{DS}$,	$BMS_{RE} \xrightarrow{p'_{RE \to DS}\downarrow} BMS_{DS}$,
where $p'_{TH \to DS} \downarrow = p'^{\max}_{TH} \cdot \lvert \sin(\pi t)\rvert$,	where $p'_{RE \to DS} \downarrow = p'^{\max}_{RE} \cdot \lvert \sin(\pi t)\rvert$,
$BMS_{TH} \xrightarrow{p'_{TH \to RE}\uparrow} BMS_{RE}$,	$BMS_{RE} \xrightarrow{p'_{RE \to AW}\uparrow} BMS_{AW}$,
where $p'_{TH \to RE} \uparrow = p'^{\max}_{DS} \cdot \lvert \sin(\pi t + \pi/2)\rvert$.	where $p'_{RE \to AW} \uparrow = p'^{\max}_{AW} \cdot \lvert \sin(\pi t + \pi/2)\rvert$.

Note By default, $p'^{\max}_{DS} = p'^{\max}_{AW} = p'^{\max}_{TH} = p'^{\max}_{RE} = 0.05$, however, the maximum values of the circadian in-between probabilities can be assigned for each state separately and vary in the interval $[0; 0.1]$

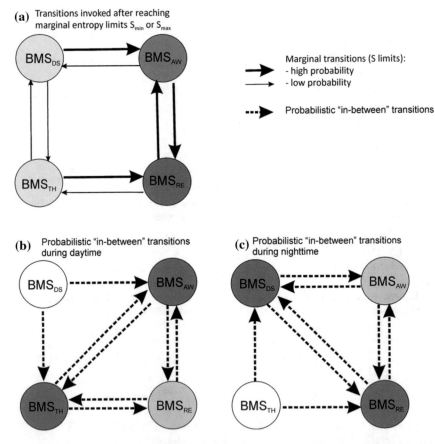

Fig. 7.14 Directed graph for the depiction of nodes (*BMS*) and the transitions between them (*directed lines*), see Table 4.2. The more frequent states (*shaded nodes*) indicate the average time spent in the corresponding BSM and the thickness of the directed lines indicates the probability level of the transitions. Part **a** indicates marginal transitions when the S_{min} or S_{max} limit is reached. Parts **b**, **c** indicate probabilistic in-between transitions during daytime and nighttime, respectively

entropy limits are indicated using solid lines and transitions invoked by the circadian in-between probabilities are indicated using dashed lines.

As can be inferred from Fig. 7.14, all the BMS participate in the daily dynamics. However, according to the model setup, some states naturally dominate. For instance, marginal transitions prevail between BMS_{AW} and BMS_{RE}. Whereas during the daytime, in-between transitions between BMS_{AW} and BMS_{TH} naturally prevail. In the nighttime, in-between transitions between BMS_{DS} and BMS_{RE} naturally dominate. In the next section, the concrete simulation results of the proposed model are provided.

7.6 Simulation Results

This section provides a description of the OAM simulation results, which were basically obtained following the simulation setup described in the previous section. Only some minor changes were implemented regarding the formation of the entropy space network curves $S(t, n)$, see the earlier version at Eq. 7.35 and Fig. 7.12. The need for a new version of the entropy space network curves stemmed from the (i) inconclusive initial simulation results regarding the poor density of the entropy curves in the entropy space (every curve covered a 24 min time period during the daytime and nighttime periods), (ii) coincidental matching of the entropy curves for the same repeating states during a day, (iii) for some states, the marginal entropy values $S(t \rightarrow 0) \ll 1$ or $S(t \rightarrow 1) \gg 0$, i.e., the simulated marginal entropy values did not follow the OAM conceptual setup well. It clearly indicated a need to apply a different $c(t)$ function in order to have a better entropy space coverage during the daytime and nighttime periods.

For this reason, a new homeostatic-circadian function $c_n(t)$ was employed, which generated a continuous and dense enough family of 72 curves $S(t, n)$ for each state depending on the period of the day (daytime and nighttime separately), see Fig. 4.3. For each state, a corresponding family of 72 curves covers a daytime or nighttime period separately. For instance, during a daytime period of $1440/2 = 720$ min, 72 curves, each covering a $720/72 = 10$ min interval of a daytime period are generated. A similar principle holds for the nighttime period too.

Thus, each 10 min time period during a day is represented by four specific $S(t,n)$ curves. Each BMS has its own family of 72 specific day time period dependent entropy curves. These curves cover the daytime and nighttime periods sequentially in the order indicated in Fig. 7.15. That is, depending on the states, the daytime curves start from one end and the nighttime curves from the other end of the same set of 72 entropy curves. In this way, the OAM is able to distinguish how (1) the BMS-related entropy processes gradually change during a day, (2) the daytime-and nighttime-related entropy curves move in the opposite directions.

Let us see an example for the BMS_{DS}: during the nighttime, the occasionally repeating deep sleep state gradually becomes shorter as the mind alternately shifts from the NREM to REM and awaking states, (this is a well-known fact, well

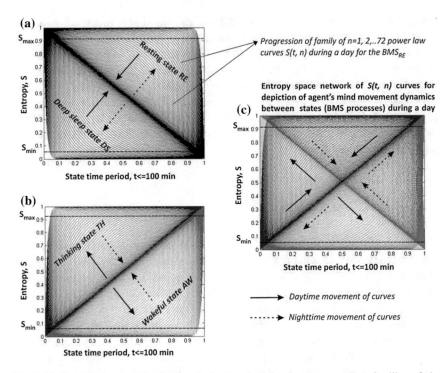

Fig. 7.15 Entropy space network $S(t, n)$ characterized by the corresponding families of the BMS-related curves (72 curves for each state), which are used for the simulation of the S (t) dynamics for the daytime and nighttime periods. Diagram **a** denotes the families of curves for the deep sleep (*DS*) and resting (*RE*) states. Diagram **b** denotes the families of curves for the wakeful (*AW*) and thinking (*TH*) states. Diagram **c** denotes entropy space network of $S(t, n)$ curves for all states (BMS processes)

reported, and documented in the mainstream literature (Booth and Diniz Behn 2014; Borb and Achermann 1999; Daan et al. 1984; Fulcher et al. 2008; Phillips et al. 2011)); meanwhile, during the daytime, the occasionally repeating deep sleep periods become longer as the mind gets tired and naturally tends to stay longer in the deep sleep state.

Hence, a new set of proposed mathematical $S(t, n)$ formulas that generate the needed sets of entropy curves for each state, depending on the period of daytime or nighttime, are depicted below

$$\begin{cases} S_{\mathrm{RE}}(t) = 1 - t^{n^{(n/72)^2}}, \\ S_{\mathrm{DS}}(t) = \varPhi(R(S_{\mathrm{RE}}(\bar{t}))), \\ S_{\mathrm{AW}}(t) = t^{n^{(n/72)^2}}, \\ S_{\mathrm{TH}}(t) = \varPhi(R(S_{\mathrm{AW}}(\bar{t}))), \end{cases} \qquad (7.37)$$

where rotation R and shifting Φ operators are used to transform the $S(t)$ function

$$\begin{cases} R = \begin{bmatrix} \cos\theta & -\sin\theta \\ \sin\theta & \cos\theta \end{bmatrix}, \\ \Phi = 1 + R(S_{\mathrm{RE}}(t)), \\ \bar{t} = 1 - t, \\ t = [0, 1], \end{cases} \tag{7.38}$$

here, the parameter $\theta = \pi$ (in radians) denotes an angle for the rotation operator R. The other variables, namely, t, $\bar{t}$ and n contribute to the formation of the homeostatic and circadian rhythms during a day:[45]

(j) homeostatic individual curves $S(t)$ for each BMS, where t denotes the state-related time period ($t < 100$ min)

(k) circadian movements of the homeostatic curves $S(n)$, depending on the period of the day.

In sum, these formulas generate a dense entropy space network $S(t, n)$, which captures the basic entropy-related processes and their specific alterations during a day following the OAM premises, see Fig. 7.15. Actually, Eqs. 7.37 and 7.38 serve much better, compared to Eq. 7.35, because now:

1. There is basically only one main formula $S(t) = t^{n^{(n/72)^2}}$, which is used to obtain the required $S(t)$ dynamics for all four BMS, using rotation and shifting operators.
2. The families of the curves obtained are smooth, dense, and cover almost all the entropic space.
3. The marginal entropy $S(t)$ values at the marginal points (when $t \to 0$ or 1) reach their intended levels as was planned, according to the OAM.

The proposed entropy space network of the $S(t, n)$ curves restricts the entropy dynamics in certain pathways, described in Eqs. 7.37 and 7.38. The mind travels through these pathways during a day. For each time moment, the mind is involved in one or another entropy-related process, which draws (actualizes) an associated curve's section in the entropy space chart, see Fig. 7.15. In each simulation, the probabilistic nature of the transitions between mind states (BMS) makes a unique pattern for the $S(t, n)$ curves.[46]

---

[45]In fact, the $\bar{t}$ variable denotes the inverse of t, as after the use of rotation operator R, the function is rotated by π, rendering $t = [1, 0]$, and consequently the points of the corresponding curves are drawn in the reverse order, see Eq. 7.37 and Fig. 7.15. In order to restore the correct order, the time flow from 1 to 0 has to be corrected using the inverse variable $\bar{t}$.

[46]Admittedly, the law of succession of entropy should hold. It states that there are no gaps between the entropy levels when the transitions between states take place. That is, the continuity of entropy level holds during transitions, and the next state proceeds from where last state ends. The proposed entropy based BMS modeling approach follows this law in a mathematical sense; however, due to some programming approximations an attentive reader can notice slight shifts between S levels during transitions on the charts below. Notwithstanding such drawback, stemming from the

Depending on the curves position (see Fig. 7.15), we can easily recognize (a) the BMS process it represents and (b) the time of day. In this way, similarly like in the phase space charts, we can observe the dynamics of states (attractors), and the main trends of the transitions between states.[47] Thus, the proposed entropy space network visualization approach is a useful tool for fast analyses of the simulation results.

Following the considerations above and the earlier sections of this chapter, some OAM simulation results are provided below using the OSIMAS V-lab simulation platform. The V-lab is available at http://vlab.vva.lt (User name: Guest, password: guest555), where the OAM simulation model can be found among other OSIMAS models on the left side menu (see for the OAM_simulation). Documentation, instructions, algorithms, and the data files needed for interactive simulations and testing are also provided in the model's page, see Fig. 7.16.

The set of the previously mentioned basic OAM parameters and some brief comments are depicted in Fig. 7.17. The importance of the parameters is listed in a descending order from top to bottom. On the left side, the parameters are listed with some comments and on the right side, their default values are provided.

For instance, selection of the first individual parameter—agent chronotype— defines the type of agent in terms of its behavioral manifestation of the underlying circadian rhythms (night owl or early bird), see Fig. 7.13. The second and third parameters from the top of the list above define the max and min entropy values for the marginal transitions to occur. The next five parameters are used to define the probabilities of the in-between transitions, etc.

Figure 7.18 shows the selection of the optional parameters used to define the probabilistic marginal transitions between states. The values shown are given by default, however, they can be chosen according to the experimentation needs. All the basic and optional parameters were discussed in terms of their meaning and feasible values earlier in this chapter. Basically, the interplay of these parameters generates OAM simulation results.

As discussed in the previous section, the proposed OAM model is capable of simulating the circadian rhythms for two different types of people chronotypes —"night owls," and "early birds," see Fig. 7.13. Let us recall that night owls tend to feel most energetic just before they go to sleep at night. Early birds as opposed to night owls, feel more energetic early in the daytime. An example of the corresponding OAM simulation results of the BMS dynamics for these two chronotypes is provided in Fig. 7.19.

The OAM generated patterns that are a bit different in their BMS dynamics for these two chronotypes, which can be discerned in Fig. 7.19. That is, at the beginning of the day, night owls tend to stay a bit longer in the resting state

---

(Footnote 46 continued)

programming approximation, the mathematics of the model remains firmly in accordance with the law of succession of entropy.

[47]The phase space of a dynamic system depicts all the possible states of a system (represented in terms of the system's main parameters) when the system's evolving state over time traces a path (a phase space trajectory).

Fig. 7.16 Title page of the OAM simulation model setup at the online OSIMAS V-lab simulation platform

(RE) and less in the wakeful state (AW), whereas, at the end of the day, early birds tend to stay a bit longer in the RE state and less in the AW state. Statistical averages are provided below for the reiterated estimates of the BMS dynamics during the daytime and nighttime periods, see Tables 7.4 and 7.5.

Table 7.4 provides statistical averages for the marginal and in-between transitions between states. As we can see, the total number of marginal and in-between transitions for early birds and night owls are similar. The daytime and nighttime number of transitions between the BMS are also similar, which presumably corresponds with the neurophysiological observations (Booth and Diniz Behn 2014). Due to the chosen setup, the marginal transitions occur around 7 times more often compared to the in-between transitions. However, depending on the chosen values of the basic and optional parameters (see Figs. 7.17 and 7.18), the total number of transitions during a day can vary within an interval of 14–50, see the previous section.

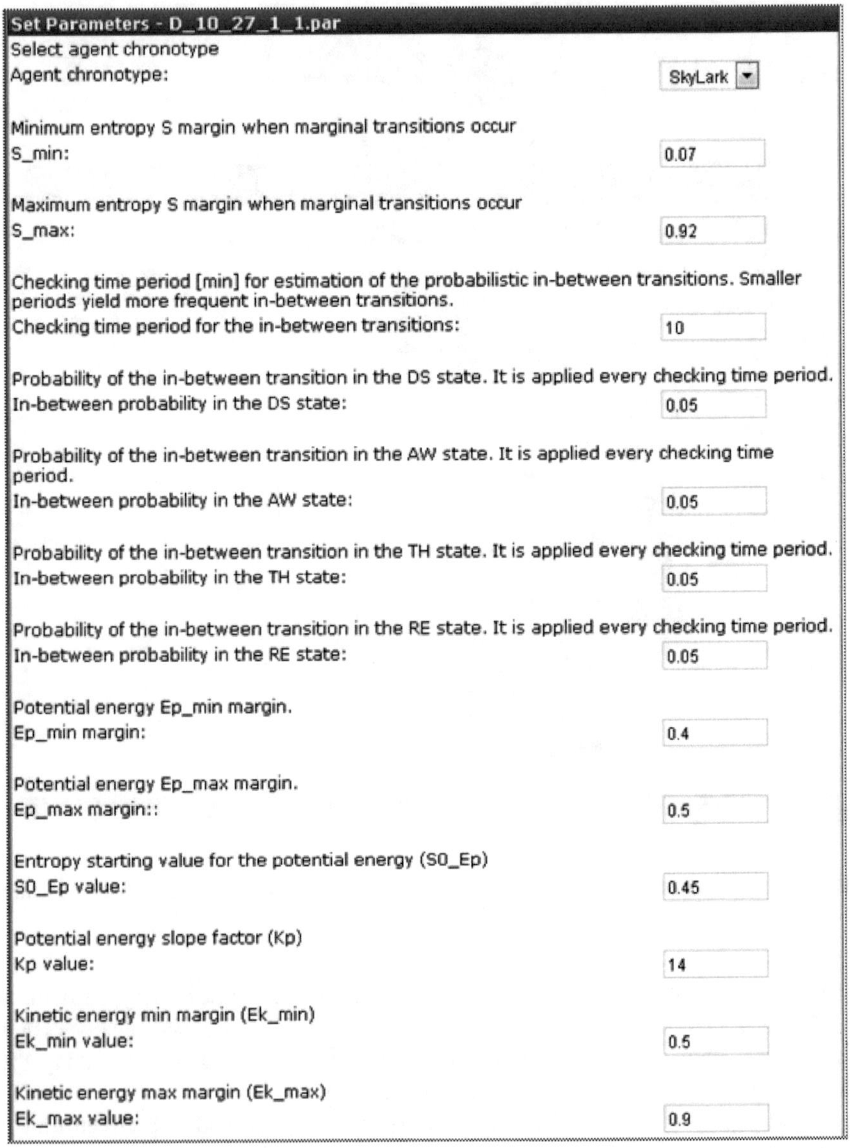

Set Parameters - D_10_27_1_1.par

Select agent chronotype
Agent chronotype: SkyLark ▾

Minimum entropy S margin when marginal transitions occur
S_min: 0.07

Maximum entropy S margin when marginal transitions occur
S_max: 0.92

Checking time period [min] for estimation of the probabilistic in-between transitions. Smaller
periods yield more frequent in-between transitions.
Checking time period for the in-between transitions: 10

Probability of the in-between transition in the DS state. It is applied every checking time period.
In-between probability in the DS state: 0.05

Probability of the in-between transition in the AW state. It is applied every checking time
period.
In-between probability in the AW state: 0.05

Probability of the in-between transition in the TH state. It is applied every checking time period.
In-between probability in the TH state: 0.05

Probability of the in-between transition in the RE state. It is applied every checking time period.
In-between probability in the RE state: 0.05

Potential energy Ep_min margin.
Ep_min margin: 0.4

Potential energy Ep_max margin.
Ep_max margin::   0.5

Entropy starting value for the potential energy (S0_Ep)
S0_Ep value: 0.45

Potential energy slope factor (Kp)
Kp value: 14

Kinetic energy min margin (Ek_min)
Ek_min value: 0.5

Kinetic energy max margin (Ek_max)
Ek_max value: 0.9

Fig. 7.17 The list of basic OAM parameters with their values given by default. The list is obtained from the V-lab simulation platform

Using the same data sample, we can also analyze the average duration of staying in each BMS, see Table 7.5. Such analysis reveals that during the day, early birds, and night owls spend a very similar total amount of time in the corresponding BMS. Naturally, most of their time is spent in the mutually opposite resting (RE) and

Selection of the optional parameters (given values by default)

Bottom margin out of interval [0,1; 0,9] of the RND values allocated for the marginal
transition DS->TH
P_min DS->TH: 0.1

Upper margin out of interval [0,1; 0,9] of the RND values allocated for the marginal
transition DS->TH
P_max DS->TH: 0.3

Bottom margin out of interval [0,1; 0,9] of the RND values allocated for the marginal
transition DS->AW
P_min DS->AW: 0.3

Upper margin out of interval [0,1; 0,9] of the RND values allocated for the marginal
transition DS->AW
P_max DS->AW: 0.9

Bottom margin out of interval [0,1; 0,9] of the RND values allocated for the marginal
transition AW->DS
P_min AW->DS: 0.1

Upper margin out of interval [0,1; 0,9] of the RND values allocated for the marginal
transition AW->DS
P_Max AW->DS: 0.3

Bottom margin out of interval [0,1; 0,9] of the RND values allocated for the marginal
transition AW->RE
P_min AW->RE: 0.3

Upper margin out of interval [0,1; 0,9] of the RND values allocated for the marginal
transition AW->RE
P_Max AW->RE: 0.9

Bottom margin out of interval [0,1; 0,9] of the RND values allocated for the marginal
transition TH->DS
P_min TH->DS: 0.1

Upper margin out of interval [0,1; 0,9] of the RND values allocated for the marginal
transition TH->DS
P_Max TH->DS: 0.3

Bottom margin out of interval [0,1; 0,9] of the RND values allocated for the marginal
transition TH->RE
P_min TH->RE: 0.3

Upper margin out of interval [0,1; 0,9] of the RND values allocated for the marginal
transition TH->RE
P_Max TH->RE: 0.9

Bottom margin out of interval [0,1; 0,9] of the RND values allocated for the marginal
transition RE->TH
P_min RE->TH: 0.1

Upper margin out of interval [0,1; 0,9] of the RND values allocated for the marginal
transition RE->TH
P_Max RE->TH: 0.3

Bottom margin out of interval [0,1; 0,9] of the RND values allocated for the marginal
transition RE->AW
P_min RE->AW: 0.3

Upper margin out of interval [0,1; 0,9] of the RND values allocated for the marginal
transition RE->AW
P_Max RE->AW: 1

◀ **Fig. 7.18** The list of optional OAM parameters used for definition of the probabilistic marginal transitions between the BMS (basic mind states: *DS* deep sleep, *AW* wakeful, *TH* thinking, *RE* resting). The list is obtained from the V-lab simulation platform

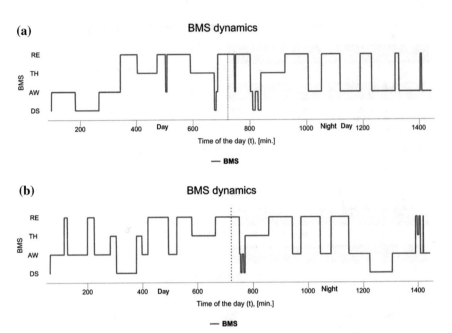

Fig. 7.19 BMS dynamics for two chronotypes—early birds and night owls. The graphs are obtained from the V-lab simulation platform, using the lists of basic and optional OAM parameters, see Figs. 7.17 and 7.18

Table 7.4 Number of transitions between the BMS (reiterated for 10 agents), using the lists of basic and optional OAM parameters, see Figs. 7.17 and 7.18

Number of transitions	Early bird		Night owl	
	Average	St. dev.	Average	St. dev.
Total number of transitions	31.9	2.18	31.8	2.66
– At daytime	15.6	2.17	14.7	1.16
– At nighttime	16.3	1.89	17.1	2.81
Total number of marginal transitions	28.4	1.65	27.3	1.64
– At daytime	13.4	1.43	13.1	1.85
– At nighttime	15	1.49	14.2	2.66
Total number of in-between transitions	3.5	1.90	4.5	1.65
– At daytime	2.2	1.55	1.6	1.51
– At nighttime	1.3	1.06	2.9	1.20

The data is obtained from the V-lab simulation platform

Table 7.5 Duration of staying in a BMS (reiterated for 10 agents)

Duration in states, t (min)	Early bird		Night owl	
	Average	St. dev.	Average	St. dev.
Total in DS	245.7	74.15	255.4	88.16
Total in AW	477.2	93.88	485.6	104.59
Total in TH	172.6	101.48	172.5	69.29
Total in RE	543.6	92.43	525.5	60.44
Average in DS	69.4	24.76	61.7	12.39
Average in AW	37.9	4.22	40.0	4.16
Average in TH	43.8	19.39	45.0	7.13
Average in RE	46.9	2.98	47.2	6.20

The data is obtained from the V-lab simulation platform, using the lists of the basic and optional OAM parameters, see Figs. 7.17 and 7.18

wakeful (AW) states, and less time is spent in the other two mutually even more extreme pairs—deep sleep (DS) and thinking (TH), see Table 7.5. This corresponds well with the neurophysiological observations of the real data, where the RE and AW states dominate during the day, whereas the DS and TH states only occur in the extreme regimes of the brain's functioning. For instance, DS usually occurs when the mind-brain system is completely exhausted and needs deep rest, whereas, thinking is the opposite extreme state of intensive mental concentration, during which the mind-brain system becomes exhausted.

From Table 7.5 we can also analyze the average time spent in each state. In the simulated case, the average time spent in the DS state is the largest, and shortest for the AW state. These simulation results are also not far from the truth. For instance, only 3–5 NREM (non-rapid eye movement, i.e., deep sleep) are usually reported for periods throughout an entire day (Phillips and Robinson 2007, 2008; Phillips et al. 2011, 2010), whereas in the simulated case, we have four DS periods on average.

It is worth noting that in the Tables 7.4 and 7.5, there are relatively high values of standard deviation. This indicates a high variance of agent state dynamics, which is due to the heterogeneity of the agent population in terms of the mind-state behavioral patterns. However, in statistical terms, it also depends on the population size. Hence, an increase in the population size from 10 to 50 agents would decrease the standard deviation values.

We also have to discuss cases where the DS state occasionally occurs during the daytime. In short, this is a consequence of the simulation of real-life situations when the mind reaches the marginal entropy limit after intensive mental activity (e.g., concentrated thinking efforts) and "switches off" to recover in the DS state, see Fig. 7.19. However, such probabilistic transitions can be ruled out using an appropriate set of parameters.

Additionally, the concerted circadian OAM model is also capable of simulating the empirically observed sleep–wake cycles during the nighttime. Depending on the OAM parameters, we can obtain a different number of REM (i.e., RE states that can also be associated with the dreaming state), NREM (DS) and AW/TH cycles during the nighttime. The results obtained are relatively close to the numerous neuro-physiological research results for sleeping stages during the nighttime (Booth and Diniz Behn 2014; Phillips and Robinson 2007, 2008).

For instance, right at the end of the nighttime period (t ∼ 1400 min) we can observe (see Fig. 7.19b) a few characteristic oscillating transitions AW → RE → TH → RE → AW → RE → AW, which are also frequently reported in the experimental EEG measurements of sleep state (REM/NREM/AW/TH) dynamics right before end of the sleep. This distinctive simulation result alone surpasses the capabilities of many other models. Moreover, a similar pattern of rapid oscillations also sometimes occurs at the beginning of the nighttime, when the mind, as often reported empirically, repeatedly wanders between various states, see Fig. 7.19b. However, additional research needs to be done to examine, in detail, the issues and criteria that will help identify the appropriate set of parameters that produce the above-mentioned patterns and other behavior exhibited in BMS dynamics.

The patterns of the BMS dynamics described above are consequences of the fundamental entropy/negentropy and stylized potential/kinetic energy-related processes, which were conceptually reasoned and algorithmically described in the earlier sections of this chapter. It is imperative to observe that it is the interaction of these fundamental processes that produces the oscillations that are essential in the OAM setup. Consequently, the ability to simulate these fundamental processes is a unique feature of the proposed approach.

In the designed simulation setup, each agent's mental behavior patterns (BMS) are uniquely represented by the oscillatory entropy/negentropy and stylized potential/kinetic energy-related processes, see Fig. 7.20 Oscillatory nature of an agent's behavior revealed in terms of the unique oscillatory patterns of entropy and stylized energy (potential and kinetic) dynamics. To the author's knowledge, there is no close modelling in this field of research that focuses on such a simulation framework, using oscillatory principles and fundamental variables that realize the oscillatory mechanisms in such a virtual setting.

Hence, in Fig. 7.20, the characteristic curves $S(t)$, $E_p(t)$, $E_k(t)$ are depicted for the previously mentioned early bird and night owl agent chronotypes throughout the entire period of 1 day. The relationships between these fundamental variables were described in the previously mentioned Sects. 7.3–7.5. As we can observe, during a day, entropy and energy oscillates in periodic, semiperiodic and at times, chaotic periods.

It is quite natural that prospective investigation of these patterns in terms of frequency, amplitude and relative phase should characterize different types of agents and their behavioral patterns. However, at this stage of the research, we only analyzed the ratio [daytime/nighttime] of entropy S, kinetic E_k, and potential E_p energy values for the early birds and night owls as shown in Table 7.6.

Even without an in-depth analysis of the frequency, amplitude, and relative phase of the agent oscillations, we can observe some minor differences in Table 7.6 in terms of the estimates of the ratio [daytime/nighttime] averaged over all states for entropy S, kinetic E_k, and potential E_p energy. For instance, for both early birds and

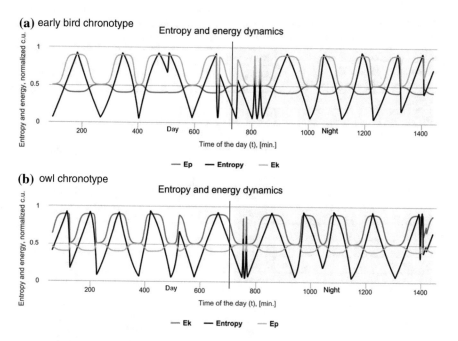

Fig. 7.20 Oscillatory nature of an agent's behavior revealed in terms of the unique oscillatory patterns of entropy and stylized energy (potential and kinetic) dynamics. The graphs are obtained from the V-lab OAM simulation platform, using the lists of basic and optional OAM parameters, see Figs. 7.17 and 7.18

night owls, the above-mentioned ratio is higher for S and E_k compared to E_p. This suggests that the levels of S and E_k (measured in c.u.) averaged over all states during the daytime (range above 1 c.u.) are higher compared to the night time (range below 1 c.u.). These results correspond well with the real life observations—during the daytime, active mind states occur a bit more frequently when entropy and stylized kinetic forms of energy are higher, whereas, during rest at nighttime, the mind states occur a bit more frequently when negentropy and stylized potential forms of energy are higher (see Sect. 7.1: Eq. 7.6 and Fig. 7.1). The standard deviation values do not invalidate this result.

In this regard, concerning the particular BMS, we also observe that:

– Exceptionally high ratios are obtained for the early birds (in the range of 4 c.u.) and night owls (in the range of 2 c.u.) in the DS. This indicates big differences between the S, E_k and E_p averaged levels during the daytime and nighttime for the deep sleep state; one possible explanation is due to the marginal transitions to the DS, which occur during the daytime when S_{max} and $E_{k\ max}$ are reached and the mind stays for some time at a high entropy level; however, it does not quite account for the unusually high differences between the E_p ratios for the early birds and night owls (3.3 vs. 1.7 c.u.), which might be due to the very high

Table 7.6 Ratio [daytime/nighttime] of entropy S, kinetic E_k and potential E_p energy values (statistical reiteration results for 10 agents)

Ratio [daytime/nighttime] of entopy S, kinetic E_k and potential E_p energy values	Early bird		Night owl	
	Average	St. dev.	Average	St. dev.
For all states				
S	1.037	0.080	1.073	0.064
E_k	1.013	0.035	1.033	0.025
E_p	0.995	0.014	0.987	0.009
For DS				
S	4.357	4.856	2.352	1.169
E_k	3.652	3.546	1.983	0.883
E_p	3.330	3.102	1.681	0.686
For AW				
S	1.013	0.597	0.970	0.243
E_k	1.007	0.589	0.922	0.209
E_p	1.005	0.570	0.885	0.203
For TH				
S	0.995	0.603	0.683	0.758
E_k	0.978	0.586	0.673	0.695
E_p	0.979	0.577	0.668	0.675
For RE				
S	0.885	0.318	1.254	0.624
E_k	0.863	0.290	1.215	0.494
E_p	0.828	0.248	1.167	0.382

The data is obtained from the V-lab simulation platform, using the lists of basic and optional parameters, see Figs. 7.17 and 7.18

standard deviation of the results obtained. Therefore, some further investigation with a larger population size might improve the statistical reiteration estimates.
– The early bird and night owl ratios also differ in the AW, TH, and RE states. For the early birds, the active AW and TH states tend to have higher ratios, whereas the night owls have higher rates in the RE state; such a result naturally indicates that during the daytime, the early birds are more active compared to the night owls, whereas, the night owls are spend more time during the day in the RE state.

In the proposed OAM simulation approach, entropy S (or negentropy N) is the major underpinning factor, which governs all the BMS-related processes. In this connection, during the day, the mind in terms of the mind state processes probabilistically travels through an a priori set entropy map $S(t, n)$. According to the proposed conceptual approach, we have introduced the entropy map, to cover all the BMS processes and the transitions between them in terms of entropy dynamics, see Eqs. 7.37 and 7.38. In this way, at each moment in time, the mind is involved in

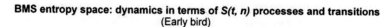

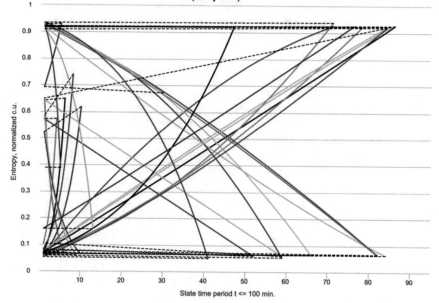

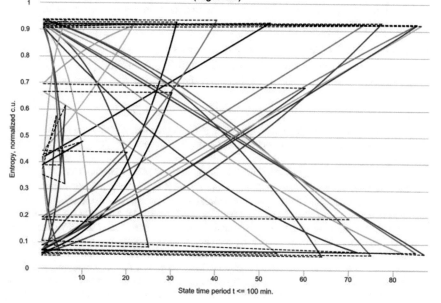

◀ **Fig. 7.21** Samples of the BMS entropy space charts depicted to illustrate the daily $S(t, n)$ dynamics for the early bird (*upper chart*) and night owl (*lower chart*) in terms of their characteristic entropy processes and probabilistic marginal and in-between transitions between the BMS. The results were obtained from the V-lab OAM simulation platform, using the lists of basic and optional OAM parameters, see Figs. 7.17 and 7.18

one or another entropy-related BMS process, which is represented by the associated curve's section in the entropy space chart, see Fig. 7.15. In each simulation, the probabilistic nature of the marginal and in-between transitions between the mind states (BMS) creates a unique pattern for these actualized curves.

Thus, in a nutshell, the daily dynamics of an agent's BMS in terms of the $S(t, n)$ processes and associated probabilistic marginal and in-between transitions can be represented using entropy space charts, see Fig. 7.21.

Let us remember that the BMS entropy space charts presented here were originally developed in the OAM framework. In the charts, each curve represents a corresponding BMS process in terms of the entropy $S(t, n)$ dynamics.[48] Here, entropy $S(t, n)$ curves are plotted in a line style. Meanwhile, transitions are marked in a dashed straight line (they return the state related time counter to zero). It follows the OAM, where each BMS process (curve) ends with a marginal or in-between probabilistic transition to another BMS process.[49] In this way, a new BMS process begins again from $t = 0$. Hence, in the BMS entropy space charts, time indicates the duration of a particular BMS process. Such an approach not only helps to visualize each BMS process, but it also helps to see the progression and overall picture of the BMS dynamics.[50] For instance, in the dynamics of the BMS entropy space charts, we can observe

- repeating BMS processes as dense curve areas
- marginal transitions occurring at the S_{min} and S_{max} margins
- in-between transitions occurring between the S_{min} and S_{max} margins

---

[48]Here, t does not denote a daytime period, but a particular state time period, and n denotes the number of the associated S curve, depending on the day period, see Eqs. 7.37 and 7.38 and Fig. 7.15.

[49]Entropy S margins $[S_{min}; S_{max}]$ are defined in the parameters setup, see Fig. 7.17.

[50]In the on-line V-lab simulation platform, colors are used to distinguish each entropy $S(t, n)$ process. Plus, users can choose which entropy process/es they want to see on the screen. It helps a great deal to visualize and analyze all the dynamics. Additional note: following the OAM setup, horizontal dashed lines denote transitions between state processes. That is, when one state's process $S(t, n)$ ends, then a transition occurs, and another state process starts from the last S value at a resettled time moment ($t = 0$). Therefore, the entropy conservation law should be in place and transitions should be denoted as dashed horizontal lines. However, due to programming approximations, we have not avoided some small biases in the S levels during transitions in the on-line OAM simulation. Therefore, some dashed transition lines are not strictly horizontal, i.e. during such transitions, a small difference in S values can be observed. This is due to a programming approximation problem for the function $S(t, n)$ between sequential BMS. However, the mathematics of the OAM model does not have this fault. Therefore, the visualization shows the approximate $S(t, n)$ values between transitions.

- DS and RE state processes as descending curves
- AW and TH state processes as ascending curves
- duration of each BMS, etc.

Next, in order to show the soundness and robustness of the OAM setup, we will briefly illustrate how the simulation results depend on a few basic parameters, see Fig. 7.17:

(i)   entropy S margins $[S_{min}; S_{max}]$
(ii) checking time period used for recalculation of the in-between transitions.[51]
(iii) probability of the in-between transition in each BMS applied every checking time period.

Hence, the dependency of the number of transitions during a day resulting from the shift of entropy margins $[S_{min}; S_{max}]$ for the early bird and night owl are depicted in Fig. 7.22.

In general, the simulation results obtained are in accord with the OAM assumptions and quite naturally follow common sense. That is, the levels of the entropy margins control probabilistic, unconditional marginal transitions. As these levels decrease, marginal transitions tend to happen more often, as the BMS $S(t, n)$ processes reach these margins more often. Hence, agents with wide entropy margins tend to have a smaller number of marginal transitions and BMS during a day, whereas agents with narrow entropy margins tend to have a higher number of marginal transitions and BMS during a day.

For illustration, charts of the BMS, energy dynamics and BMS entropy space for the early birds and night owls are depicted in Figs. 7.23 and 7.24 respectively. As we can see, the entropy margins clearly change:

- the number and duration of BMS.
- the oscillation frequency of the associated entropy and energy dynamics.
- the BMS entropy space patterns.

In sum, the latter simulation results provide more observations regarding how the S margins regulate:

(i)   the various experimentally observed physiological mind state behavioral patterns, e.g. repeating REM (BMS_{RE}), NREM (BMS_{DS}), and waking (BMS_{AW}) rhythms during the nighttime and the repeating rhythms between BMS_{AW}, $BMS_{TH,}$ and occasionally BMS_{RE} during the daytime.
(ii) the amplitudes of the entropy and energy oscillations (wider allowable S margins produce higher amplitudes).

---

[51]The time period for the recalculation of the in-between transitions is a parameter, which sets (checking) the time periods when the probabilistic algorithm is employed to define (a) whether a transition at the particular time period should take place, (b) the next BMS. It is a purely probabilistic process, where the probabilities for transitions to other states are defined in the same list of the basic parameters, see Fig. 7.17.

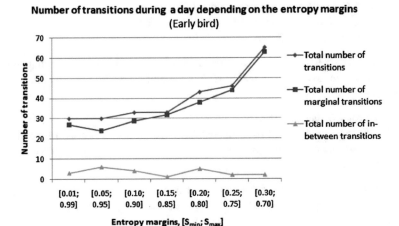

Fig. 7.22 Graphs depicting the averaged dependency of the number of transitions during a day subject to change of entropy margins $[S_{min}; S_{max}]$

(iii) the differences between early bird and night owl behavior, e.g., for the night owls, a decrease in the S margins tends to comparatively increase the DS state duration and frequency during the day compared to the night (night owls tend to be less active during the daytime period).

In fact, deeper comparative analysis is required to examine the effects created by change in the S margins, e.g., number and duration of the BMS or frequency, amplitude, and phase of the associated entropy and energy oscillations for the early birds and night owls. The same applies to the comparative analysis of the corresponding changes in the BMS entropy space patterns. However, even this brief analysis provides some insights about the importance of the S margins $[S_{min}; S_{max}]$,

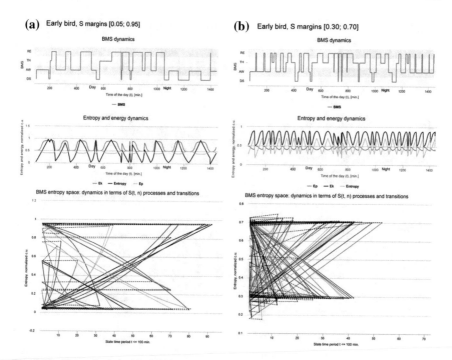

Fig. 7.23 Charts depicting the BMS, energy dynamics and BMS entropy space for the early birds subject to different values of the S margins $[S_{min}; S_{max}]$: $[0.05; 0.95]$ in the A case, and $[0.30; 0.70]$ in the B case respectively

basic parameters in the overall behavior of the simulated agents' main characteristics, see Figs. 7.23 and 7.24.

Next, we should examine how the simulated agents' behavior changes subject to another basic parameter—the time period for the recalculation of the in-between transitions, see Fig. 7.17, where it is called the checking time period of the in-between transitions. Let us remember that this is the parameter that sets the length of the recalculated periods, which define whether a transition at the particular (checking) time period should take place. These checking periods take place during a current BMS process before it reaches its entropy or energy margins. Therefore, we can call such transitions "in-between." Similarly, like in the case of marginal transitions, they can probabilistically initiate transition to another BMS process. Consequently, the occurrence of the probabilistic in-between transitions is settled for the particular time periods when the recalculation of the in-between probabilities is taking place. The transitions are purely probabilistic, and are defined in the same list of the basic parameters for each BMS, see Fig. 7.17.

Following the basic parameters setup listed in Fig. 7.17, the simulation results of the in-between transitions for seven different recalculation (checking) time periods (3, 5, 7, 9, 11, 13, and 15 min) are provided below, see Fig. 7.25.

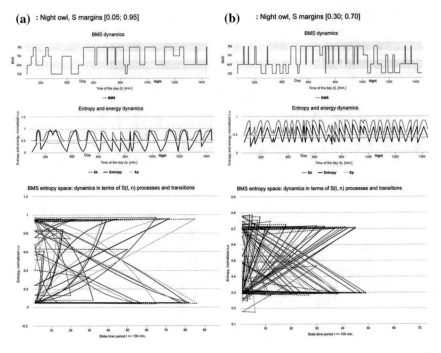

Fig. 7.24 Charts illustrating the BMS, energy dynamics and BMS entropy space for the night owls subject to the different values of the basic parameter—S margins [S_{min}; S_{max}]

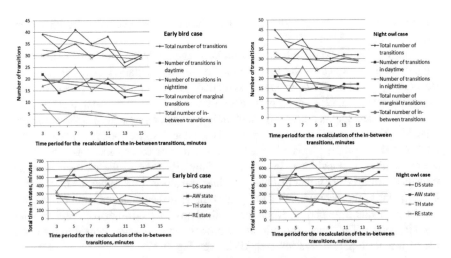

Fig. 7.25 Illustration of a number of transitions (*upper left* and *right graphs*) and the total time spent in the corresponding BMS (*lower left* and *right graphs*) for the early bird and night owl cases subject to the time period used for the recalculation of the probabilities of the in-between transitions. Linear trendlines have been added to accentuate their averaged slopes

Thus, the analysis depicted in Fig. 7.25 can be shortly summarized as follows:

- As expected, the number of total (daytime and nighttime), marginal and in-between transitions gradually diminishes for the early bird and night owl cases subject to the increase in the time period used for the recalculation of the probabilistic in-between transitions (see top left and right graphs).
- For both the early bird and night owl cases, a similar slight increment was observed in the total time spent in the less extreme, but opposing AW (active wakefulness) and RE (resting) states, subject to the increase in the time period used for the recalculation of the probabilistic in-between transitions (see the bottom left and right graphs).
- For both the early bird and night owl cases, a similar slight reduction was observed in the total time spent in the most extreme and opposing TH (thinking) and DS (deep sleep) states, subject to the increase in the time period used for the recalculation of the probabilistic in-between transitions (see the bottom left and right graphs).
- The latter two results indicate that in the proposed model, a decrease in the number of in-between transitions leads to the dominance of the less extreme BMS, i.e., RE and AW. It also points to the fact that an increase in the number of the in-between transitions comparatively increases the total time spent in the more extreme, but opposing TH and DS states.
- Thus, an increase in the time period used for the recalculation of the probabilistic in-between transitions not only diminishes the total number of (daytime and nighttime) in-between transitions, but also differentiates the total time spent in the less extreme and more extreme BMS.
- The above-mentioned behavioral dynamics for the early bird and night owl cases were similar.

Now let us examine the influence of one more chosen parameter on the BMS dynamics—the probability of the in-between transitions as applied equally for each BMS in every time period used for the recalculation of the probabilities of the in-between transitions, see Fig. 7.17. Since each BMS has its own corresponding parameter (probability), all the corresponding parameters were equally changed in the interval [0.01; 0.31] using increments of 0.03; the common results are depicted in Fig. 7.26.

As we can notice from Fig. 7.26, there is a direct link between the increasing probabilities of the in-between transitions and the number of transitions (total, during daytime, during nighttime, marginal, and in-between) for early birds and night owls (see the upper left and right charts, respectively). This is a natural outcome following the model setup. Plus, we can also see that the total time spent in the corresponding states (DS, AW, TH, RE):

(a) does not change for the early birds.
(b) may have two differentiated tendencies for the night owls: a direct dependency for the DS and TH states and an inverse dependency for the RE and AW states (however, this has to be verified because the variations are very high).

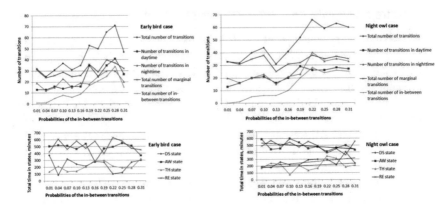

Fig. 7.26 Illustration of the number of transitions (*upper left* and *right graphs*) and total time spent in the corresponding BMS (*lower left* and *right graphs*) for the early bird and night owl cases, subject to the probabilities of the in-between transitions in each BMS

The charts depicted in Fig. 7.27 clearly illustrate sharp differences between the BMS, energy dynamics and BMS entropy space patterns observed for the chosen extreme probabilities of the in-between transitions (0.01 for the left side and 0.28 for the right side charts). As we can see, the probability of the in-between transitions effectively controls the total number of states and transitions during a day (see the top two charts). The entropy and energy patterns (see the middle two charts) also confirm this—higher probabilities result in a increasing number of in-between transitions, occurring before entropy and the corresponding marginal energy values reached (see the small ripples in the middle right chart). Meanwhile, the entropy space charts depict another outcome—the segregation of transitions in the time periods used for the recalculation of the probabilities of the in-between transitions, which are recalculated every 10 min period; see the transitions at the 10, 20 and 40 min time intervals in Fig. 7.27, lower right chart. In short, these results follow the intended OAM simulation setup and provide rich opportunities to tune the agent model according to design needs.

In the example above, the probabilities of the in-between transitions were applied equally for all the BMS. That is, the corresponding probabilities of the in-between transitions were equally changed for every BMS in the interval [0.01; 0.31]. However, the OAM model can be also used to explore the impact of changing the probabilities of the in-between transitions of particularly chosen BMS, while the other states' probabilities of in-between transitions remain unchanged.

In Fig. 7.28, an example of this is depicted, where only the probability of the resting (RE) state was changed, whereas, the other BMS probabilities for in-between transitions were equal to 0.05. For illustration purposes, only the extreme probability values are depicted, 0.05 (chart A1) and 0.50 (chart B1).

Let us keep in mind that the right side charts, A2 and B2 (see Fig. 7.28), were generated using a statistical gamma distribution function $\Gamma(k,\mu)$ with two

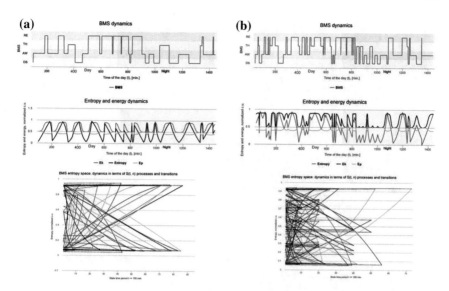

Fig. 7.27 Charts illustrating two extreme examples of the BMS, energy and BMS entropy space dynamics for the early bird case, depending on the probabilities of the in-between transitions: **a** *Left side charts*—probability 0.01, **b** *Right side charts*—probability 0.28

controlling parameters (k,μ), associated with the entropy S and total internal energy U, respectively (see Sect. 7.3.).[52] The gamma distribution function, depending on the values of the controlling parameters (S,U), approximates the corresponding EEG PSD distributions for all the BMS (see Fig. 7.10 and Eq. 7.25). In this way, each point in the BMS, entropy and energy dynamic charts can be transformed into the characteristic EEG PSD chart (see Chap. 4 for more details about the relevant EEG experiments and analysis). Hence, in Fig. 7.28, the A2 graphs demonstrate the active wakeful (AW) and resting states (RE) for points 134 and 368 from the A1 BMS dynamics chart. The corresponding PSD distributions were generated using the appropriate S and U values at these points. In a similar way, the PSD distributions are generated (see B2 graphs) for points 385 and 459 (see BMS dynamics chart B1). As the simulation results reveal, the PSD activation wave continuously moves throughout the EEG spectral range for each state and between states. In this manner, the simulated PSD distributions approximately follow the empirical EEG observations of the real PSD distributions typical for each BMS (see Chaps. 3 and 4).

Coming back to Fig. 7.28, the higher probabilities of the in-between transitions (see the entropy and energy dynamics in chart B1) reveal yet another OAM property: the entropy S oscillations appear to be modulated by the less frequent oscillations of the stylized kinetic energy E_k. Such a property follows from the

---

[52]Shape parameter k and scale parameter μ can be naturally associated with the relative levels of entropy S and potential field type energy E_p respectively.

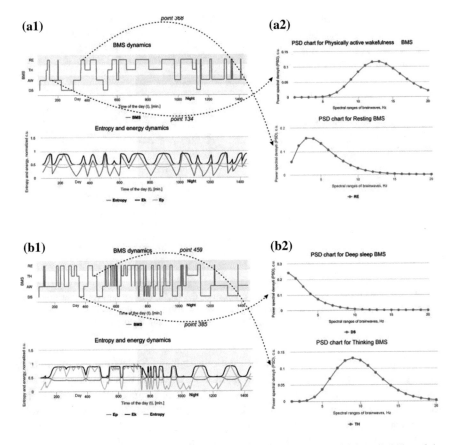

Fig. 7.28 Charts **a1** (probability of the in-between transition = 0.05) and **b1** (probability of the in-between transition = 0.50) depict an example of extreme BMS, entropy and energy dynamics for the early bird case in the RE state, while the other states' probabilities remain constant (0.05). The right side charts, **a2** and **b2**, illustrate the stylized EEG power spectral density (*PSD*) distribution graphs, generated by the OAM, for the chosen BMS points (see the *dashed lines*)

OAM setup, where the frequency of the S oscillations is slightly affected by the marginal and in-between transitions, and the energy oscillations are much more sensitive to the marginal, and less sensitive to the in-between transitions.

In the proposed OAM approach, the key driving factor of the BMS processes is entropy, whereas, energy a secondary factor of dependency (see Eqs. 7.3–7.6 and 7.10–7.13), which follows the trend of the S oscillations. From another viewpoint, the in-between transitions can be understood as a depiction of probabilistic tunneling between the BMS processes, which bypass an a priori set of marginal transition conditions. We can assume that the probabilistic nature of the in-between transitions is an intrinsically inherent property of some entropy (negentropy) driven, complex neurodynamic processes.

At this time, due to objective reasons, we cannot fully understand the complexity of the neurodynamic processes that create the dynamics observed between the states. Therefore, the author has proposed an approximation by using a model of the marginal and in-between transitions between the entropy and stylized energy-driven mental processes (see previous sections) called BMS. The validity of the proposed simulation approach stands on firm ground, as it employs the most fundamental physical, although stylized, terms of entropy, and energy.

To the author's knowledge, the closest and most prospective alternative theoretical approach at this point in time can be attributed to the 'Orch OR' (orchestrated objective reduction) theory of consciousness proposed by Hameroff and Penrose (2014), see Sect. 7.2., where observed EEG wave patterns are treated as interference beats taking place within a much higher frequency range (MHz). However, at this point of development, even this theory cannot provide conclusive answers regarding states of consciousness and their observed dynamics.

Thus, we have consistently explored just a few basic parameters of the OAM simulation approach. However, according to the lists provided in Figs. 7.17 and 7.18, there are also other basic and optional parameters, which have their own influence on the simulation results. They should be explored in prospective research too. In this sense, future research of the OAM simulation framework could reveal more complex properties and opportunities for use of the proposed simulation approach.

In fact, the OAM setup was targeted to simulate agent circadian rhythms based on the BMS processes and probabilistic marginal and in-between transitions. However, the probabilistic nature of transitions requires further study and more in-depth modelling. In essence, to the author's point of view, probabilistic transitions should be substituted with parametric transitional models based on, for instance, Orch OR theory, the Kuramoto oscillators' model, neurophysiological and other biophysical or empirical data driven models. In this way, the proposed probabilistic transitions setup would serve as a ready to use simulation framework to be enhanced with a sound parametric or empirical transitions model.

In the next chapter, based on the OAM described above, an OAM-based MAS (multi-agent system) simulation framework is presented along with simulation results. Hence, the OAM serves an important role in the larger scheme of MAS development. Essentially, we will explore how an oscillations-based paradigm may be employed in a coherent multi-agent system setup. In this way, we lay the foundations for a principally novel approach to social system simulation.

Chapter 8
OAM-Based MAS: Collective Circadian Rhythms' Simulation Setup and Results

8.1 OAM-Based MAS Construction Principles

In this chapter, multi-agent system (MAS) simulation setup is essentially based on the previously described oscillating agent model (OAM, see Chap. 7). In general, OAM-based MAS construction is aimed at simulating group-wide, i.e., collective mind-brain states and correspondingly emerging coherent phenomena like social circadian rhythms, collective synergy effects, agents' spatial, or virtual features based clustering in the Hilbert space, etc. The OAM-based MAS simulation is oriented to reveal the basic conditions and parameters for the emergence of some coherent group-wide behaviors. In turn, the modeling of these conditions and parameters should add a conceptually novel understanding of social phenomena and new ways to model MAS.

Hence, this chapter is aimed toward the simulation setup for a group of agents, which coordinate (synchronize) to some degree in their internal BMS processes not by using pair-to-pair interaction, but a field-like coordination that pertains to (i) the common external circadian rhythms[1] and (ii) some basic collective synergy effects, which save the systems' total energy and decrease entropy.[2]

---

[1]According to the literature, daily rhythms in sleep and waking performance are generated by the interplay of multiple external and internal oscillators. These include the light–dark and social cycles, a circadian hypothalamic oscillator oscillating virtually independently of behavior, and a homeostatic oscillator primarily driven by sleep-wake behavior. Both internal oscillators contribute to variation in many aspects of sleep and wakefulness (e.g., sleep timing and duration, REM sleep, non-REM sleep, REM density, sleep spindles, slow-wave sleep, electroencephalographic oscillations during wakefulness and sleep, and performance parameters, including attention and memory). The relative contribution of the oscillators varies greatly between these variables. Sleep and performance cannot be predicted by either oscillator independently, but critically depend on their phase relationship and amplitude (Dijk and Schantz 2005; Skeldon et al. 2014).

[2]The latter effects are due to observations that the simulation of natural coherent and/or synchronized behavior (e.g., bacterial colony optimization, swarm optimization, ant colony optimization, etc.) can also be applied for human "swarming", which, via control systems, enable real-time groups of human participants (agents) to behave as a unified collective intelligence (mass collaboration). There are plenty of applications in sociobiology, political science, crowdsourcing, social media, voting systems, etc.

© Springer International Publishing Switzerland 2016
D. Plikynas, *Introducing the Oscillations Based Paradigm*,
DOI 10.1007/978-3-319-39040-6_8

Admittedly, in the proposed simulation scheme, there are two major mutually related levels of self-organization, namely, individual (earlier described OAM) and group-wide (MAS or multi-agent system). To begin with, we have the following tasks:

1. to adapt a single agent's OAM simulation setup for the group-wide MAS case and afterwards
2. to define the group-wide constraints and optimization goals for the MAS simulation level.

Starting from the first task, we admit that agents can be homogeneous or heterogeneous in terms of their characteristic set of parameters. In the heterogeneous agents' case, their individuality can be differentiated using OAM parameters (see the previous chapter) such as:

– entropy margins S_{max}, S_{min} (see Fig. 7.8)
– energy margins $E_{k\ min}$, $E_{k\ max}$, $E_{p\ min}$, $E_{p\ max}$ (see Fig. 7.8)
– $E_p(S)$ and $E_k(S)$ curves' logistic parameters S_0, k, L_1, L_2 (see Eq. 7.19 and Fig. 7.7)
– freely chosen energy curves' parameters C_1 and C_2 (see Eqs. 7.20 and 7.22, Fig. 7.9)
– constraints of probabilities for the marginal transitions between states, see Table 7.3 and Fig. 7.14a
– constraints of probabilities for the in-between transitions like p'_{max} (see Fig. 7.14 b, c)
– night owls or early birds options of transitions between states (see Fig. 7.13).

Depending on the simulation case, a researcher chooses some of the above listed parameters for the individual needs of the agents in a concrete OAM setup. Next comes the mathematical description of the interaction mechanism between agents. Admittedly, we are not looking for the commonly used peer-to-peer or matchmaking agent-based communication approach.[3] Instead, in following OSIMAS

---

[3]In the peer-to-peer based communications of ABM (agent based models) or MAS applications, (i) the sender and receiver are constrained by time and space, (ii) the agents exchange ontologically classified semantic information by using a common communication protocol or a matchmaking agent. However, such communication is not efficient enough for multitasking, parallel processing, broadcasting, congested traffic control, conflict resolution, etc. In short, the current approaches lag behind digital reality, where we often observe one-to-many, many-to-one and many-to-many interactions, which can be attributed to the nonlocal distributed cognitive systems. In the context of modern ICT (information and communication technologies), each agent can be understood not only as individual, but also as a distributed cognitive system, which unconsciously internalizes and, therefore, shares social norms, behaviors and, more broadly, the cognitive environment, see Appendix C. In short, in modern information societies (1) each agent can instantly send and receive information simultaneously through multiple communication channels, (2) information flows are locally managed by the agent's preferences as if having the ability to "tune" to different broadcasting channels, (3) agents become producing, processing, storing and retransmitting information nodes in the social networks (network economy).

premises (Plikynas et al. 2014), we base our reasoning on the pervasive information field (PIF) theoretical concept actualized in the form of a potential energy field.[4]

However, for practical modeling reasons, we adapt a simple approximation, using a slightly changed thermodynamic free energy term, see Eqs. 7.8 and 7.10

$$A = U - TS + \tau \breve{N}, \tag{8.1}$$

where the additional external energy term $\tau \breve{N}$ can be expressed as

$$\tau \breve{N} = \frac{\partial U}{\partial \breve{N}} \breve{N} = \frac{\partial (E_k + E_p)}{\partial \breve{N}} \breve{N} = \left(\frac{\partial E_k}{\partial \breve{N}} + \frac{\partial E_p}{\partial \breve{N}} \right) \breve{N} \xrightarrow{\frac{\partial E_k}{\partial \breve{N}} \to 0} \frac{\partial E_p}{\partial \breve{N}} \breve{N}, \tag{8.2}$$

here, $\breve{N} > 0$ denotes external negentropy (in short, order), which pertains to the conceptual OSIMAS idea of the PIF (pervasive information field); $\tau \breve{N}$ actualizes the coherent collective mind field in terms of the potential energy field.

Partial derivative $\tau = \partial U / \partial \breve{N}$ denotes the sensitivity of an agent's internal energy to the small change in the external superposed collective mind state denoted as $\partial \breve{N}$. For some agents who are not sensitive to the collective mind field, $\partial E_p / \partial \breve{N} \to 0$ and consequently $\tau \breve{N} \to 0$. Thus, this sensitivity term is an additional important intrinsic characteristic of each agent. In essence, it determines the degree of "atonement" or coherence with the collective mind field.

There is one more important consideration concerning the term $\partial E_p / \partial \breve{N}$—it depends on the function $E_p(\breve{N})$, which presumably takes a common form of a logistic function, see Eq. 7.20. In this way, a gradual increase in $\breve{N}$ evokes a nonlinear change in the E_p, see Fig. 7.7.

In fact, the additionally included external energy term $\tau \breve{N}$ is constructed in a way that is similar to the traditional heat waste term $-TS$. However, both terms work in the opposite direction—they have different signs and different origins. That is, the energy term $-TS$ is an internal product of the agent's mind-brain processes,

---

[4]A PIF concept stems from the mean-field theory (MFT), also known in physics and probability theory as the self-consistent field theory, which models the effect of all the other individuals (which interact with each other) on any given individual using an approximation of a single averaged effect, thus reducing a many-body problem to a one-body problem with a chosen good external field. The external field replaces the interaction of all the other individuals to an arbitrary individual. Then it naturally follows that if the individual (agent) exhibits many interactions in the original system, the MFT will be more accurate for such a system. This is true in cases of high dimensionality, or when the Hamiltonian includes long range forces (Bensoussan et al. 2013). In the OAM-based MAS, agents can also be placed in the virtual high dimensional Hilbert space of mental states and be affected by long range forces. Actually, in the proposed OSIMAS paradigm, the pervasive information field (PIF) concept can technically be modeled using the MFT approach.

whereas, the energy term $\tau\breve{N}$ reflects the agent's mind-brain reaction to the PIF, i.e., the superposed collective mind field.

When $\Delta\breve{N}(t) < 0$, the coherence of the collective mind field decreases and for $\Delta\breve{N}(t) > 0$ it increases. Admittedly, the collective mind state has a direct influence on an individual agent's state. The agent's mind state reacts to environmental (collective mind) changes depending on the individual sensitivity term $\partial U/\partial\breve{N} = \partial E_p/\partial\breve{N}$, see Eq. 8.2.

In general, the collective mind field $\breve{N}$ not only fluctuates in time, but also has a dynamic distribution in Euclidean space that can be denoted as the gradient of a scalar function

$$\nabla\breve{N}(x,y,z,t) = \frac{\partial\breve{N}(t)}{\partial x}i + \frac{\partial\breve{N}(t)}{\partial y}j + \frac{\partial\breve{N}(t)}{\partial z}k, \qquad (8.3)$$

where i, j, and k are the standard unit vectors in Euclidean 3D space. Gradient $\nabla\breve{N}(x,y,z,t)$ depends on the spatial distributions of the agents who are actually creating the collective mind field $\breve{N}$.

In the presented application case, we are concerned about construction and simulation of the circadian daily BMS rhythms of the OAM-based MAS, however, in general, a space metric can be arbitrarily adapted according to the spatial application case. That is, not only 2D or 3D spatial Euclidean space, but abstract Hilbert vector spaces can also be used with any finite or infinite number of dimensions. After all, mental functions (sometimes called cognitive functions) determine the things that individuals can do with their minds. These include perception, memory, thinking (such as ideation, imagination, belief, reasoning, etc.), volition, and emotion. Each of these cognitive functions is capable of formatting high dimensional mental spaces that can be modeled, e.g., using Hilbert vector spaces.

For instance, in the case of emotional mental space, there are several well recognized emotional states,[5] which can be modeled (similarly like the BMS in the OAM setup) using specially designed virtual Hilbert space in order to represent the agents' movements between emotional states. In this sense, the virtual space opportunities for the simulation of various cognitive functions are immense. In prospective research, the simulation of the spatial dynamics of clustering patterns of mental states in Euclidean and also virtual Hilbert spaces is also foreseen.

---

[5]For instance, positive and negative emotions: curiosity, alarm, panic, attraction, desire, admiration, aversion, disgust, revulsion, surprise, amusement, indifference, familiarity, habituation, gratitude, thankfulness, anger, rage, joy, triumph, sorrow, grief, relief, frustration, disappointment, love, hate, etc. All these emotional states can together constitute a high dimensional emotional space that could be modeled in virtual Hilbert space.

Coming back to the external energy, it can be interpreted as the superposed mind fields of other agents, as each agent's open mind-brain system emits some small part of its internal field energy into the environment via the term −TS, see Eqs. 7.2, 7.3 and 8.2

$$-TS = \frac{\partial(E_k + E_p)}{\partial S} S = \frac{\partial E_k}{\partial S} S + \frac{\partial E_p}{\partial S} S,$$

$$\frac{\partial E_p}{\partial S} S = \frac{\partial \widehat{E}_p}{\partial(S_{max} - N)}(S_{max} - N) + \frac{\partial \breve{E}_p}{\partial(S_{max} - N)}(S_{max} - N), \quad (8.4)$$

$$\partial E_p = \partial \widehat{E}_p + \partial \breve{E}_p, \ \partial \widehat{E}_p \gg \partial \breve{E}_p,$$

where $\widehat{E}_p$ and $\breve{E}_p$ denote an agent's intracranial and extracranial mind fields.[6] We assume that $\breve{E}_p$ takes part in the superposition process with other agents' extracranial mindfields. Aside from the complicated quantum superposition notation, the classical superposition of the extracranial mindfields stemming from the set of agents $\{A_i\}$ can be denoted in a linear way as[7]

$$\sum_{A_i} \left[\tau(t)\breve{N}(t) \right]_{A_i} = \sum_{A_i} \left[\frac{\partial \breve{E}_p(t)}{\partial \breve{N}(t)} \breve{N}(t) \right]_{A_i}. \quad (8.5)$$

With regard to superposition, it can be described as interference (for a few oscillating sources) and diffraction (for many sources). Let us first see how interference works. In the most simple case, if we denote an oscillating agent as the most basic wave (a form of a plane wave)

$$u(x,t)_{A_i} = [A(x,t) \cdot \cos(kx - \omega t + \varphi)]_{A_i} \quad (8.6)$$

or using Euler's formula

$$\psi(x,t)_{A_i} = \left[A(x,t) \cdot e^{i(kx - \omega t + \varphi)} \right]_{A_i},$$
$$\text{Re}\{\psi(x,t)_{A_i}\} = A(x,t) \cdot \cos(kx - \omega t + \varphi), \quad (8.7)$$

then an ideal property of oscillations that enables stationary (i.e., temporally and spatially constant) interference occurs when oscillating sources, i.e., simulated

---

[6]The extracranial mind field part, $\breve{E}_p$ can be registered on the cranium using an EEG.

[7]According to the superposition property, for linear sources of two or more stimuli, the net response at a given place and time is the sum of the responses which would have been caused by each stimulus individually. That is, the homogeneity and additivity properties together are called the superposition property.

agents A_i in this case, are perfectly coherent (synchronized) with the same frequency ω and constant phase difference $\Delta\varphi$.

However, in the OAM case, the frequency term ω is multimodal, i.e., associated with the experimentally measured dominating EEG brainwave frequencies in the delta ω_Δ, theta ω_θ, alpha ω_α, beta ω_β, and gamma ω_γ spectral ranges. Following the OSIMAS paradigm, these spectral ranges have a specific distribution of spectral energy patterns depending on the BMS. That is, each mind state has a specific spectral energy distribution pattern in these spectral ranges. Hence, the set of five wave functions corresponds with the experimental EEG observations of characteristic spectral brainwave distributions that describe the states. Transitions between the BMS actualize the redistribution of energy between the spectral ranges (see Chaps. 2 and 3).

For that reason, instead, of just one wave function, there is a combination of five wave functions for each frequency range

$$\left[\psi(x,t)\big|_{\omega_\Delta,\omega_\theta,\omega_\alpha,\omega_\beta,\omega_\gamma}\right]_{BMS}. \tag{8.8}$$

For convenience, the spectral power of each state is defined as a sum of squared values of the real part of the wave signals in each brainwave frequency

$$\begin{aligned}
P_{A_i} = &\left(\mathrm{Re}\{\psi(x,t)_{A_i}\}^2\right)\big|_{\omega_\Delta} + \left(\mathrm{Re}\{\psi(x,t)_{A_i}\}^2\right)\big|_{\omega_\theta} \\
&+ \left(\mathrm{Re}\{\psi(x,t)_{A_i}\}\right)^2\big|_{\omega_\alpha} + \left(\mathrm{Re}\{\psi(x,t)_{A_i}\}\right)^2\big|_{\omega_\beta} + \left(\mathrm{Re}\{\psi(x,t)_{A_i}\}\right)^2\big|_{\omega_\gamma}.
\end{aligned} \tag{8.9}$$

In order for extracranial synchronization between agents to take place, a set of agents at least in some degree, has to meet the previously mentioned interference conditions. That is, following Euler's formula (see Eq. 8.7), the real parts of the appropriate $\psi\big|_{\omega_\Delta,\omega_\theta,\omega_\alpha,\omega_\beta,\omega_\gamma}$ functions have to be synchronized in phase. In this regard, the phase $\varphi\big|_{\omega_\Delta,\omega_\theta,\omega_\alpha,\omega_\beta,\omega_\gamma}$ lends itself well for the computational modeling of the interference between the agents' brainwave oscillations represented in terms of the spectral power distributions in the five main brainwave frequencies.[8]

In other words, the phases (initial offsets) of individual brainwave frequencies have to be in sync with a collective mind field (PIF), which can technically be modeled in terms of the mean-field theory (MFT). According to the MFT, the effect of all the other individuals on any given individual can be approximated by a single averaged effect, thus reducing a many-body problem to a one-body problem (Bensoussan et al. 2013). A simple version of such an approach is employed in the next section. However, in the prospective research, a bit more sophisticated models,

---

[8]Synchronization can also deal with the cooperative behavior of the chaotic subsystems of agents under noisy and excitable social conditions, where chaotic phase synchronization can take place (Osipov et al. 2007).

based on the Kuramoto model (used to describe synchronization for the behavior of a large set of coupled oscillators) can be created.

Now let us briefly discuss two levels of OAM-based MAS simulations. The first level can be called the "bottom-up" as it uses the OAM approach described in the previous chapter. Here, agents do not exchange energy or information with the surrounding environment. They experience some synchronization via common (i) circadian daily functions, which govern the BMS processes and, (ii) the probabilistic transition rules between the BMS.

The second level can be called the "top-down" as it employs a meta-layer of self-organization, which optimizes the performance of the MAS in terms of the meta-layer laws:

1. It is a truism in this research field that social systems, similarly to other biological systems, naturally tend to increase their total negentropy, i.e., the order or constraints of chaotic behavior (this property is also observed at the cellular and multicellular self-organization levels).[9] We imply that for a system composed from social agents, the law of maximization of total negentropy applies

$$N^{\text{MAS}} = \max\left(\sum_i N^i\right), \tag{8.10}$$

following Eq. 7.3, agent's A_i negentropy $N_i = S_{\max} - S_i$.

2. Hand in hand with the negentropy maximization law, the law of potential energy maximization[10] also applies

$$E_p^{\text{MAS}} = \max\left(\sum_i \breve{E}_p^i\right), \tag{8.11}$$

here, $\breve{E}_p^i$ denotes an agent's A_i extracranial potential energy, which is emitted through the surface of the cranium in the form of an electromagnetic field, see Eq. 8.3. Whereas, E_p^{MAS} pertains to the energy part of the PIF (pervasive information field; see Chaps. 2 and 3).

For both of the mutually related maximization terms, we can simply use one integrated optimization term

$$f_{\text{opt}}\left[N^{\text{MAS}}, E_p^{\text{MAS}}\right]. \tag{8.12}$$

---

[9]However, for nonliving matter, the universal laws of negentropy and energy minimization hold, as they do not have to maintain biological metabolism, self-preservation and organization.

[10]Actually, both maximization terms are two sides of the same coin, where N represents information and E_p, the energy parts of the collective mind field (earlier referred as the PIF).

This integral optimization function represents the meta self-organization level. However, the question is how this optimization function operates for the OAM-based MAS if the number of agents is fixed and holds to the previously described free energy terms and entropy conservation law.

In fact, biological and social systems as well, via natural evolution, have found an efficient way out. They employ some coherent processes to withstand the environment's chaotic pressure via the rhythmic synchronization of the activities of their constituent parts (Fingelkurts et al. 2006).[11] The emerging rhythms or oscillations are themselves a form of a negentropy at the meta-organizational level, which differs substantially from purely chaotic behavior. The rhythms and oscillations are directly related to the emergence of the coherent superposition states at the meta-level. *In sum, the optimization function (see Eqs. 8.10 and 8.11) operates via the emergence of social rhythms that are directly associated with the increment of negentropy and potential energy at the social organization (collective mind) level.*

Admittedly, every order can be assigned to one or another form of rhythms or oscillations, which create synergistic effects, when the otherwise counteracting individual oscillations are coordinated to maintain order at the system level (Osipov et al. 2007). This universal principle is also valid at the social level of self-organization, where the agents' oscillations in the mental space are coordinated to sustain social order (the social level of negentropy).

Hence, the answer to the earlier posed question—how optimization function operates for the OAM-based MAS—*is related with the simulation of emergent social rhythms. Social order and consequently emerging social rhythms can only occur if there is a long ranging feedback mechanism between each agent and the rest of the MAS. Such coordination can be achieved via negentropy and associated potential energy exchange mechanism.*

In this regard, in the proposed OAM-based MAS, the information and associated energy[12] exchange mechanism is explored between the individual agent and social meta-levels, i.e., between agent A_i and the MAS. Let us assume, in following the OSIMAS assumptions, that the energy exchange mechanism works via the change in the potential energy field $\Delta E_p'^i(t)$

$$\Delta E_p'^i(t) \longrightarrow A_i \leftrightarrow MAS \tau^i(t)\, \breve{N}^i(t) - T^i(t)S^i(t),\ \tau^i(t)\, \breve{N}^i(t) > 0, \qquad (8.13)$$

where here, the term $T^i(t)S^i(t)$ indicates agent's A_i time t dependent potential field emitted to the MAS and $\tau^i(t)\, \breve{N}^i(t)$ denotes the potential field absorbed by the same agent from the MAS. The sign of the difference $\Delta E_p'^i$ between both potential fields indicates whether an agent emits more than it absorbs from the MAS environment:

---

[11]The formation of complex, sometimes chaotic structures, where the interacting parts exhibit long range correlations, were first introduced by the physical chemist Ilya Prigogine in his dissipative systems research (Glansdorff and Prigogine 1971).

[12]Energy works as a carrier of information.

if $\Delta E_p^{\prime i}(t) < 0$, then the agent emits more than it absorbs from the MAS,
if $\Delta E_p^{\prime i}(t) > 0$, then the agent emits less than it absorbs from the MAS.

In the first case, the agent emits more than it receives. The agent's inner potential energy feeds the collective mindfield of the MAS. In the second case, an agent shares less than it receives from the surrounding MAS environment. Both of these cases are possible, depending on the agent's properties. Namely, the value of the individual potential energy field term $T^i(t)S^i(t)$ makes a difference. In the MAS setup, we can add an individual parameter $0 < \lambda < 1$, which will denote the agent's willingness or ability to share part of its negentropy and associated potential field energy with the MAS environment

$$E_p^{\prime i}(t) = \lambda \cdot T^i(t) \cdot S^i(t). \tag{8.14}$$

In this way, an agent can share or contribute three major inner characteristics to the MAS environment, namely,

$$\left\{ \overset{\smile i}{N}(t), E_p^{\prime i}(t), \mathrm{BMS}^i(t) \right\}:$$
$$\overset{\smile i}{N}(t) = \lambda \cdot N^i(t), \tag{8.15}$$
$$E_p^{\prime i}(t) = \lambda \cdot T^i(t) \cdot S^i(t),$$

where $\overset{\smile i}{N}(t)$ denotes the shared proportion λ of the agent's A_i inner negentropy $N^i(t)$; the $\mathrm{BMS}^i(t)$ denotes the agent's A_i state at time moment t.

At the same time, each agent receives the average feedback values from the MAS environment, namely

$$\left\{ \overline{N^{\mathrm{MAS}}(t)}, \overline{E_p^{\prime\mathrm{MAS}}(t)}, D_{\mathrm{BMS}}^{\mathrm{MAS}}(t) \right\}:$$
$$\overline{N^{\mathrm{MAS}}(t)} = \sum_{i=1}^{n} \frac{\overset{\smile i}{N}(t)}{n},$$
$$\overline{E_p^{\prime\mathrm{MAS}}(t)} = \sum_{i=1}^{n} \frac{E_p^{\prime i}(t)}{n}, \tag{8.16}$$
$$D_{\mathrm{BMS}}^{\mathrm{MAS}}(t) = \frac{n_{\mathrm{BMS}}(t) - \bar{n}_{\mathrm{BMS}}}{n},$$

where $D_{\mathrm{BMS}}^{\mathrm{MAS}}(t)$ denotes the normalized absolute deviation from the mean number of agents $\bar{n}_{\mathrm{BMS}}$ in each BMS.[13]

---

[13]For instance, if the agent population $n = 100$, the number of agents at time moment t in the thinking state $n_{\mathrm{TH}}(t) = 28$ and $\bar{n}_{\mathrm{TH}} = 100/4 = 25$, then $n_{\mathrm{BMS}}(t) - \bar{n}_{\mathrm{BMS}} = 3$ and $D_{\mathrm{TH}}(t) = 0.03$.

It is important to note that the MAS environment, where an agent operates, communicates its state to the agents via the PIF (so-called pervasive information field), which has the nature of a common information field. In fact, the PIF carries information about the results of the superposition of all the agents' states. Hence, the MAS environment operates via the PIF and the associated superposed potential energy field.

This provides an important clue concerning the nature of the agents' feedback mechanism with the MAS environment. It follows from the above considerations that the superposed MAS environment operates via negentropy terms, which affect the individual agents' mind states in a probabilistic way. That is, the MAS states that via the negentropy terms, it affects the probabilistic transitions between the BMS in the individual agent's level. Energy is associated with the negentropy terms. In this manner, it follows the negentropy terms accordingly. Thus, we can assume that the probabilistic feedback mechanism operates in more of a quantum nature, as it operates at the quantum states level (see Chap. 5). Consequently, the agents' feedback reaction to the MAS environmental stimuli is due to the probabilistic in-between transitions.

Following the above-described general principles of the OAM-based multi-agent system (MAS) setup, we can construct a simple simulation model in the following section.

8.2   Simulation Setup: Coherent Collective Circadian Cycles

This section discusses OAM-based MAS modeling issues and results pertaining to the simulation of the collective circadian rhythms. For the modeling of the coherent collective mind states (CMS), the harmonization mechanism between the individual and CMS is employed, i.e., we use the previously mentioned mean-field approach. We assume that each agent A_i emits information about its state to the environment and at the same time, receives information back about the state of all other agents in the form of an averaged mean field.

In this simple model, we assume that the agents are capable of exchanging information about their basic mind states (BMS) in an appropriate form of negentropy N. The general scheme of the mean-field approach is depicted in Fig. 8.1.

According to the proposed scheme, an agent A_i interacts with the mean field generated by the rest of the agents $\{A_i\}$. In this way, the so-called difficult problem involving multiple body interaction is reduced to the much less problematic interaction of each agent with the averaged mean field produced by the rest of agents. Such an approach ensures a feedback mechanism, which shapes the synchronization process of the agents' mind states.

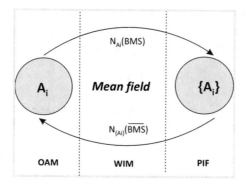

Fig. 8.1 General scheme of the information (negentropy) exchange between an individual agent, A_i and the mean field of a set of agents $\{Ai\}$. The agent is represented in terms of the OAM, and the interaction mechanism is based on the principles of *WIM* [wavelike interaction mechanism (Plikynas et al. 2014)], and the set of agents that $\{A_i\}$ produces creates a mean field called the *PIF* (pervasive information field) in the OSIMAS paradigm

The agent's synchronization with the mean field is realized via the imposed additional probabilistic transitions, which increase correlations between the individual and prevailing CMS. For further reference, we shall call them *harmonizing transitions*. They are similar in kind to the in-between transitions, but have a different origin and higher priority.[14]

For the sake of clarity, let us briefly review the OAM-based MAS model algorithm below. First, for each time moment t there is an estimated number of agents in each basic mind state. This allows knowing a prevailing basic mind state with the maximum number of agents in it

$$\Omega(t)|_{\text{BMS}} = \max(n(\text{BMS}, t)) = \begin{cases} n_{\text{DS}}(t) = \sum \text{BMS}_{\text{DS}} \\ n_{\text{AW}}(t) = \sum \text{BMS}_{\text{AW}} \\ n_{\text{TH}}(t) = \sum \text{BMS}_{\text{TH}} \\ n_{\text{RE}}(t) = \sum \text{BMS}_{\text{RE}} \end{cases}. \qquad (8.17)$$

The prevailing state's $\Omega(t)|_{\text{BMS}}$ relative and normalized intensity $\Lambda(t)$

$$\Lambda(t) = \frac{\max(n(\text{BMS}, t)) - \bar{n}(t)}{\bar{n}(t)}, \qquad (8.18)$$

where $\bar{n}(t)$ denotes the average number of agents across all states.

---

[14]Marginal probabilistic transitions have the highest priority, then the harmonizing transitions follow, and after them, the in-between transitions.

Each agent's probability of a harmonizing transition $p_h(t)$ to the prevailing common state $\Omega(t)|_{BMS}$ is estimated as a product of the prevailing common states's intensity Λ and agent's individual sensitivity parameter ξ_{Ai}

$$p_h(t)|_{A_i} = \Lambda(t) \cdot \xi_{A_i}, \tag{8.19}$$

here, a sensitivity parameter ξ_{A_i} is individually assigned for each agent in a random fashion. However, if a transition is not allowed following the OAM model, then it does not take place.

Next, we want to have (i) individual (for each agent) and (ii) common (for the system of agents) measures of synchronicity with the prevailing collective state $\Omega(t)|_{BMS}$ in each time moment t and for the chosen period of time $T = t_2 - t_1$. Hence, for measuring the individual synchronicity at each time moment, we simply use the binary values of 1 or 0, depending on whether the agent's state $A_i(t)|_{BMS}$ coincides or does not coincide, respectively, with the collective state $\Omega(t)|_{BMS}$

$$c_{A_i}(t) = \begin{cases} 0, \ A_i(t)|_{BMS} = \Omega(t)|_{BMS} \\ 1, \ A_i(t)|_{BMS} \neq \Omega(t)|_{BMS} \end{cases}. \tag{8.20}$$

Similarly, the individual normalized measure of synchronicity for the period $T = t_2 - t_1$ is

$$c_{A_i}(T) = \frac{\sum_{t_1}^{t_2} c_{A_i}(t)}{t_2 - t_1}. \tag{8.21}$$

In this way, $c_{A_i}(T)$ varies in the interval [0; 1]. That is, the synchronicity is very low when $c_{A_i}(T) \to 0$ and the synchronicity is high when $c_{A_i}(T) \to 1$.[15]

In a similar fashion, we estimate the synchronicity level (with the prevailing common state) of the system of agents for the time moment t and for the time period T

$$\Omega_{MAS}(t) = \sum_i c_{A_i}(t)/n, \tag{8.22}$$

$$\Omega_{MAS}(T) = \sum_{t_1}^{t_2} \Omega_{MAS}(t) = \frac{\sum_{t_1}^{t_2} \sum_i c_{A_i}(t)/n}{t_2 - t_1}, \tag{8.23}$$

---

[15]Value $c_{A_i}(T) \to 0.25$ is related with a random walk, because the random probability of the coincidence of the individual and collective states (when we just have four BMS) is equal to ¼. That is, at each time moment, the chance of the coincidence of the individual and collective states is equal to 0.25. Let as also remember that the collective states are the averaged individual states throughout the entire population of agents.

where n denotes the total number of agents in the system. Whereas, $\Omega_{MAS}(T)$ varies in the interval $[0; 1]$. Similarly, like for the $c_{A_i}(T)$, the system's synchronicity is very low when $\Omega_{MAS}(T) \rightarrow 0.25$, and the synchronicity is high when $\Omega_{MAS}(T) \rightarrow 1$.

All the above-mentioned temporal dynamics can be seen for the chosen agent or their system in the corresponding charts, which are generated during simulation sessions in our interactive modeling V-lab environment (http://vlab.vva.lt; User name: Guest, password: guest555). Some charts also depicted a few paragraphs below for illustrative purposes.

However, in order to improve the capacity to monitor the dynamics and coherence patterns of the individual and collective states, we have additionally adapted the Kuramoto oscillators visualization approach (Kuramoto 2012), which uses polar coordinates (r, φ) for the depiction of the oscillators' phases and their collective coherence level. That is, according to the Kuramoto visualization approach, the oscillators are represented as dots on a unitary circle with a radius $r = 1$. Their position on the circle depends on their phases φ.

Using polar coordinates, the collective state of the entire system of oscillators is represented as a central dot (phase centroid) with the polar coordinates $(R_\Omega, \varphi_\Omega)$. Radius R_Ω varies, depending on the coherence level of the entire system as $R_\Omega = \Omega_{MAS}(t)$, see Eq. 8.22. The greater the coherence, the larger the radius R_Ω, where $R_\Omega(max) = 1$, whereas the central dot's inclination φ_Ω denotes the averaged phase, see Fig. 8.2.

In our adapted Kuramoto oscillators' visualization approach, all agents are interpreted as oscillators too. In a similar way, they are depicted on the radius of the unitary circle. However, their inclination in the polar coordinates, denoted as the phase $\varphi_{A_i}(t)$, has a slightly different meaning. That is, in the proposed approach, the time dependent inclination $\varphi_{A_i}(t)$ represents an agent's mind state taking place in the appropriate quarter of the circle that is assigned to represent a specific BMS process. Hence, four quarters are allocated to the four BMS. Depending on the agent's progress in the particular state process, its inclination is gradually shifting anticlockwise in the appropriate quarter, see Fig. 8.3. An agent's transition to another BMS process is represented by the instantaneous shift of its phase $\varphi_{A_i}(t)$ in the polar coordinates. That is, once the next BMS process is known, an agent instantaneously shifts to the appropriate quarter. After the transition, an agent starts the next BMS process in the appropriate quarter and its position gradually rotates with time.

Similarly, like in the Kuramoto oscillators' visualization approach, we assume that the central dot (phase centroid) with the polar coordinates $(R_\Omega, \varphi_\Omega)$ represents the collective prevailing mind state of the entire system of agents (oscillators), see Fig. 8.3. Its radius R_Ω is the measure of coherence and varies depending on the coupling strength and the distribution of the intrinsic phases.[16] The large coherence

---

[16]Let us remember that phases are understood here as $\varphi = \omega t + \varphi_0$, where ω represents the natural frequencies $(\Delta, \theta, \alpha, \beta, \gamma)$ of brainwaves.

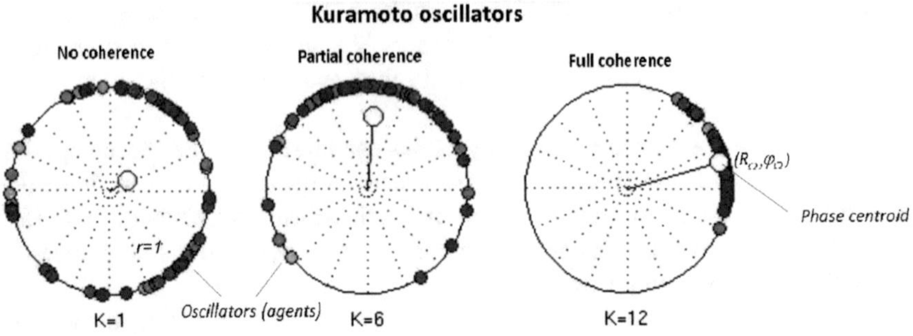

Kuramoto oscillators

Fig. 8.2 All-to-all coupled network of Kuramoto oscillators depicted for various degrees of phase-locked coherence, which is characterized by the coupling strength K and distribution of natural frequencies. In polar coordinates, the phase centroid's radius R_Ω measures the system's coherence and φ_Ω prevailing phase

is represented by the larger radius R_Ω, which is estimated as the synchronicity (coherence) level $0.25 < \Omega_{\mathrm{MAS}}(t) < 1$, see Eq. 7.22, while the central dot's inclination at the appropriate quarter φ_Ω denotes the prevailing common mind state and its anticlockwise rotation in time, see Fig. 8.3. The angular quarters denote the appropriate mind states in a following manner:

$$\varphi_{A_i}(t), \varphi_\Omega(t) = \begin{cases} [0, \pi/2] : \mathrm{BMS_{DS}} \\ [\pi/2, \pi] : \mathrm{BMS_{AW}} \\ [\pi, 3\pi/2] : \mathrm{BMS_{TH}} \\ [3\pi/2, 2\pi] : \mathrm{BMS_{RE}} \end{cases} \qquad (8.24)$$

For instance, in the polar coordinates, an agent A_i starts a $\mathrm{BMS_{DS}}$ process from the phase angle 0 and rotates counterclockwise to the phase angle $\pi/2$, where a marginal transition takes place and the agent starts from the beginning of another process in another circular quarter, see Fig. 8.3. However, if an in-between or probabilistic harmonization transition occurs before the agent reaches the marginal entropy limit, then another state starts, not from the beginning, but from the appropriate $S(t)$ level, which can be linearly associated with the phase $\varphi(t)$

$$S(t) = [0, 1] \Rightarrow \varphi(t)$$

In this way, we associate time dependent entropy values with the appropriate phase values. For instance, for the $\mathrm{BMS_{AW}}$, these values range between $\pi/2$ and π, see Eq. 8.24. The starting angular position in another quarter (state) depends on the entropy value $S(t)$, which determines the appropriate phase $\varphi(t)$ values, see Fig. 7.15 and Eqs. 7.37 and 7.38. This works in a similar fashion for all the state processes and the transitions between them.

In sum, the polar representation of the individual and collective temporal dynamics of the mind states is intuitive and well suited for the analyses' purposes.

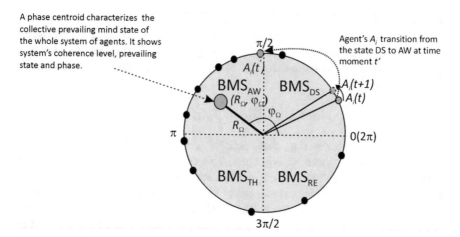

A phase centroid characterizes the collective prevailing mind state of the whole system of agents. It shows system's coherence level, prevailing state and phase.

Agent's A_i transition from the state DS to AW at time moment t'

Fig. 8.3 The adapted Kuramoto oscillators' visualization approach, where the oscillators represent OAM-based agents (see *dots* on the circle). *The central dot* (R_Ω, φ_Ω) represents the coherence level of the collective prevailing mind state of the entire system of agents

However, the crux of developing such a model rests with our interpretation of the simulation results in terms of how an individual agent interacts with the mean field of the entire system of agents. For that reason, the essential question arises—How can such interactions lead to some basic social phenomena, e.g., synergy effect, which, according to the Synergism Hypothesis, are the drivers of cooperative relationships of all kinds and at all levels in living systems? The thesis, in a nutshell, is that synergistic effects have often provided functional advantages (economic benefits) in relation to survival and reproduction that have been favored by natural selection (Corning 1983).

Hence, synergy relates to human interaction and teamwork. In general, the most common reason why people cooperate is that it creates a synergy. In the context of organizational behavior, following the view that a cohesive group is more than the sum of its parts, this synergy is the ability of a coherent group to outperform the summed performance of the incoherent individual members. The synergy effects are often called "synergistic benefits", representing the direct and implied result of the developed/adopted synergistic actions.

In a technical context, its meaning is a construct or collection of different elements working together to produce results not obtainable by any of the elements alone. The elements, or parts, can include people, hardware, software, facilities, policies, or documents: all the things required to produce system-level results. The value added by the system as a whole, beyond that contributed independently by the agents, is created primarily by the relationship among the agents, that is, how they are interconnected. In essence, a system constitutes a set of interrelated components working together with a common objective: fulfilling some designated need.

It is important to observe that a coherence mechanism is needed to share and direct some of the social systems' negentropy and energy correspondingly, which would otherwise be diminished or canceled as the incoherent agents would act in a counterproductive way. In this regard, following OAM-based MAS model, agents A_i participate in the stylized potential energy (mind field)-related mental processes (see Fig. 7.8), which synchronicity can be uniquely represented using the notation of phases $\varphi_{A_i}(t)$

$$\text{BMS}_{Ai}(t) : [S(t) \Rightarrow E_p(t) \Rightarrow \varphi(t)]_{Ai}. \tag{8.25}$$

If the agents' phases (i.e., state processes, see Eq. 8.23) are randomly distributed, then such a system does not produce a coherent collective behavior. In this regard, following the extended free energy expression (see Eq. 8.1), we imply that coherence is a result of the potential field-like energy exchange between individual agent and the system as a whole. This exchange takes place at the agent level because of the terms $-TS$ and $\tau\eta$. The first term $-TS$ represents an agent's emitted field-like internal energy, while the second term $\tau\eta$ represents the incoming energy from the collective mean field. The aforementioned mean field, denoted as $\overline{MF(t)}$, is created by that part of agents from the prevailing state, who are above the average distribution of agents across all states

$$\overline{MF(t)} = \tau\eta(t) = \frac{1}{\breve{n}(t)} \sum_{1}^{n} TS(t), \tag{8.26}$$

here, $\breve{n}(t)$ denotes the difference between the number of agents in the prevailing state and the average number of agents in all states

$$\breve{n}(t) = \max(n(\text{BMS}, t)) - \bar{n}(t). \tag{8.27}$$

The mean field $\overline{MF(t)}$ is created by these $\breve{n}(t)$ agents. We assume that each agent interacts with the $\overline{MF(t)}$ as if with another agent in a pair-to-pair fashion, see Fig. 8.1. According to the proposed model, the synergy effects occur mainly because of the synchronization (coherence) mechanism between the individual agents and the mean-field states. The coherence mechanism itself can be modeled using the interference principle when harmonic oscillators with similar phases produce resonance. In our case, these phases for agent A_i and the entire system of agents Ω are accordingly $\varphi_{A_i}(t)$ and $\varphi_{\Omega}(t)$, see Eq. 8.24 and Fig. 8.3. The values of the phases $\varphi_{A_i}(t)$ are obtained from the OAM simulation setup, using a simple linear relation

$$\varphi(t) = f(S(t)), \tag{8.28}$$

where the values of the function $S(t) = [0; 1]$ are linearly mapped to the phase values in the appropriate $\pi/2$ intervals depicted in Eq. 8.24.

However, in a more advanced OAM-based MAS setting, we do foresee phases as discriminated functions of the sort

$$\varphi(t) = \omega t + \varphi_0, \qquad (8.29)$$

where ω denotes the natural frequency of the oscillating entity (agent) and φ_0 denotes its initial phase. In fact, both parameters can be intrinsically related with the agent's properties and can vary in the MAS setup following the Kuramotto model used to describe the synchronization of a large set of coupled oscillators (Kuramoto 2012). Such an approach will extend the current OAM-based MAS setup in prospective research.

However, in the simple OAM-based MAS version presented here, the synergy effect only depends on the nondiscriminated form of the difference between the individual $\varphi_{A_i}(t)$ and the mean-field $\varphi_\Omega(t)$ phases

$$\Delta\varphi(t)_{A_i} = \varphi_\Omega(t) - \varphi_{A_i}(t). \qquad (8.30)$$

When the phases are opposite, the harmonic oscillators cancel each other and when they are close, resonance occurs and the synergy effect takes place, see Fig. 8.4. It can be expressed mathematically in terms of agent's synergetic contribution

$$(E_s)_{A_i} = \left| (TS(t))_{A_i} \cdot \cos(\Delta\varphi(t))_{A_i} \right|, \qquad (8.31)$$

here the term $(TS(t))_{A_i}$ is derived from the free energy expression, see Fig. 7.8 and Eq. 7.14. The total synergy effect $(E_s)_\Omega$ depends on those $\breve{n}(t)$ agents in the prevailing state who are above the average number distributed across all states (see Eq. 8.17)

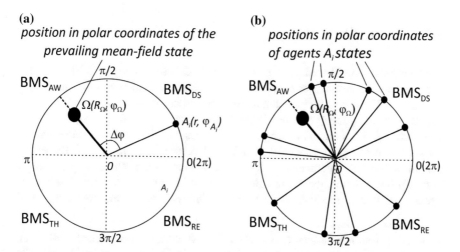

Fig. 8.4 The prevailing mean field (**a**) and individual (**b**) phases are depicted in the polar coordinates. The phase difference $\Delta\varphi$ indicates the interference angle between the prevailing mean-field state and an individual agent A_i state. Mean-field (MAS) prevailing state's coordinates $(R_\Omega, \varphi_\Omega)$ indicate strenght and phase of the MAS synergy effect accordingly

$$(E_s)_{\Omega} = \sum_{n(t)} \left(\left| (TS(t))_{\underset{n(t)}{\smile}} \cdot \cos(\Delta\varphi(t))_{A_i} \right| \right). \tag{8.32}$$

For such agents the individual synergy term is positive $(E_s)_{A_i} \geq 0$ as coherent agents contribute accordingly some inner energy depending on their phase difference with the mean field of the prevailing state.

In Fig. 8.4, according to the model assumptions, the coherent agents create the synergy effect. We have to emphasize that the coherence is governed by the probabilistic negentropic principles that were discussed a few sections earlier. In this way, negentropy drives synergy effect.

The synergy effect shows the coherent potential energy (mean field) in the agents' system, which is obtained from the individual agents staying in the prevailing state. In fact, in living systems, the coherence mechanism helps to share and direct synergy for maintaining order in various forms. In the social systems context, the synergy is exploited to maintain social order (a form of negentropy in the mesoscopic scale of living systems). Due to the positive or negative feedback between the system and the agents, the synergy level may increase or decrease accordingly. It depends on whether the social system's synergy is effectively used to coordinate individual agent behavior and it also depends on the agents' sensitivity to the coherent mean field.

In this way, in the proposed OAM-based MAS model, just few fundamental simulation principles and parameters are outlined. In the next section, there are demonstrated some simulation results, which pay the ground for the prospective possibilities for modeling the behavior of complex social systems in the extended versions of the current setup.

8.3 Simulation Results

Following the OAM-based MAS descriptions in the earlier sections of this chapter, some simulation results are provided below using the OSIMAS V-lab simulation platform.[17] The simulation session starts with the creation of a chosen number of agents with individually selected OAM features, see Sect. 7.6 and Fig. 8.5.

Following the simulation setup, a set of basic individual parameters can be chosen for each agent using the 'Configure' button, see Fig. 8.5. A list of the default individual parameters with brief comments about them is provided in Fig. 8.6. For instance, the second and third individual parameters listed define the agent's reaction to the MAS state (the agent's sensitivity to the prevailing MAS

---

[17]The V-lab simulation platform is available at http://vlab.vva.lt (Guest: guest555), where the OAM-based MAS simulation model can be found among other OSIMAS models in the menu on the left side (OAM_based_multiagent_system_MAS_1). Documentation, instructions, algorithms and the data files needed for interactive simulations and testing are also provided in the model's page.

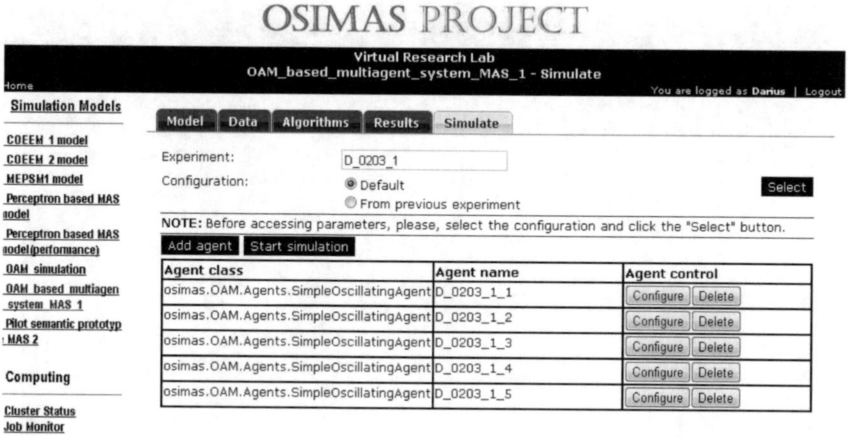

OSIMAS PROJECT

Fig. 8.5 V-lab window used for the generation of agents

state and share of energy that can be used for synergy with the MAS). The other parameters define internal characteristics that do not depend on the MAS. They are described in Sect. 7.6. (see Figs. 7.17 and 7.18).

After the parameter setup is over we can start simulation. The simulation results obtained for each agent and for the entire MAS are in the form of charts and *.csv data sheets. Next, we are going to briefly explore the results for a MAS consisting of 10 agents, using the default set of parameters, see Fig. 8.6. Afterwards, we will explore how changes in some of the basic MAS-related and individual parameters influence overall MAS performance.

Thus, an example of the dynamics of the total number of agents in each state is depicted in the appropriate chart, see Fig. 8.7. It shows that the entire MAS starts from the DS (deep sleep) state early in the day, which is quickly changed by the AW (active wakefulness) state that alternates for some time with the RE (resting) state. In the middle of the day, the RE state mostly alternates with short periods of the TH (thinking) state. Whereas, at the end of the day majority of agents are either in the AW or TH states. Such tendencies are typical for the default set of agents' parameters. These simulated tendencies well correspond with the ordinary dynamics of human mind states during a day. However, simulated dynamics during a day depends, for instance, from the relative proportion of agents with the particular chronotype in the MAS.

In fact, the total number of agents staying in the same dominant state indirectly indicates the MAS synchronicity level. In this regard, in the previous section we introduced a special synchronicity measure, see Eq. 8.21. In Fig. 8.8, some examples of the individual normalized synchronicity curves are depicted for a few agents. Let us recall that synchronicity indicates how well an individual agent's state coincides with the collective (MAS) state. In the OAM-based MAS model we can regulate, using the appropriate parameters, the probabilities of individual

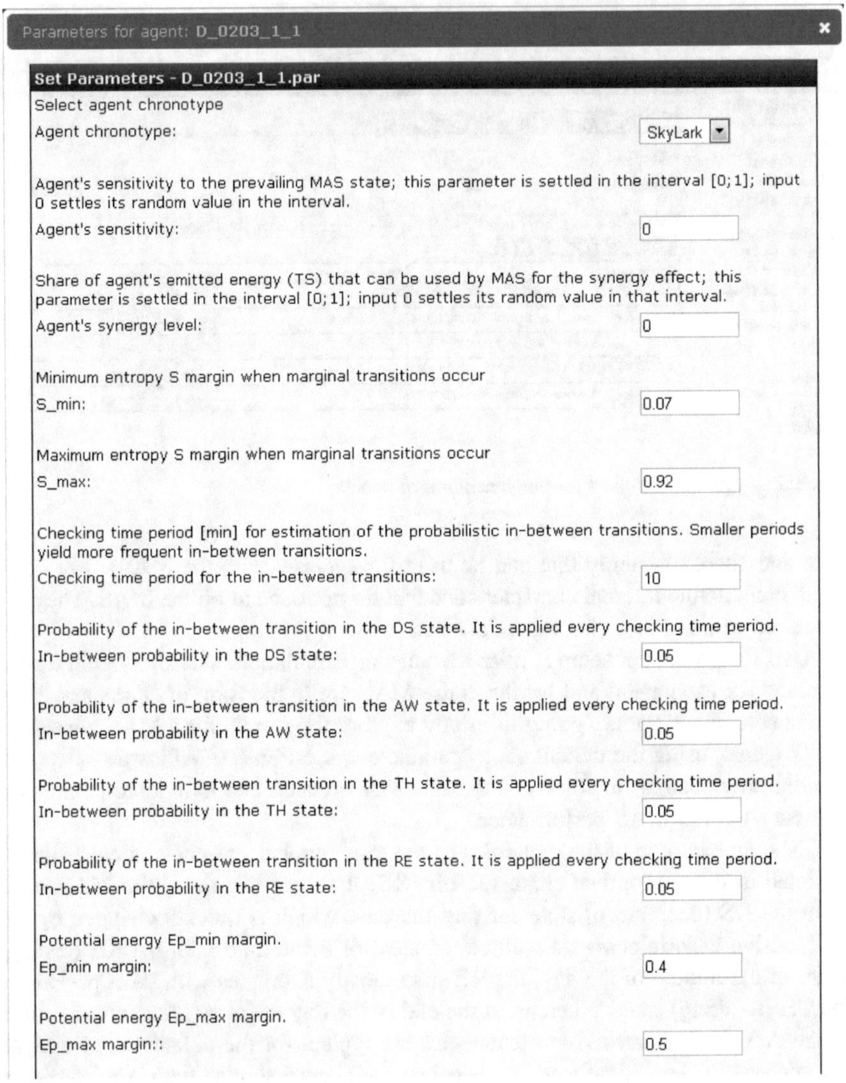

| Parameters for agent: D_0203_1_1 | ✕ |

Set Parameters - D_0203_1_1.par

Select agent chronotype
Agent chronotype: SkyLark ▾

Agent's sensitivity to the prevailing MAS state; this parameter is settled in the interval [0;1]; input
0 settles its random value in the interval.
Agent's sensitivity: 0

Share of agent's emitted energy (TS) that can be used by MAS for the synergy effect; this
parameter is settled in the interval [0;1]; input 0 settles its random value in that interval.
Agent's synergy level: 0

Minimum entropy S margin when marginal transitions occur
S_min: 0.07

Maximum entropy S margin when marginal transitions occur
S_max: 0.92

Checking time period [min] for estimation of the probabilistic in-between transitions. Smaller periods
yield more frequent in-between transitions.
Checking time period for the in-between transitions: 10

Probability of the in-between transition in the DS state. It is applied every checking time period.
In-between probability in the DS state: 0.05

Probability of the in-between transition in the AW state. It is applied every checking time period.
In-between probability in the AW state: 0.05

Probability of the in-between transition in the TH state. It is applied every checking time period.
In-between probability in the TH state: 0.05

Probability of the in-between transition in the RE state. It is applied every checking time period.
In-between probability in the RE state: 0.05

Potential energy Ep_min margin.
Ep_min margin: 0.4

Potential energy Ep_max margin.
Ep_max margin::   0.5

Fig. 8.6 List of the basic OAM-based MAS parameters with their default values

transitions occurring in the prevailing collective states. In this way, we can obtain
various individual synchronicity levels.

Following Eq. 8.22, an example of the collective (MAS) synchronicity curve for
10 agents is depicted in Fig. 8.9. As we can see, the collective synchronicity level
tends to fluctuate rapidly. The overall MAS synchronicity level depends on the
values chosen for the individual agents' sensitivity to the collective MAS state
parameter. This dependency has been also investigated a few pages below.

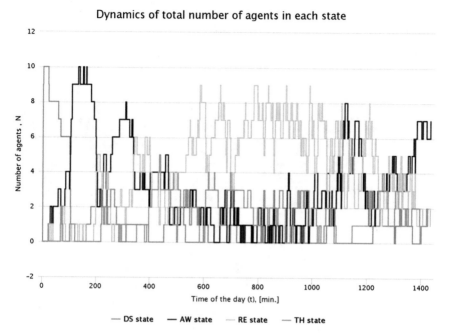

Fig. 8.7 Dynamics of the 10 agent states during a day (default simulation setup)

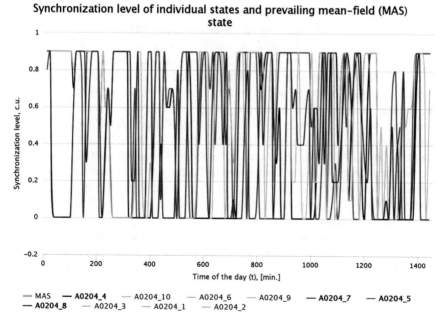

Fig. 8.8 The dynamics of individual synchronization levels for agents #4, 5, 7 and 8

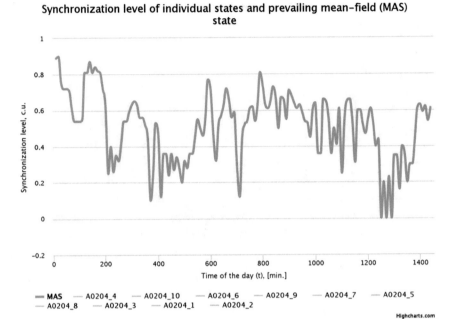

Fig. 8.9 The dynamics of the collective (*MAS*) synchronization level

The dynamics of the collective synchronization level indicate the coherence of the collective mind state. High levels of coherence are indicative of a well attuned collective mind state, which is an essential prerequisite for an effective team, organization or society. As mentioned in the previous section, the coherence mechanism is needed to share and direct some of the social systems' negentropy and energy correspondingly, which would otherwise be diminished or cancelled, as the incoherent agents would act in a counterproductive way.

In this way, the coherence of the collective mind state relates to the synergy of human interaction and teamwork. In general, the most common reason why people cooperate and become coherent, even in the sense of common mental states, is that it creates a synergy. Let us remember that in the context of organizational behavior, following the view that a cohesive group is more than the sum of its parts, this synergy is the ability of a coherent group to outperform the summed performance of individual incoherent members.

In the OAM-based MAS version presented here, the synergy effect only depends on the nondiscriminated form of the difference between the individual and the mean-field (MAS) phases.[18] An example of the synergy dynamics for a few agents is depicted in Fig. 8.10.

---

[18]Using a simple linear relation (see Eq. 8.28), the phases are related with the entropy $S(t)$.

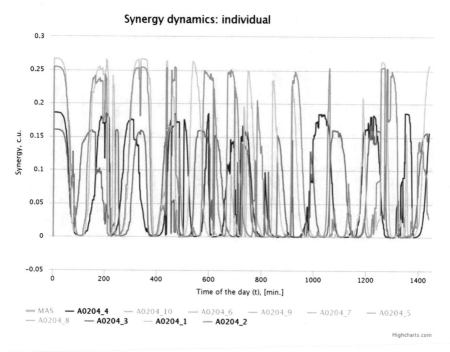

Fig. 8.10 Synergy dynamics for a few agents (#1, 2, 3, and 4) during a day period

As we can observe from Fig. 8.10, synergy dynamics can exhibit fluctuations that are quite volatile. When the phases of an individual agent and the prevailing MAS state are opposite, they cancel each other, and when they are close, resonance occurs and the synergy effect takes place, see Eq. 8.31. In this way, depending on an agent's coherence with the prevailing MAS state, he is able to contribute some internal energy to the MAS. In the model, it is called the agent's synergetic contribution.

According to the model, the total synergetic contribution depends on those agents in the prevailing state who are above the average distribution calculated for all states, see Eqs. 8.17 and 8.32. An example of the total synergy dynamics for the entire MAS is depicted in the Fig. 8.11.

Hence, the synergy effect shows the coherent potential energy (mean field) in the agents' system, which is obtained from the individual agents' interference with the prevailing MAS state. Let us remember that such synergy is a form of energy at the social level, which not only serves as an attribute of social organization, but is also actively involved in maintaining itself, i.e., the social order that produces such synergy.[19]

---

[19]The key ideas contained in the approach presented here are also very close to the well-known autopoietic theory, which deals with the dynamics of the transformation processes of mind states, which through their interactions, continuously regenerate and reorganize the network of processes (relations) that produced them (Maturana 1980). In this regard, it is quite natural to conclude that the idea of the cyclical nature of the self-referential mindfield presented earlier is very close to the autopoietic theory.

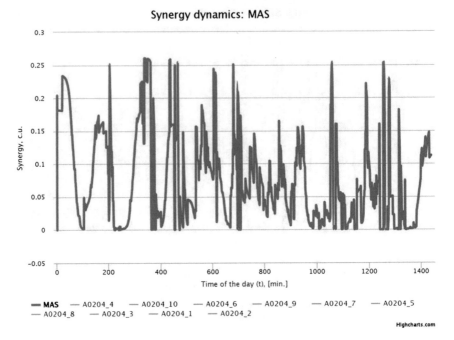

Fig. 8.11 Total synergy dynamics for the MAS consisting of 10 agents (five owls and five early birds)

In order to visualize the state related synchronicity of the entire MAS, the prevailing phases of the mean field and individual agents were depicted in polar coordinates, see Figs. 8.4 and 8.12. Such live animation of the agent's state dynamics well manifests the time periods of high and low coherence within the entire MAS. That is, when the agents are coherent, their phase-related beams are closely aligned together and the MAS beam's length is close to one, see Fig. 8.12a. Otherwise, the agents' beams are dispersed and the MAS beam's length is close to zero, see Fig. 8.12b. Thus, this representation depicts the overall MAS state at each moment in time, its dynamics over time, MAS coherence and the corresponding synergy level.

The synchronization effects occurring in the OAM-based MAS (see Figs. 8.8 and 8.9) are mainly influenced by two parameters, i.e., the agents' sensitivity to the prevailing MAS state and the amount of energy emitted (TS) by the agents that can be used by the MAS to produce the synergy effect. Both of these controlling parameters are designed to influence the individual mind state dynamics, depending on the collective MAS behavior.

For instance, an agent's sensitivity parameter has a slight effect on the transition probabilities for switching to the prevailing MAS state, see Fig. 8.13. That is, values of the agents' sensitivity parameter ranging from [0; 1] increase the overall MAS synchronization level in the interval [0.5; 0.6]. The MAS synchronization level's confidence interval is ± 0.22 for $\alpha = 0.05$. We have to admit that the

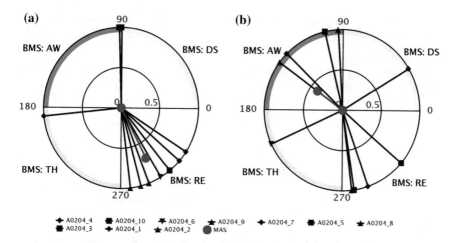

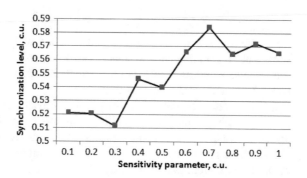

Fig. 8.12 A snapshot of the phase dynamics for 10 agents (five owls and five early birds) in polar coordinates for two different moments in time (see (**a**) and (**b**) charts). The phases of the individual agents and prevailing mean field (MAS) are depicted as beams located in the appropriate mind state sectors (quarters). The length of the MAS beam denotes the strength of the MAS synergy effect

Fig. 8.13 Dependence of the averaged synchronization level on the sensitivity parameter for the MAS composed of 10 agents (five owls and five early birds), using the default set of parameters

OAM-based MAS synchronization effect obtained could be regulated according to the modeler's needs by changing the values of each individual agent's sensitivity parameters ξ_{A_i}, see Eq. 8.19.

Now, let us examine the influence of another OAM-based MAS parameter, i.e., the energy emitted by the agents $(T \cdot S)$ that is used for the attainment of the MAS synergy effect, see Eqs. 8.30 and 8.31. As we can observe in Fig. 8.14, the averaged synergy is in direct proportion to the synergy parameter.

However, investigation also revealed a very high variation of the MAS synergy effect, i.e., the confidence varies in the interval [0.005; 0.06] accordingly. It does not change much even if the sensitivity parameter remains constant. Such an OAM-based MAS setup fosters volatile dynamics and the observed chaotic oscillations, see Figs. 8.10 and 8.11.

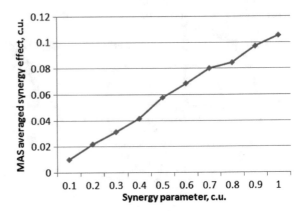

Fig. 8.14 Dependence of the averaged synergy effect on the values of the synergy parameter for the MAS composed of 10 agents (five owls and five early birds), using the default set of parameters

Thus, we have consistently explored just a few of the basic parameters of the OAM-based MAS simulation approach. Future research could reveal more complex properties and opportunities for use of the proposed simulation approach. In fact, analysis of the chaotic oscillations is foreseen in the prospective research. For instance, depending on the OAM-based MAS application domain, the character of the simulated chaotic oscillations can be investigated, depending on the setup of the agents' individual parameters. Analysis of the chaotic dynamics could reveal some characteristic chaotic invariants, phase attractors and fractal structures. Consequently, it can be compared and matched with the behavioral dynamics of some empirical biological and social systems. In this way, the proposed simulation setup could serve as a ready-to-use simulation framework to be enhanced with sound parametric and empirical data.

In the last two chapters of the book, another macro level simulation framework is presented using neural networks (NN) for modeling the fluctuating behavior of agents in volatile social media situations and automated trading systems. At first glance, these might appear to be a totally different area for research. However, the methical differences may actually complement each other, as the state dynamics of NN-based agents can be modeled using OAM and OAM-based MAS methods. In this way, the universality of the possible applications of the OSIMAS paradigm is demonstrated.

Chapter 9
Agent-Based Modeling of Excitable Social Media Fluctuations

A macro model is needed for the simulation of the system of agents. At this stage of our research, though, it is premature to present a fully functional large scale-based MAS (multi-agent system). Hence, we have started from simple simulations, employing nature inspired 2D cellular automaton (CA) model, to analyze information processing and transmission properties in the heterogenous excitable media. The suggested model is located between CA with discrete outputs and differential equation-based models.

The characteristic peculiarity of the excitable media is that, not the physical signal, however, the information transmitted from one media element (agent or group of agents) to other ones is very important. An objective is to consider the excitation-wave initiation, propagation, and breakdown from a point of view of information transmission. Hence, granular artificial neural network-based models were employed. Contrary to standard CA approach with discrete outputs of the automaton, we utilize a single layer perceptron (SLP) with smooth sigmoid function that gives continuous outputs. Such an approach allows to obtain many wave-propagation patterns observed in real-world experiments and known simulation studies.

Our prior analysis hints that the wave breakdown and daughter wavelet bursting behavior possibly is inherent peculiarity of excitable media with weak ties between the cells (agents), short refractory period and granular structure. It helps to understand wave bursting, propagation and annihilation processes in homogeneous and nonhomogeneous media too. Information propagation in the simple CA media allows not only to study excitatory waves, but also provides an opportunity to substitute simple SLP with the earlier described oscillations-based agent model (OAM).

The content of this chapter is based on the adapted material from our previous publication conducted together with co-authors (Plikynas et al. 2014), who have substantially contributed in the OSIMAS research. In particularly, I am most grateful to prof. S. Raudys for his kind contribution.

In essence, the AI (artificial intelligence)-based modeling of the fluctuations in volatile social media situations presented below, using SLP as a proxy for the OAM, provides a proper understanding of and a useful tool for OAM integration into the AI-based macro-level MAS simulation framework. While SLP serves here only as a useful approximation for the OAM, nevertheless, its related information processing and transmission properties reflect some of the basic features of the OAM approach. Thus, the SLP-based MAS simulation framework presented below provides a much better understanding of some of the prospective applications of the proposed OSIMAS paradigm.

9.1   Overview

This chapter investigates the propagation of excitation information in artificial societies. We use a CA approach, in which it is assumed that social media is composed of tens of thousands of community agents, where useful (innovative) information can be transmitted to the closest neighboring agents. The model's originality consists of the exploitation of artificial neuron-based agent schema with a nonlinear activation function to determine the reaction delay, the refractory (agent recovery) period and algorithms that define mutual cooperation among several excitable groups that comprise the agent population. In the grouped model, each agent group can send its excitation signal to the leaders of the groups. The novel media model allows a methodical analysis of the propagation of several competing novelty signals. The simulations are very fast and can be useful for understanding and controlling excitation propagation in social media, planning, and social and economic research.

An exponential increase in the amount of information exchanged between individuals or economic or social units makes it necessary to understand the ways and regularities of detailed information-transmission processes in complex social media. Analyses of questions such as how people, organizations and social media in general react to novelties or new information, play a leading role here. This sort of knowledge allows the actions of individuals and social groups to be predicted and partially controlled.

Social life is increasingly characterized by interdependencies among actors (individuals, companies, or institutions), which give rise to self-organization phenomena, feedback loops, and unpredictable and sometimes counterintuitive interaction patterns (Helbing 2013). Agent-based models can simulate and help to understand macro-level phenomena from the bottom–up. The rule-based approach to agent-based modeling is simple, flexible, and intuitive. Agent-based modeling promises to be a major tool for providing a better understanding of the complexities that fundamentally challenge the way social phenomena are approached. Cognitive architectures may serve as a good basis for building mind- and brain-inspired, psychologically realistic cognitive agents for various applications that require or prefer human-like behavior and performance (Sun and Hélie 2013). Examples of this type of information-transmission research include adopting new technology (Rogers 2010) and the diffusion of innovations in social networks (Acemoglu et al. 2011; Keller-Schmidt and Klemm 2012; Valente 1996; Young 2006). To our knowledge, the major area of virtual wave-based social simulations and applications are related to emerging research in social-networking, agent-oriented, and multi-agent systems. As a matter of fact, wave- or field-based modeling is usually applied in studies inspired by nature. Such analysis is necessary in the search for a better understanding of very complex, large and highly dynamic physical, biological and social networks. Unfortunately, these studies use the wave-like coordination approach as a technical solution without a deeper exploration of the basic nature of such an approach.

For instance, in social-networking research, because of its large scale and complexity, attempts are frequently being made to simulate social networks using wave-propagation analysis. Some of the applications deal with message-broadcasting and rumor-spreading problems (Wang et al. 2012) and simulations of rapidly changing social and financial phenomena. In mainstream research papers, individuals and social units are represented by agents that are affected by their social and physical surroundings and produce cognition-based behavior patterns (Perc et al. 2013; Wijermans et al. 2013). Comprehension of social behavior is sought by relating intra- and inter-individual levels of behavior generation to emerging behavior patterns at the individual or group level. In social-networking research, interactions are mostly realized through connections between pairs of nearby agents.

As an example of the necessity for wave-propagation research in social spheres, we look at the analysis of some time series of high-dimensional financial data used in automated trading systems (ATS) (Raudys 2013). In Fig. 9.1, we depict two time series of profit/loss financial time series. In ATS, we use hundreds or even thousands of investments for decision-making. To develop and train such a complex trading system, we need the same or higher numbers of training vectors.

Economic and financial markets are changing very rapidly nowadays. Two years' (less than 500 trading days') history becomes lengthy. For this reason, in the development and testing of diverse versions of trading schemas we include artificial data generated by means of the wave-propagation model. Figure 9.1 illustrates an example of two synthetic time series.

The data generation was based on the hypothesis that the economic and financial world is composed of a large number of mutually communicating social actors that exchange novel information in a similar way to how we have described this in the social media model. We modeled ATS as sets of neighboring nodes in an excitable media network. Sums of excitations of groups of social agents modeled the profits and losses of the investment. We also introduced an investment risk, where the ATS refuse to make investments if the risk is too high. A change in the parameters of the data-generating rule allows diverse economic activities to be investigated. Moreover, we can generate a large number of principally diverse wave-propagation patterns. In this way, the developer of decision-making algorithms can have long time series sufficient for reliable performance evaluations of algorithms. Our analysis shows that synthetic data are useful for modeling chaotic economic

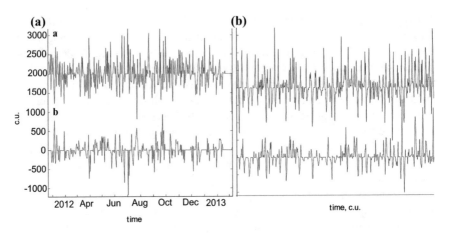

Fig. 9.1 Fluctuations of profits generated by algorithmic trading systems during 15 months period (on the *left*) and fluctuations of profits simulated by means of an excitable media model (on the *right*)

behaviors observed in real life and developing automatic decision-making algorithms that can withstand unexpected changes in the real world.

In this chapter, we try to use an almost universal methodology: the stimulation signal propagation in an excitable media. It takes the bottom–up approach, which is used successfully in many research disciplines, such as particle physics, micro and cell biology, chemistry, medicine, meteorology, and astrophysics (Ben-Jacob et al. 2000; Martin and Martin-Granel 2006; Moe et al. 1964; Spach 1997). We propose that these kinds of models can also be used in the analysis of information-propagation processes in social media, and can lead to a deeper conceptual understanding of the fundamental nature of information dissemination in societies of virtual agents. The objective of this chapter is to demonstrate this statement.

Theoretical and subsequent simulation analyses in this area belong to methods based on systems of nonlinear differential equations. This approach investigates phenomena in continuous space and time. Many researchers claim that real objects and processes have a discrete nature. In this way, CA research has been adopted in many research fields (Bandini et al. 2007; Chua et al. 1995; Mehta and Gregor 2010). In CA-based models, the excitation signals are transferred to neighboring knots (agents and cells). CA-based models are much simpler, easier to understand, and faster to simulate using computers. Computer simulations were first performed by Moe et al. (1964), who laid out a virtual neural network (NN) on a rectangular grid, used 0 or 1 outputs for each cell (node of the network) and measured time in discrete periods. A plethora of subsequent researchers developed the CA approach further. In these models, the number of excited states is greater than 1, and more than one neighbor must be active in order to induce a resting cell to an excited state (see the short review in Raudys 2004).

The propagation of information, new technologies, and cultural matters are nonlinear phenomena, such as wave propagation and chaos origination. The differential equations and cellular automata methodologies both allow us to model a number of signal propagation patterns that have been observed experimentally in physical, chemical, and social media, e.g., regular uniform wave propagation outwards from a point of starting excitation; the generation of spiral waves after their initiation; waves surrounding a nonconductive obstacle; disorganized activity; and broken wave propagation in one or several directions. These phenomena have been actively investigated in numerous disciplines, including chemistry, biology, meteorology, ecology, medicine, engineering, physics and astrophysics. Some researchers believe that these models are universally inspired by nature because they exhibit oscillating cells, cooperative self-organization, and collective behavior (Ben-Jacob et al. 2000; Chua et al. 1995; Mehta and Gregor 2010; Steele et al. 2008). Some of the authors started using similar models in a much wider sphere of research disciplines, including ecology, epidemiology and, possibly, the economic and social sciences (Berestycki et al. 2013; Litvak-Hinenzon and Stone 2007).

In classical media models, the similarity between the agents is measured geometrically as a distance in two-dimensional (2D) or three-dimensional (3D) space. In research into economic and social phenomena, similarities are typically characterized by the larger number of characteristics of a dissimilar nature. In this regard, in the present chapter, for practical applications we assume that the distance or similarity between agents can be defined freely (see the next section). Measures of similarity may be applied to define match and connectivity between agents (nodes of excitable economic or social media). Like in the grid-based models, each node sums the input signals transmitted from neighboring nodes. In the new model, however, the magnitudes of the output signals and their release times depend on the sum of the accumulated inputs.

This chapter is organized as follows. For a simpler explanation of the virtual agent population model and for visualization purposes, we describe and extend a 2D cellular, grid-based, social media model and its modifications in next section. Diverse excitation-signal propagation patterns are demonstrated. In next sections, we introduce a group-based agent model in which the agent population is split into groups. Inside a single group, the excitation signals are transmitted only to neighboring nodes. Limited interaction of groups, as well as propagation of diverse informational signals, is allowed. We also consider the propagation of two competing excitations in the grouped-agent population. Finally we summarize the simulation results and discuss the potential for further studies.

9.2 The Basic Model

Hexagonal 2D CA Model

To understand the main properties of the model of signal propagation in excitable economics or social media, we will first describe its simplified versions: 2D rectangular and hexagonal models of an excitable medium. The basic model consists of

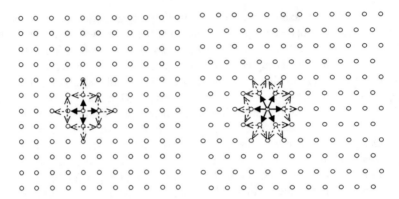

Fig. 9.2 Rectangular and hexagonal grid-based models and directions of the transmitted signals. Excitation originates from a randomly chosen point in the regular spaced grid

nodes (cells and agents) spaced on regular grids (see Fig. 9.2). It is assumed that an excitable medium is composed of tens of thousands of community agents, where useful (innovative) information can be transmitted to the closest neighboring agents (see the arrows in Fig. 9.2).

Each element of the grid is represented by a single layer perceptron, which has a number of inputs (say p), that uses weights (connection strengths between the nodes) $w_1, w_2, \ldots, w_p$ and x_i (denotes signals received via inputs) to calculate a weighted sum $arg = \sum_{i=1}^{p} w_i x_i$, and produces an output, $o = f(arg)$ by using sigmoid nonlinearity, $f(arg) = 1/(1 + \exp(-arg))$ habitually used in artificial NNs (Haykin 1998; Raudys 2001). We adopted this function for feasible needs:

$$f_s(arg) = \frac{\gamma}{1 + \exp(-\eta \times arg - \theta)} \quad \text{if } arg \leq \Delta^*; \ f_s(arg) = 0 \text{ otherwise,} \quad (9.1)$$

where $\Delta^* \leq 0$ is a priori-defined sensitivity threshold. In simulations, we used $\gamma = 1.333$, $\eta = 5$, $\Delta^* = 0.4$, $\theta = -1.333$. The constants were selected to have the weights w_i, and outputs o, between 0 and 1.

After obtaining excitation signals from neighbors, the agent calculates the weighted sum and after some delay fires out transformed sum of input signals, $f_s(\sum_i w_i x_i)$, accumulated during two previous time periods. In our cellular automata model, the signal transfer time, t_{transf}, is discrete, $1, 2, \ldots, m$. This time depends on the strength of the accumulated output signal, o. For that reason, the output signal interval (0, 1) is split into m equal intervals, corresponding to time periods, $1, 2, \ldots, m$ (see Fig. 9.3).

In our social media model, we have a negative feedback loop: *the larger the excitation signal*, arg, *the later the jth agent will fire out its output signal*. To define t_{transf}, we followed observations from CA and differential equation-based excitable media models (Spach 1997): the longer the delay in transferring the excitation, the greater the node-to-node strength of the signal. The minimal transmission time can

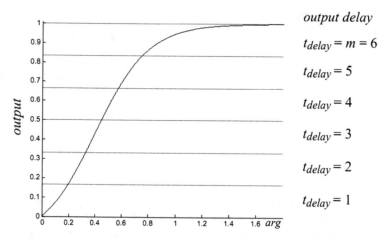

Fig. 9.3 Dependence of the output, $o = f_s\left(\sum_i w_i x_i\right)$, the signal-transmission delay, t_{delay}, on the input, the weighted sum, arg

be equal to 1, i.e., the cell does not transmit a signal if $f(\text{arg}) < 1/m$. The minimal transmission time is also affected by sensitivity threshold, Δ^*. The proportionality of the signal-transmission delay to the strength of the signal makes the CA-based model more similar to differential equation-based models. This is a very important new feature of the model. Both the discrete time and the negative feedback intro-duce stochastic (chaotic) components. Comparative experiments showed that the utilization of fixed or variable time delays essentially changes the characteristics of the model. To speed up computer calculations, a look-up table was used to find $o = f(\text{arg})$. The use of the look-up table introduces additional stochastic compo-nents into the model. Accumulated signals are transferred only to non-excited neighboring agents.

An important parameter traditionally used in signal propagation models is *the refractory period*, t_{refr}, the number of elementary time periods; after excitation, the node cannot be excited again. In our model, the refractory period is determined by the saturation, sat_j, of the jth agent after its excitation and the strength of potential new excitation, o_{new}. Just after excitation, $\text{sat}_j = o$. The saturation exponentially decreases with time, t:

$$\text{sat}_j(t) = o \times \exp(-\alpha_{\text{refr}} \times (t - t_{\text{excitation}})). \qquad (9.2)$$

When saturation falls below a threshold,

$$\Delta^* - \beta \times (o_{\text{new}} - \Delta^*), \qquad (9.3)$$

the refractory period terminates, and the jth agent can be excited by a new excitation signal, o_{new}, if $o_{\text{new}} \geq \Delta^*$. In the above equations, $\alpha_{\text{refr}} = 1/(\text{refr} \times m)$, and β is a small positive constant. The parameter refr is the a priori determined time constant,

Fig. 9.4 Hexagonal shape of
a 2D excitable media with
$N_y = 7$ nodes in each side

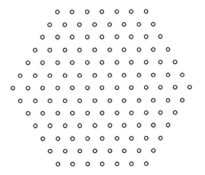

the refractory time parameter. This parameter can be common to all agents or individuals. Equation 9.3 shows that powerful excitations of neighboring agents, o_new, can shorten the refractory period by a small amount. For simplicity, in experiments reported in this chapter, $\beta = 0$.

The size and the shape of the media are also important characteristics. In this chapter, for 2D illustrations we used a hexagonal shape, where we have N_y nodes along each edge (Fig. 9.4). The border nodes may transfer their outputs only to adjacent cells. If the excitation spreads from the center of the media towards the edges, after reaching the edge the excited cells can find unexcited cells to transmit a signal to. In such a case, the propagation of the signals closest to the media edges fades.

We emphasize again that the single layer perceptron is a model of information processing and transmission inspired by nature (Haykin 1998). It has a number of universal properties (Raudys 2004). The introduction of nonlinearity into the excitable media model perceptibly influences the wave-propagation patterns. A very important peculiarity of excitable social media is not the physical signal, but that the information is transmitted from one media element to others. The following lists, the parameters that can be used to specify concrete social media models:

- N_y, the number of nodes on each edge of the hexagonal media in 2D space;
- p, the number of neighbors, and the shape of the grid;
- $w_1, w_2, \ldots, w_p$, the connection's weights;
- the concrete parameters of the transfer function, $o = f(\text{arg})$;
- Δ^*, the sensitivity threshold;
- arg_start, the starting excitation, and the position of the cell (or cells) in the grid that transmit the starting excitation;
- m, maximal time-delay periods in the signal-transmission process;
- t_refr, the rule and its parameters that determine the refractory period;
- t_transf, the rule used to determine the excitation time (Fig. 9.3); and
- the rules used to determine the similarity and neighborhood of the media nodes.

The parameters and required rules can be the same for all nodes in the social media; however, they can also be fixed for each node (agent) individually. In principle, the parameters can be learned during the simulation of media evolution in changing environments (for a similar example, see (Raudys 2008)). Finding a correct set of parameters is a subject for future research. In a strictly hexagonal 2D media model, we can select either 6 or 18 neighbors. In a number of simulation studies, we investigated nonhomogeneous media. For that reason, we shifted the nodes randomly to left–right and up–down directions. We were then able to specify the neighborhood according to new distances. In principle, it is possible to define the nearest neighbors arbitrarily. Such attributes of excitable media are more natural and more desirable for the analysis of economic and social systems.

Interpretation of the Model's Parameters

In this section, we provide more details of how to use and interpret the model's parameters in the applied social media analysis tasks. In the analysis one needs to bear in mind that the *interpretation of the model's parameters depends on the particular social problem to be solved*. In different tasks, interpretations can differ in essence.

In general, parameter N_y describes the size of the agents' population in the social media simulation. In a virtual experimental setting, agents are spatially distributed in either hexagonal or rectangular grids. In a social analysis, the agents can be spatially distributed in spaces of higher dimensionality.

Depending on the type of grid, parameter p determines the size of the neighborhood, i.e., the number of closest neighbors surrounding an agent. The agent sends an excitation signal (new information, or novelty) to those neighbors only. In general, the neighborhood can be determined in terms of common business activity, shared interests, religion, family relations, family relationships, organizational infrastructure and scheme of social network.

The parameters $w_1, w_2, \ldots, w_p$ denote agents' weights of input connection with their neighboring agents; in other words, the strength of the social relationships between neighboring agents. Strong connection weights cause regular signal propagation. Weak weights are associated with chaotic behavior.

The function of the parameter for transferring the excitation signal to neighboring agents $o = f(\text{arg})$ is determined in Eq. 9.1. Mathematically, it describes the nonlinear output transmitted from a particular agent to its closest neighbors. It cannot be smaller than 0 or larger than 1. Depending on the application in question, this function can be changed in order to reflect specific needs of the wave-propagation set-up process. However, in this chapter the output function we have presented was selected by trial and error after numerous attempts to obtain different wave-propagation patterns.

The parameter Δ^* denotes the sensitivity threshold. It is related to the minimum value of input excitation that can excite a neighboring social agent. The agent does not get excited if the sum of input signals is smaller than $\Delta^* - a \times (o_{\text{new}} - \Delta^*)$; see Eq. 9.3. It must be remembered that humans live in a noisy environment and strive to filter and adapt important information only. In the prospective research, we foresee introducing a dependence of the sensitivity threshold on the type of research task. Determining a suitable threshold value could enhance an agent-specific reaction to dissimilar types of information, which is so common in social domains.

Parameter $\arg_{start}$ is the magnitude of the signal (novelty, etc.) which begins excitation (triggering) in the simulated social media. It is related to the position of the starting cell (or cells) in the grid. If the magnitude of the starting signal is too small or too powerful, in some cases this causes the signal to either vanish or overwhelm the population. In fact, the origin of the source of excitation is task-dependent and requires thorough further investigation. We infer that the triggering signal is produced by an innovator whose individual parameters allow useful information to be accumulated and tested outside of the noisy environment. This line of research, though, requires yet another layer of sophistication in our model.

Parameter m characterizes the model complexity (smoothness of nonlinear transfer function). It denotes a maximum elementary time-delay period, t_{transf}, in the signal-transmission process ($t_{transf} = 1, 2, \ldots, m$; see Fig. 9.2). If m is too large, more computer time is required; if it is too small, the CA-based social media model is too simplified. We have established its value by trial and error after several attempts to obtain consistent wave-propagation patterns.

The parameters of activation function (1) define the rule used to calculate the excitation delay, t_{transf}. After receiving the excitation signal, the social agent needs time to distinguish between useful and erroneous information. A number of elementary time periods, t_{refr}, characterize a refractory (rest) period after the agent fires out the excitation information to its neighbors. The refractory period plays a key role in all excitatory systems, including social ones. After becoming excited, the agents become involved in the process of mastering the new information, performing necessary changes in their own activity (behavior, manufacturing, etc.). This is why some modern societies have better means to participate in the high-speed information economy than others. Determination and the use of excitation states are essential in social and business analyses, such as advertising media scheduling (Sissors et al. 2010), epidemic dynamics (Martin and Martin-Granel 2006), and diffusion of innovations (Alkemade and Castaldi 2005). In each specific application domain, one needs to find an explanation of formal parameters of the universal model by a constituent set of observable variables.

Excitation Signal Propagation Patterns in 2D Space

In economic and social systems analysis, three-, four- or higher-dimensional models are preferable. If the space has more dimensions, this enhances the visualization of the wave propagation. 2D illustrations show interesting patterns of excitation-wave propagation.

In simulation studies we performed hundreds of experiments using hexagonal grids. We found that some of the parameters affect wave propagation only in a very narrow interval of their variation. Outside this interval, the wave propagation ceases. In some cases, adequate selection of other parameter values can restore the wave-propagation process. To introduce nonhomogeneity, in some simulations the coordinates of the model's nodes were affected by a small noise. In our experiments, the triggering of the social media ($\arg_{start} = 0.7$) was carried out on the node at the very center of the media model. If excitation is sufficiently powerful, a strong

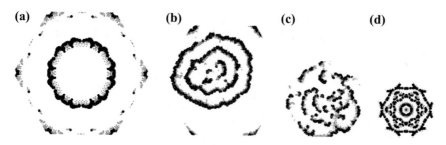

Fig. 9.5 Diverse wave propagation patterns after **a** 160 and 300; **b–d** 215 wave propagation steps in four social media examples differing in size and other model parameters

signal with some delay is transferred to $p = 6$ neighboring nodes. Then, each of the p already excited nodes transfer their output signals to a further 1, 2, or even 3 non-excited nodes (see Fig. 9.2, right). In Fig. 9.5a, c, d, we show four wave-propagation patterns. In the media on the far left and right (a, d), the nodes were placed strictly hexagonally, while in the two central ones (b, c) the nodes were placed in a nonuniform way. To make the distribution of the nodes nonuniform and ensure bursting, a noise was added to the coordinates of the nodes. The other parameters of the four media models were as follows: $N = [184\ 146\ 124\ 104]$; weights $= [0.7\ 0.8\ 0.7\ 0.73]$; refractory period refr $= [29\ 5\ 4\ 2.5]$; excitation threshold $= [0.5\ 0.6\ 0.45\ 0.55]$; noise level $= [0.0\ 0.5\ 0.5\ 0.0]$. More examples of excitation-wave patterns can be found in Fig. 9.5.

If the weights and initial excitation are small and the excitation threshold is too high (or very small), the wave propagation may cease. In intermediate situations, gaps (non-excited nodes) appear in a circle of excited nodes. If the refractory period is long, the gap is quickly filled by excitations from neighboring nodes. If the refractory period is short, after the excitation the nodes can be excited again without a lengthy delay. In such cases, the number of non-excited nodes increases quickly and the wave can start to propagate backwards. Such a situation is observed in the propagation patterns depicted in Fig. 9.5b–d.

If the values of the weights are close to 1, almost all the incoming signals can be transferred on. We then have a regular hexagonal-formed signal propagation, even in situations where the refractory period is short. This situation is depicted in Fig. 9.5a, d. Freshly (for the two most recent time periods) excited nodes are shown in black, and cells in the refractory period are shown in gray. The edge of this media $N_y = 37$, refractory period $t_{refr} = 4$, and weight $w = 0.8$.

In Fig. 9.5a we have a strictly hexagonal model with a comparatively long refractory period. The wave-propagation pattern is almost circular (see the wave after $t = 160$ time steps). Later, the wave reaches the hexagonal edges of the media. After 300 steps in six corners, we see only a few exited nodes. We can also observe a small number of gray nodes close to the end of the refractory period. After two time periods, all the excited nodes disappear. Only the nodes in the refractory period remain. After a couple more time periods, they disappear as well. In short, novelty for the social media investigations approach allowed us to obtain diverse wave-propagation patterns observed in real-world experiments (Litvak-Hinenzon and Stone 2007).

9.3   Social Media Model Composed of Several Agent Groups

The analysis of cell behavior in biology and social organizations of human beings and enterprises shows their clustered structure and even their synchronized behavior. The same is confirmed by theory and by simulations of wave-propagation processes (Chua et al. 1995) and other approaches (Keller-Schmidt and Klemm 2012). In general, many aspects of nature, biology, and even society have become part of the techniques and algorithms used in computer science, or have been used to enhance or hybridize techniques through the inclusion of advanced evolution, cooperation or biology (Cruz et al. 2010).

With this knowledge in mind, we developed *a structured schema of agent society*. The agents (the nodes in the model) are split into groups with diverse characteristics (parameters of the social media model). It is assumed that the groups in the single populations are partners and share the novel information they get from outside. Inside a single group, the wave excitation and propagation are performed as described in previous section. If a large number of nodes in the single media (group) exceed *a certain level*, the group leader transmits excitation (novelty) signals to leaders of other groups of the agent populations (the central nodes in the groups). To determine this level, we selected a fraction $T_{excitation} = 0.02$ of the recently excited nodes in a central area of the media as a criterion to transmit excitation information to other partner groups. The central area was characterized by a fraction of the total number of agents found excited in that area. In simulations, we used $N_{closest} = 0.8$. In Fig. 9.6 we present the wave-distribution patterns after $t = 361$ and $t = 516$ wave-propagation steps. In this experiment, we have one group (c) in which the signal does not fade, and three groups in which the signals vanished after reaching the media edges. Because the model parameters of the groups are different, the disappearance of the excitation in diverse groups occurs at different moments in time. For that reason, diversity of the groups is an almost essential condition for the survival of the excitation waves in the agent population.

Mathematical notation aside, the motivation behind the group approach is very simple. To simulate complex social agent-based systems, explicit information encoded in the properties of individual agents is not sufficient, as social agents are not actually so individual. Social agents are open systems influenced by the external (i.e., not only local, but also regional and global) environment at large (Bandini et al. 2007). The new social media model provides new perspectives on observing the phase-dependent field-like reality of social media excitations (De Paoli and Vizzari 2003). These findings lead us to the idea that contextual (implicit) information via the mechanism of dissipating excitation waves is distributed in fields and

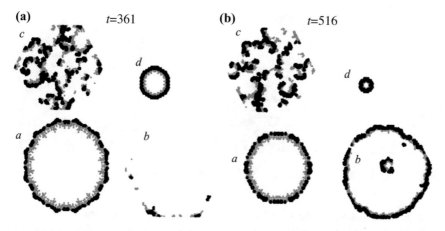

Fig. 9.6 Wave patterns in four (NG = 4) agent groups after **a** 361 and **b** 516 wave propagation steps. Parameters: sizes N_y = [128 120 108 74]; weights = [0.7 0.8 0.7 0.8]; refractory period = [20 16 4 4]; excitation threshold = [0.5 0.6 0.45 0.5]; noise level = [0.0 0.6 0.5 0.0]

that those fields, although expressing some global information, are locally (mostly unconsciously) perceived by the agents. Based on the earlier considerations, we foresee the further expansion of the present information-transmission model. One needs to introduce the global variables and states of the whole system. In such a simulation, a global implicit as well as local explicit level of contextual information could be captured and effectively exploited (Bandini et al. 2007).

9.4   Propagation of Two Competing Signals in Populations of Agents

Two Colliding Waves and Their Breakdown

Of course, bimodal simulation is a mere simplification of social multimodal reality. Nevertheless, such reductionism helps us to observe the basic features of excitation-propagation dynamics. In fact, two competing excitation modes (originating from two different excitation centers) are often perceived in social media; for instance, two competing major parties, presidential candidates, opposing opinions, genders or states of faith. This happens naturally, even in complex heterogeneous agents' environments where extreme competition (or cooperation) boils down to just two dominating states of social excitation. Hence, bimodal modeling of such states of social excitation reveals from bottom to top the agents' parameters, which foster observed social behavioral patterns.

Up until now, we have considered the propagation of innovation (excitation) signals in unexcited social media. In reality, everybody—human beings, the economy and social organizations—is already excited by previous innovations. In fact, the

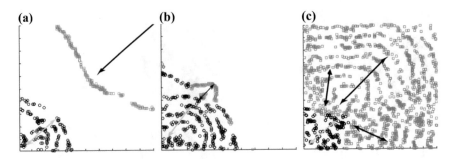

Fig. 9.7 Annihilations of two colliding waves and their breakdown

competitive dissemination of innovations takes place; that is, one innovation is replaced by another. In order to meet such a multi-excitatory reality, we have to enrich a single excitation model with multi-excitatory properties, where several competing signals can be modeled.

First, let us consider a single excitable media model where two signals begin propagating from different nodes. In the 2D example in Fig. 9.7a (after $t = 100$ propagation time steps), we see two propagating excitation (novelty) waves. Both excitation signals are propagating in strictly hexagonal media. A relatively short refractory period usually causes chaotic but periodic patterns for both propagating signals. After the waves collide (Fig. 9.7b, after $t = 170$ propagation time steps), almost all the nodes near the collision area are in their refractory period. Due to a short refractory period, the excited nodes of both signals are able to find some nodes whose refractory period has just ended. In Fig. 9.7b we can see that the signal represented by black squares is dominant and the signal represented in gray has almost died; however, after one hundred time steps, the gray signal recovers and begins to dominate almost all areas of the media space. The character of the wave propagation remains the same during the subsequent 10,000 time steps (Fig. 9.6c).

By changing the parameters of the social media, one can observe an enormous variety of wave-propagating patterns, with one signal outperforming another and the other dying, and slow or rapid movement of borders between two differently excited areas. An analysis of a large number of simulations with a variety of model parameters can lead us toward a better understanding of wave-colliding mechanisms.

Propagation of Two Signals in the Grouped Population

Like (Keller-Schmidt and Klemm 2012), the authors of this chapter assumed at first that innovations were rare. In previous sections we considered the propagation of novelty signals in empty, non-excited social media. In previous grouped-agent population models, the novel excitation from outside arose just after the end of the refractory period in the central area of the nodes. Nowadays, scientific, technological, political, and economic novelties appear fairly frequently. In reality, we do not have an unexcited economy or social media. They are all full of activity due to previous fruitful ideas.

In order to analyze the propagation of two (EXS = 2) or more excitation signals that appear at different moments in time, we expanded the grouped structure of the agent population. In the analysis of two, the old and new, innovations, we considered a population composed of NG = 4 agent groups. Each group controls two excitable social media: one for the propagation of each single innovation signal. In total, we had to examine the wave-propagation patterns in 2NG = 8 groups of media. To bring the model closer to problems we face in reality, we introduced the mutual influence of two groups of parameters allocated to each solitary agent group. We supposed that the propagation of alternative signals (first four groups of media) would influence the media parameters of the remaining four groups: the excitation threshold refractory period and the magnitudes of the weights. In such a way, one excitation signal disturbed the propagation of another signal. We investigated a variety of scenarios and found a wide range of innovative information-propagation patterns:

1. both signals propagate in the population of the groups; and
2. new signal wins or loses, etc.

A variety of possibilities requires a separate large-scale study. A two-fold analysis of information propagations in the grouped excitable social media model, however, shows that the wave-propagation approach can be useful for understanding the distribution of a variety of innovations and controlling their dissemination processes.

Concluding Remarks

In this chapter, we have presented a new information-propagation approach developed to simulate excitations in large social media. Our model provides a bottom–up method of simulating and investigating information-diffusion properties in social media based on the characteristics of the simulated simple agents. Before concluding, we interpret the social context and reality a little more broadly. Hence, meaningful information can be understood as an economic resource or intangible commodity in modern societies, where a large share of gross national product is produced by the information-related service sectors. In this regard, depending on the properties of the social networks, some societies perform better than others in the new information economy.

Following the line of these investigations, we should observe that the majority of agents act as information processing, storing, and transmitting nodes. Models like the ones presented in this chapter help to reduce investigation problems to the most fundamental constituent parts. In this way, bottom–up and wave-like excitation-propagation research assists us to identify the most basic properties of social agents, which produce observed phenomena such as information broadcasting, destruction, competition between several sources of excitation, and excitation of transfers to other social domains. The main results of our analysis are as follows.

1. To make the wave propagation approach useful in excitable social media, we expanded CA- and artificial NNs-based schema (Raudys 2004) and suggested a novel one in which the node excitation and recovery times depend on the

strength of the signal accumulated during two time periods for a single node. Our results indicate that the nonlinearity of the activation function plays an especially important role.

2. This chapter provides insights into, and an overview of, the interpretation of basic parameters; for example in the applied case of advertising media scheduling we have shown a concrete way to exploit the excitation saturation decay function.

3. Broadening the homogeneous media model to make it heterogeneous revealed population-clustering dynamics and conditions of cooperation between different groups. We showed that a clustered and diverse nature of the population is vital for the overall performance of a society.

4. We investigated a large number of wave-propagating patterns in differently structured multigroup populations of agents and found propagation patterns frequently observed in the case of the diffusion of innovations, the spread of behavior in online social network experiments, and the diffusion of messages in social groups, etc. To view these and other simulation results, visit our online virtual lab (V-lab) at http://vlab.vva.lt/; MEPSM1 model, login: Guest; password: guest555.

5. We suggested a structure and model for population groups in order to analyze the dissemination of two or more competing innovation signals. This provides new possibilities to understand what kinds of social media characteristics influence the dominance of one or another of the signals.

Our analysis supports the conclusion concerning the bursting branching process of evolution observed in databases and theoretically explained in (Keller-Schmidt and Klemm 2012). In summary, this chapter presents a mechanism to explore and set parameters for uniform, chaotic, spiral, and rhythmic oscillations, as well as, bursting excitation-propagation patterns that originate from essentially similar basic properties; namely, agents' propensity to adopt a new behavior depending on the proportion of a neighboring reference group that has already adopted it.

Based on the observed oscillation dynamics in the grouped population simulations, we infer our models in accordance with chaos theory, where rhythmic oscillations are typical indicators of (i) coherence between the constituent parts (i.e., agents); and (ii) self-organization on the whole-system scale (Osipov et al. 2007). There are also such implications for self-organized social systems, as coherence between agents is mainly established through the constant recalibration (i.e., propagation of excitation waves) of agents' states. This is why waves of excitation occur rhythmically in self-organized social systems. Hence, simulations of oscillations (waves of excitation) in social systems from the bottom–up (i.e., agents' perspectives) could yield new knowledge and surprising results concerning conditions for the occurrence of coherent social order. It therefore appears to be essential to direct our prospective research efforts towards the simulation of oscillations in social systems.

Analysis of the synthetic data generated by modeling wave propagation can help to distinguish the most important factors that affect the social phenomena under investigation. For instance, in the absence of reliable empirical data, wave-propagation modeling becomes a useful instrument to determine the attributes necessary for carrying out, for example, effective sociological surveys, political campaigns and news distribution. We also intend to bring into play 2D and 3D visualizations of excitation-propagation patterns in the analysis of time series of high-dimensional financial data used in automated trading (Raudys 2013).

The agent groups and wave-propagation-based models can be expanded in a number of ways. For instance, we can use dynamical system theory to develop a mechanistic mean-field model for neural activity to study the abrupt transition from consciousness to unconsciousness as the concentration of an anaesthetic or other type of agent increases (Hou et al. 2015).

Various agent-proximity measures can be applied, including OAM-based or other approaches, and the simultaneous propagation of several dissimilar information waves can be considered. In following the OSIMAS paradigm, there could also be the introduction of global variables and the ability of the states of the entire system to transmit excitations to distant neighbors, using the broadband one-to-many information-transmission approach. In such models, we can adjust the size, refractory period, and importance of the agents. The agent groups can be joined together with subpopulations. We will then obtain multilevel, hierarchical agent population structures. An important task, which has not yet been touched on, is an automated evaluation of the model's parameters. The simulation of media evolution during long periods of innovation becomes a useful tool here (Raudys 2008). In prospective research, however, the simulation and examination of concrete social phenomena follows, and seeking for task-dependant interpretations of the parameters (or sets of them) in the applied OAM-based MAS models.

Chapter 10
Agent-Based Modeling of Fluctuations in Automated Trading Systems

This applied chapter uses large scale social media excitation wave propagation model for the simulation of automated investment strategies. Some OAM, PIF, and WIM principles are integrated into large scale CA (cellular automata) and AI-based MAS modeling. Changes in the model's parameters allow periodic, vibrant, and turbulent excitation wave propagation behaviors to be obtained in the large scale simulated social media. The type of signal behavior can change overtime. Without external intervention, these behavioral characteristics can reverse. The spectral representation of the multidimensional time series allows specific features to be generated that are necessary for recognizing the type of chaos-generating model. This observation suggests a novel evolvable portfolio-calculation scheme. The novel portfolio design schema notably outperforms three benchmark methods in automated securities trading, where one deals with thousands of investment strategies.

10.1 Introduction

The rapid development of information processing and transmitting technologies initiated the birth of automated trading systems (ATS) in financial markets, and increased their number and interdependence (spatial contagion Araújo and Castro 2010; Durante and Foscolo 2013). This increase in number, as well as the interactions between economic and financial units, became one of the causes of chaotic

The content of this chapter is based on the joint research together with Prof. Habil. Dr. S. Raudys in the framework of the OSIMAS project.

and unpredictably changing environments (Plikynas et al. 2015; Raudys 2004; Raudys et al. 2014). This chapter considers the turbulent, oscillating behavior of ATS in the investment sector. In automated trading term refers to any collection of financial assets or computerized trading strategies (robots) and a set of coefficients (weights) necessary to define the investment proportions. In financial markets, each participant is attempting to design its own portfolio that is capable of competing successfully. To design the optimal portfolio, one needs to choose optimal criteria and know the probability distribution of portfolio inputs (components and investment proportions).

An objective of portfolio design is to create the most profitable proportions for p assets, $x_1, x_2, ..., x_p$. Instead of assets, one can employ trading robots (profit and loss generated by trading robots). In this chapter, we consider daily trading. In this case, x_{ij}, is the jth investment's profit or loss during the ith day. In portfolio design, for the p-dimensional (p-D) input vector, $x_i = [x_{i1}, x_{i2}, ..., x_{ip}]$, one calculates the weighted sum

$$P(x_i, w) = \sum_{j=1}^{p} w_j x_{ji} = x_i w^T = q_i, \qquad (10.1)$$

where the coefficients of the portfolio's weight vector $w = (w_1, w_2, ..., w_p)$, fulfill the requirements: $w_j \geq 0$ and $\sum^p w_j = 1$ If L days' profit and loss history of vectors x_i is used as a training set, one can seek for optimal w where a profit and risk are taken into account. A *mean/variance approach* is often used here. One maximizes the Sharpe ratio based on the training data (Reilly and Brown 2011)

$$Sh_R = \text{mean}(P(w))/\text{std}(P(w)) = \bar{x} w^T / \sqrt{w S w^T}, \qquad (10.2)$$

where $\bar{x}$ is the sample N-D mean vector and S is the $p \times p$ sample covariance matrix (CM). We will use *an annualized Sharpe ratio modifier*, $Sh_A = Sh_R \times \sqrt{252}$, instead of Eq. 10.2.

If the number of portfolio inputs, p, is large, the distribution of sum (see Eq. 10.1) approaches the normal law. The highest losses can be characterized by a maximized probability defined as *Prob* $(P(x_i, w) < P_{max})$, where P_{max} is a risk level. If the portfolio dimensionality, p, is fixed, and the sample size, L, is very large (strictly speaking, $L \to \infty$), the maximization of ratio (see Eq. 10.2) gives the largest profit, Sh_A^{max}, for any *apriori* fixed risk value, P_{max}.

Difficulties arise when L is small and close to p. In situations with a small learning set, the portfolio profit diminishes due to the inexact evaluation of $\bar{x}$ and S. Asymptotically, it was shown that when both p and L are large, an expected profit, $Sh_A^{Expected}$, decreases as described in (Raudys 2013).

$$\text{Sh}_{\text{A}}^{\text{Expected}} = \text{Sh}_{\text{A}}^{\max}/T_{\text{mean}}/T_{\text{CM}}, \tag{10.3}$$

where

$$T_{\text{mean}} = \sqrt{1 + p/\left(L \times (\text{Sh}_{\text{A}}^{\max})^2\right)} \text{ and } T_{\text{CM}} = \sqrt{1 + p/(L-p)}.$$

Term T_{mean} shows a decrease in profit due to an inexact estimation of the means of the returns. The term T_{CM} indicates a decrease in profit due to the inexact estimation of CM. If the sample size, L, is small, different methods to regularize or reconfigure the sample estimate, S, have been developed. In the simplest regularization, one reduces the diagonal elements of matrix S:

$$S_{\text{reg}} = S \times (1 - \lambda) + \lambda \times \text{diagonal}(S), \tag{10.4}$$

where λ is the regularization coefficient. When $\lambda = 0$, we have no regularization. When $\lambda = 1$, we ignore all the correlations between the portfolio inputs. In the latter case, asymptotically, the term $T_{\text{CM}} = \rightarrow 1$ when both p and L tend to infinity. In such a portfolio design strategy, only the term T_{mean} controls the profit deterioration (Raudys 2013).

Nowadays, we have to deal with tens of thousands of portfolio inputs. The variation of the inputs overtime is often very chaotic. For this reason, we have an insufficient amount of reliable real-world data to redesign the portfolio weighting coefficients over time. It is possible that only a trading history of 500–600 days is reliable for a multidimensional portfolio design. This amount of data is clearly insufficient when the portfolio dimensionality is in excess of thousands.

Here is an example. If $\text{Sh}_{\text{R}}^{\max} = 0.5(\text{Sh}_{\text{A}}^{\max} = 7.9372)$, the number of trading robots $p = 6000$, for $L = 500$ we have $\text{Sh}_{\text{A}}^{\text{Expected}} \approx 1.13$. This means, due to insufficient data, the portfolio inefficacy decreases to one-seventh. Due to the presence of transaction costs, the equal weighting becomes ineffectual in high-dimensional situations. For that reason, automated trading becomes unprofitable as well.

One of the ways to overcome the problems related to small amounts of data is the equation-based modeling that is frequently applied in physics and chemistry. Such an approach would certainly be difficult to transfer to the social sciences, as the behavior of most systems has not been mathematically formalized. One of the exceptions would be analysis of noise in the financial time series as uncertain processes, described by uncertain differential equations (Liu et al. 2015). One method that seems to be suited for the computer simulation of socio-economic systems is agent-based computational modeling (Helbing 2013; Helbing and Balietti 2013). Depending on the problem in question, agents may, for example, represent individuals, groups, companies, or political units (Helbing 2013; Helbing and Balietti 2013; Plikynas et al. 2015; Raudys et al. 2014). Computer modeling of the agent-based social and economic mediums can become a useful tool for improving insights into interconnected economic and financial formations. Such an analysis becomes important in comprehending how people and organizations react to information about new technological, political, economic, and environmental changes.

An objective of the present chapter is to use a bottom–up approach to model semi-chaotic activities in financial markets in order to detect regularities that could suggest a more rational way to calculate the portfolio weights. Next section describes a multi-group excitable media-based multi-agent economy model and its features. We demonstrate that the synthetic time series behaves in an almost regular, vibrant, and chaotic way. We observe lasting changes in the signal behavior as well as the possibility of returning to a former state of activity without external intervention. Further on we deal with an adaptive, multilayer, multi-agent system (MAS)-like portfolio design for use in changing environments. The final section presents our conclusions.

10.2   Mimicking of Financial Times Series by Excitable Agent Media

The financial world is influenced by billions of interacting factors. It is impossible to develop "exact" dynamic models. Below we will describe, use, and examine a fairly simple model in which chaotic and vibrant behaviors have already been observed (Plikynas et al. 2015; Raudys et al. 2014). Wave propagation in excitable media is a fairly general approach. It has been used in the analysis of chaotic, spiral, and vibrant wave propagation in biology, medicine, epidemiology, chemistry, physics, and astrophysics. To explain the properties of the model more easily, we will describe a simplified version of a single agent group: a two-dimensional (2-D) model of nodes. After receiving a novel excitation (an information signal), each single node in the model can transmit an excitation (*novelty*) signal to adjacent "companion" nodes (see Fig. 10.1a). One of the nodes is excited at the very start

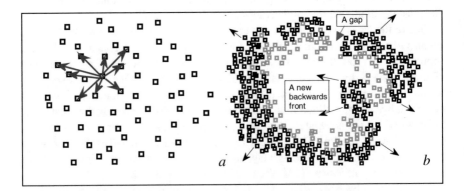

Fig. 10.1   Signal propagation: in the model of almost *hexagonally* distributed agents in 2-D space: **a** signal transmission during the first step of simulation; **b** the creation of wave gaps and the inwards front (after 40 time steps)

and transmits an excitation signal to nine other nodes (blue arrows). As with the single-layer perceptron (Haykin 1998), each of the model's nodes is stimulated by the sum of the excitations coming from other nodes.

The transmission of the signal is performed with a certain delay. The delay and power of excitations to adjacent companion nodes depend on the sum of the strength of the node's excitation. After transmitting a signal to its neighbors, each node has to take a rest (the refractory period). After this period of delay, the excitation signals are transferred to further nodes. More details about the model can be found in previous papers (Plikynas et al. 2015; Raudys 2004; Raudys et al. 2014). In principle, a wide variety of modifications can be applied to the model. In economics and finance we have to deal with multidimensional space, so the distance (similarity) should be measured in a more complex way. In Fig. 10.1 and later in Fig. 10.2, we made 2-D simplifications.

The following parameters of the model affect the excitation wave propagation: the size and configuration of the grid, the strengths and directions of signal transmissions to adjacent nodes, the method of determining the refractory period, the threshold for a node to be excited, and the type of nonlinear activation function that bounds the node's output signal. In Fig. 10.1b we see a sequence of excitation waves originating from the stimulation of a single cell 40 time steps before. The recently excited nodes are colored black. They are at the front of the wave. The nodes during the final moments of the refractory period are colored gray.

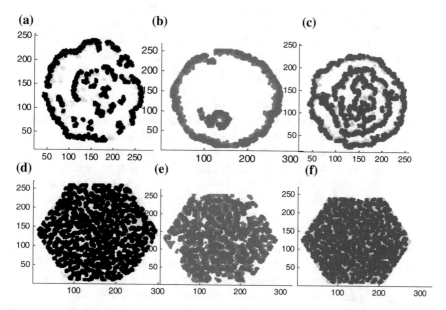

Fig. 10.2 Wave-propagation patterns in three agent clouds: **a** after ~ 200 time steps; **b** after 4400 time steps; **c** after 10000 time steps; **d** after 15000 time steps; **e** after 20000 time steps and **f** after 25000 time steps. In each wave chip, we can observe that the wave front (*darker colors*) is followed by nodes during the last moments of the refractory period (*lighter colors*)

In Fig. 10.2a, d we show the wave-propagation patterns in an almost hexagonal media model composed of 14,911 nodes. Figure 10.2a clearly illustrates the split of the excitation wave into "chips." Due to the heterogeneity of the excitable-media, gaps (marked by the letter A_{GAP} in Fig. 10.1b) are starting to develop at the front of the wave. In this way, the wave starts to propagate inward, towards the center of the grid. An inner wave in Fig. 10.2a is the product of such inward propagation. In Fig. 10.2a–c we can already see the separate wave chips. After 5000 time steps the wave chips are packed closely together. Inspections of a large number of grids (node clouds) characterized by diverse sets of model parameters demonstrates that wave-propagation patterns can become *regular*, *spiral*, *chaotic*, or *vibrant*. The wave can cease at the very beginning, after a moderate number of time steps, or after overstepping the hexagonal borders of the grid.

To avoid the excitation signals ceasing in our "economy model" composed of a number of separate groups (grids of nodes), we introduced a restricted level of cooperation between them. If a sum of node's excitations in a subgroup of nodes goes above a certain threshold, the agent group can transmit the excitation signal to another group in which excitation has already ceased. This way, we mimic team-work. If the parameters of such a team are chosen correctly, excitation does not cease.

To mimic fluctuation behavior we formed 12 nonintersecting subgroups of adjacent nodes (agents) inside each node group. Each single subgroup was composed of 200 nodes (~ 1.5 % of nodes in the cloud). We supposed that the total excitation of a solitary subgroup could mimic the fluctuations of a solitary "asset" or profit/loss of a solitary trading strategy realized in the ATS.

The assets of several subgroups in the distinct clouds turned out to be correlated. It should be noted that the model we have just considered is very simple. It is useful for understanding the chaotic behavior of the financial markets; however, the questions of mutual cooperation are worthy of further analysis and development in the future.

In Fig. 10.3a we observe fluctuations of "artificial assets" (one for each of three groups) during 256 time steps taking place once the wave-propagation process has already become well-established (~ 5000 time steps). We observe clearly periodic signal components, which hint that the Fourier transform can be useful to discriminate the clouds according to their histories of excitation wave propagation.

Figure 10.3b displays three spectra of the artificial time series. The spectra were obtained using the 2nd to 129th absolute values of the real part of the spectra x_1, x_2, … , x_{128}. The spectra confirm presence of obvious periodic signal components.

Figure 10.3b demonstrates that all three spectra are different. We analyzed multitudes of spectra calculated from relatively short time periods (256 iterations). We found that that the spectra inside a single cloud are fairly diverse. We can discern, however, differences between the spectra belonging to signals from different agent groups characterized by distinct parameters of the excitation model.

To demonstrate the differences between multitudes of spectra according to type of vibrant-chaotic behavior, we reduced the number of features from 128 to 2. The 128-D spectral data was mapped onto 2-D space. For mapping, the popular method

(a)   **(b)**

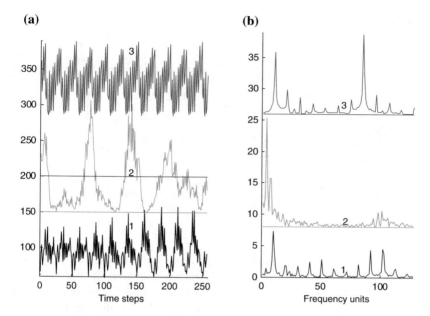

Fig. 10.3 a The oscillation of synthetic assets in each of the three agent groups; **b** the spectra of artificial assets in the three agent subgroups

of a linear discriminant analysis (LDA) (Fukunaga 2013) was used. Here, in the multi-category tasks, one maximizes the "normalized" distances between the sample mean vectors $\bar{X}_i$ ($i = 1, 2, \ldots, K$) of K classes

$$W = \underset{W}{\arg\max}\left(WS_b W^T / WS_w W^T\right), \tag{10.5}$$

where W stands for the $(K - 1) \times p$-D transformation matrix where one obtains the $(K - 1)$-D vector of new features,

$$Y = (y_1, y_2, \ldots, y_{k-1}) = XW^T,$$

$$S_w = (L - 1)^{-1} \sum_i^K \sum_j^{L_i} (X_{ij} - \bar{X}_i)^T (X_{ij} - \bar{X}_i).$$

Here S_w stands for the between and within classes sample CMs, $\bar{X}$ is the mean vector of all training data, $L = \sum_i^K L_i$, X_{ij} represents the jth training vector-row of the ith class (group), and L_i is the number of vectors of the ith class used to estimate the mean vector $\bar{X}_i$.

In this demonstration example, $K = 3$, $L_i = 155$, $p = 128$, and 5000 time steps were used to obtain a well-established wave-propagation process. Afterwards,

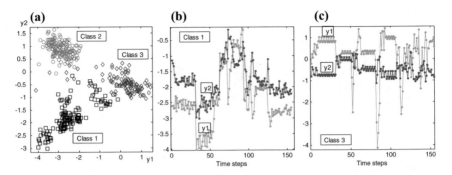

Fig. 10.4 a A scatter diagram of the spectral description distribution of 465 wave-propagation time intervals mapped onto 2-D space; **b, c** the variation of two new features, y_1 and y_2 of Class1 and Class 2

310×256 subsequent time steps were used to obtain $2K \times L$ 128-D spectra of three time series corresponding to three dissimilar node groups. The first half of the data was used as a training set to evaluate the (2×128)-dimensional data transformation matrix, $\mathbf{W}$. The second half of the data served as the test set, where the data was mapped to 2-D space, $(y_1, y_2) = \mathbf{XW}^T$. Figure 10.3a shows that the wave-propagation patterns inside the three groups used for testing can be discriminated well, even in the 2-D space. Thus, the spectral features are informative in specifying the wave-propagation style.

We observe in Fig. 10.4a that the distributions of two classes are composed from several small subclusters. This means the type of wave-propagation processes is changing over time. The scatter diagrams in Fig. 10.4a do not show the time series behaviors overtime. In Fig. 10.4b, c for two pattern classes (Class1 and Class3) we display variation of two new features, y_1 and y_2, during 155 time steps. We see that the 2-D vectors are ranging in small areas, say, Ω_1, during a certain number of time steps. After some fluctuation, the vectors in the same category (class) start moving to other areas, for example, Ω_2, Ω_3, and so on, or return near to the area Ω_1. The fluctuation of the vector in a single area continues for some time. This time period is longer than that of the changes in investment values. One can guess that a mixture of trading agents could be more stable than their investment values. It is possible that this fact is the main result of the study of chaotic behavior performed in this section. If the wandering behavior of agent2 is verified for finance and economics, this means that some parts of the old data history could be useful for forecasting the chaotic behavior of financial time series and for adaptive portfolio management. Knowledge of such signal behavior could affect the calculation schemas of portfolio weightings. The following section will be devoted to looking for such behavior in real-world financial time series and using the knowledge obtained for portfolio management in a changing environment.

10.3   Portfolio Management Rules in High-Dimensional Situations

Tens of thousands of trading strategies are candidates for selection as the portfolio components of finance ATS today. The term T_{CM} in Eq. 10.3 indicates that in such situations one is obliged to ignore correlations. The term T_{mean} says that a decrease in the true Sharpe ratio due to an inexact sample estimation of the vector of returns is huge (see the comments following Eq. 10.3 in the introduction). Our objective is to design realistic schema to achieve the task and compare it with benchmark methods.

Benchmark methods

The simplest rule is the nontrainable 1/N portfolio-management rule. Using this method, all inputs are equally weighted. DeMiguel and his colleagues (DeMiguel et al. 2009) compared several methods and suggested using the 1/N portfolio as a benchmark rule (we will refer to this as rule **BA**).

The second benchmark method (**BB**) uses L_1 previous days' history and estimates the Sharpe ratio for each of the p trading strategies (robots) individually. Then, one ranks the robots and selects p_B best of them in order to design the 1/N portfolio rule with p_B inputs and all weights, $1/p_B$.

In our previous paper (Raudys et al. 2014), the adaptive MAS scheme (benchmark rule **BC**) was designed for recalculating the portfolio weights). This schema is similar to the feed-forward multilayer perceptron (MLP) (Haykin 1998). This MAS has p inputs, H_1 hidden layer nodes, and a single output that calculates the portfolio return. The inputs of the hidden layer units are fed by different subsets of inputs (one day's values of different time series). To evaluate the schema's performance the $H_{effective}$ best hidden layer nodes (trading agents, intermediate portfolios) are fused by the weighted sum rule (the Markowitz rule, see below). Effective nodes are selected on the basis of an additional cost-sensitive classification algorithm (Raudys and Raudys 2010). Thus, a sense of the hidden layer and subsequent classification is *dimensionality reduction*: We design and select a small number of effective trading agents, specific to each time period. This algorithm is used as the benchmark rule **C**.

In the **standard mean–variance framework (the Markowitz rule)**, the finding of the vector w is performed as a calculation of an efficient frontier for F return values, $w_1\bar{X}^T = q_1$, $w_2\bar{X}^T = q_2$, ... , $w_F\bar{X}^T = q_F$ (Durante and Foscolo 2013). The standard maximization of ratio (see Eq. 10.2) is based on Lagrange multipliers, which require the fulfillment of two constraints, $w_j \geq 0$, $\sum_{j=1}^{p} w_j = 1$, and $\bar{X}w_t^T = q_t$, $t = 1, 2, ... , F$. In our experiments, we calculated the weight vectors w_1, w_2, ... , w_F analytically (Raudys 2013)

$$w_t = (q_t\bar{X} + 1)\,(S/\eta + \bar{X}^T\bar{X} + 1^T1)^{-1}, \tag{10.6}$$

where $\mathbf{1} = [1\ 1\ ...\ 1]$ stands for a row vector composed of r "ones" and scalar η controls the violations of constraints (an accuracy of fulfilling the constraints). At the same time this regularizes the solution. The term η should be sufficiently large (Raudys 2013). This algorithm is many times faster. For that reason, it is suitable for the multiple calculations that are necessary in the MAS design. In the regularized versions, instead of a conventional sample estimate of the covariance matrix, S, we use the regularized one, S_{reg}, defined in Eq. 10.3.

Adaptation to changes in the novel MAS-Based portfolio rule

To develop a novel automatic trading system we used our previous experience, where decision-making schema reminded the multilayer perceptron with a single layer (Raudys 2013; Raudys et al. 2014a, b). In all of them, the p inputs were split into groups. Diverse heuristics were used for this splitting. To generate a smaller number of more complex trading agents, weighted sums of inputs in each single group were calculated. At the final portfolio design stage, the outputs of the designed agents were fused by using the regularized Markowitz rule.

The chaos modeling performed in previous section showed that the parameters of excitation wave propagation behavior wander between small areas (say, Ω_1) for certain time periods. Then, almost immediately, the vector parameters move to other areas. After some time the parameter vector can return close to area Ω_1 again. While discussing the data used in the experiments we will show that the "wandering phenomenon" can also be observed in the financial data.

In *the novel portfolio design scheme* we assign L_1 days' data and two hidden layers aimed to generate specific agents (named "agents2"). We want them to be highly comparative (Fulcher and Jones 2014), specific, and efficient in financial trading, even for a short period of time. For the best agents2 selection we rank agents2 according to the Sharpe ratio evaluated during relatively short $N_{\text{intervals}}$ succeeding sections of training data ($L_{\text{rank}} = 100$, see the schema in Fig. 10.4). Subsequent $L_1 + 1$ to $L_1 + L_{\text{sel}}$ days' data is used for the most efficient agents2 selection and serves as input into the final calculation of portfolio weights. The weights are calculated according to the regularized Markowitz rule (Eqs. 10.4 and 10.6).

We present more details below. The first layer of the system is based on a careful clustering of p input time series data, L_1 day length (see learning and testing schema in Fig. 10.5). One of the most important aspects of clustering is the similarity measure. In a previous scheme (benchmark method **C**), the cluster analysis of the input time series was done according to the correlations between input time series during L_1 days' history. In contrast, in the present chapter we used *spectral features* obtained after the fast Fourier transformation (FFT). The justification for this approach was presented in the previous section. The time series we are dealing with have a large number of outliers (days when the input robots' profits or losses are extremely large). To reduce the influence of outliers on clustering quality, prior to FFT we performed a square root transformation of the profit and loss time series:

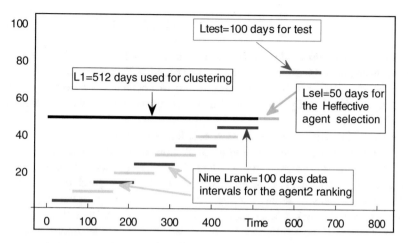

Fig. 10.5 Training and performance-testing schema for $N_{\text{intervals}} = 9$

$$y_i = \text{sign}(x_i) \times \sqrt{|x_i|}, \tag{10.7}$$

where x_i is ith value of the input time series. The transformed y_i, $(i = 1, 2, \ldots, L_1)$ data is used for only FFT, the subsequent principal component analysis-based dimensionality reduction, and the clustering of p portfolio inputs time series.

A characteristic feature of the spectra estimated from relatively short data series is the high correlation between the spectral features. To reduce dimensionality, the principal component analysis of L_1 training days' transformed data was performed. Only the first $r = 64$ new data components were used for clustering. We used the K-means algorithm. The inputs (robots' outputs) that composed a single cluster were used to form trading agent1. We require high-quality clustering. Therefore, we chose H_1, the number of clusters, to be large.

Good clustering means that within a single cluster the robots' outputs can be highly correlated. Consequently, we can make an assumption that the correlations between the time series belonging to the same cluster are approximately equal. The equality of the correlations strongly suggests that classification according to the class mean vectors is near to the optimal classification rule (Fulcher and Jones 2014). For this reason, in each cluster we used averaging. Thus, the 1/N BA rule was used in each cluster. Here, all weights were equal to $1/p_j$ (p_j is the number of time series assigned to the jth cluster).

To make trading agents *more specific* for successful trading during the time assigned to train the MAS (L_1 days) we generated $H_2 = H_1(H_1 - 1)/2$ pairs of the first-level agents1. In a single node of the second hidden layer, the outputs of two agents1, $\bar{x}_i$ and $\bar{x}_j$, were fused by the linear weighting

$$y_s = \alpha_s \bar{x}_i + (1 - \alpha_s)\bar{x}_j, s = 1, 2, \ldots, H_2. \qquad (10.8)$$

So, for $H_1 = 100$ hidden units we generated $H_2 = H_1(H_1 - 1)/2 = 4950$ composite agents2 for possible inclusion in the second hidden layer for the final portfolio weights determination. The best values of coefficients α_s were determined from L_1 training days by calculating the mean/variance ratio (2) values 101 times in interval [0 1].

We remind readers that the above simulation of the artificially generated chaotic and vibrant time series (see previous section) showed that time intervals exist where the wave signals of one class, more or less, do not change their properties. To make the second hidden layer more effective for developing a good portfolio for the subsequent L_{test} days' trading, we ought to find a small subset of the $H_{effective}$ most effective agents2 (second-layer nodes) from the large collection of H_2 nodes. In this regard, we ranked the agents2 for a number, say $N_{intervals}$, of data intervals. We used an additional L_{sel} days to identify which interval's data was most useful for the calculation of the portfolio weights. In this way, we extracted an ensemble of features for an effective portfolio weights calculation for the next time period. The schema of training, validation, and testing is presented in Fig. 10.5. In an empirical comparison of the novel and benchmark methods, the walk-forward experiments were performed many times. To obtain *Exp* performance estimates, the $L_1 + L_{sel} + L_{test}$ days' data was moved forward *Exp*-1 times with a shift of L_{shift} days.

10.4   Data Used and Its Preliminary Analysis

As before (Raudys et al. 2014a, b) we analyzed thousands of financial time series created by simulating automated trading strategies (trading robots) and recording daily profit and loss. We considered the most liquid US and European futures in global exchanges (CME, CBOT, NYBOT, ICE, COMEX, EUREX): stock indexes, energies, metals, commodities, interest-rate products, foreign-exchange products (E-mini S&P 500, E-mini S&P MidCap 400, E-mini Russell 2000, E-mini ASDAQ-100, E-mini DOW ($5), Canadian dollar, Swiss franc, Japanese yen, Australian dollar, Euro FX, British pound, sugar, coffee, soybeans, gold, silver, copper, DAX, EURO STOXX 50, natural gas, oil, 2-, 5-, 10- and 30-year US Treasury notes and others). Some time series were relatively sparse, and consisted of 50 % of zeros ("empty days" when trading robots refuse to invest and obtain zero profit on the next day).

To verify the new ideas that followed from the chaos analysis, we considered three real-world financial datasets that cover prior-to-crisis and crisis time periods, characterized by a multitude of sudden changes. The first, 10441-dimensional dataset, **DSA**, covers the period from January 20, 2008 to February 13, 2013. The second, 7480-D dataset, **DSB**, covers the period from August 30, 2009 to March 21,

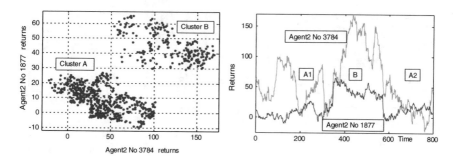

Fig. 10.6 a A scatter diagram of two agent2 returns observed in a 1024-day period; **b** variation of markedly *smoothed* returns in time

2014. Finally, the third, 14686-D dataset, **DSC**, covers period from January 2, 2009 to November 4, 2014 (1462 working days).

The chaos modeling carried out in the previous section showed that the parameters indicative of the excitation wave propagation style wander in diverse small areas for some time. After a period of time, the parameter vector can return near to its previous region. The wave-propagation patterns hint that part of the data history could be useful for future portfolio management in chaotically changing environments. We conducted a special study to identify whether similar behavior can be observed in real-world financial time series.

In an analysis of profit and loss time series used in the finance ATSs, where tens of thousands of trading strategies were candidates for selection as portfolio components, we did not succeed in finding similar behavior patterns. In a continuation of previous research , we started to analyze very specific agents produced by the first two layers of the MAS described in the previous section. In Fig. 10.6a we present a scatter diagram of the distribution of two agents2 returns during *the last 1024 working days* of financial data from **DSA**. Both agents2 (the 1877th and 3784th) where chosen out of $H_2 = 4950$ ones generated after clustering the portfolio input time series into $H_1 = 100$ groups.

In the scatter diagram Fig. 10.6a we observe two big clusters. In the left-hand cluster (A) the returns are smaller than in cluster B. Figure 10.6b shows that cluster A can be split into two parts, with cluster B between these two parts. The agent wandering in time confirms the conclusions obtained from the chaos behavior analysis in the second section: the older historical data can be useful in the portfolio design process. It seems that the data segment A1 fits better than data B for use in portfolio design during the time interval A2.

In Fig. 10.7a we present the smoothed spectra of the 4140th agent2. To obtain the smoother graph we shifted 512-day-long data 31 times and calculated the spectra each time. Then we averaged $N_{SE} = 32$ estimates of the spectra, smoothed the result slightly, and displayed it in Fig. 10.7a. We can see that the spectrum has two clearly expressed peaks. At least two sinusoidal components exist in this

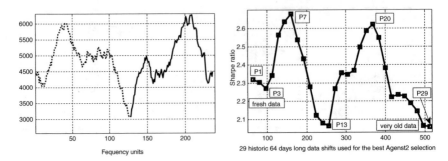

Fig. 10.7 a *Smoothed* spectra of the 4140th agent2; **b** the influence of old data on the portfolio quality

spectrum. This means the 4140th agent2 returns possess vibration properties, and the older behavior of the data can reappear from time to time.

Figure 10.7b exhibits the following phenomenon: an influence of *the selection data interval* on the portfolio quality. For a demonstration we selected the $H_{\text{effective}} = 1485$ most effective agents2 (30 % of 4950) to be used for the 1/N portfolio calculation.

The schema of this data-investigation experiment is as follows:

1. Training data: $[1:L_1]$ days for clustering, agents1 and agents2 design. The $N_{SE} = 29$ overlapping windows (**OW**), $L_{rank} = 64$ days each for ranking the agents2. Validation data (**VD**): next to each single **OW** $L_1 + [1:L_{sel}]$ data interval (in this experiment, $L_{sel} = 64$). The **VD** and **OW** datasets are $L_1 + L_{sel} = 576$ days.
2. The *H2* Sharpe ratio values are calculated for each single **OW**. The best $H_{\text{effective}}$ agents2 are selected for each single **OW**.
3. For all N_{SE} single **OWs**, the same **VD** set is used to estimate the efficacy (the Sharpe ratio for the 1/N portfolio rule) of the $H_{\text{effective selected}}$ agents2.
4. As a result, we have $N_{SE} = 29$ Sharpe ratio estimates with different data window delays used for ranking the agents2.
5. For a more exact and a smoother evaluation, the 576-day data periods are shifted forward 28 times. In this way, the experiments are repeated $Exp = 29$ times. The averages of Exp experiments are used for comparison.

The averages are displayed in Fig. 10.7b and confirm that older data can be more preferable than fresh data. For the 10441-dimensional financial dataset, the most useful appeared to be the 385th- to 448th-day-old data interval, P_7. In Fig. 10.7b we can see that the older data interval P_{20} also provides good results. Traditional statistical methods, however, use the latest history. Possibly, argumentation by chaos simulations and the experimental verification of the usefulness of older data fora very specific selection of agents for the final portfolio weights calculation is the main finding of the present chapter.

Experiments using real-world data

Experiments using several real-world data sets showed that the optimal agent ranking interval, Dat_{rank}, varies depending on changes in the real-world financial dataset and the number of agents2, $H_{effective}$. Therefore, in the development and validation of the novel MAS we decided to determine the most effective parameter values based on previous experience.

To find the most suitable Dat_{rank} and $H_{effective}$, we used the schema described in Fig. 10.6. For the FFT, subsequent clustering into $H_1 = 100$ groups, and for the design of agents2 we used a 512-day history. To rank the agents, we tested $m = 9$ 100-day overlapping intervals, Dat_{rank}. To evaluate the performance of the MAS in order to choose the most effective set of Dat_{rank} and $H_{effective}$ values, we used $L_{sel} = 50$ extra days. We tested n diverse $H_{effective}$ values, e.g., $H_{effective}^{set} = [50 : 50 \times n]$. In such a way we obtained the $(m \times n)$-dimensional matrix $\mathbf{Sh} = ((Sh_{ij}))$ of the Sharpe ratio values $i = 1, \ldots, m$ and $j = 1, 2, \ldots, n$. In this way, after finding the position of the highest Sh_{ij} value, we determined a number, P_i, of the "best" interval Dat_{rank}^{best} and $H_{effective}^{best}$.

After determining the MAS's parameters Dat_{rank}^{best} and $H_{effective}^{best}$, the 512 + 50-day data was shifted by L_{sel}. According to the weights calculation and performance testing schema presented in Fig. 10.6, the subsequent $L_{test} = 100$ days were used for the portfolio performance evaluation *in an out-of-sample regime*.

In an empirical comparison of novel and benchmark methods, an experiment was run many times by moving the $L_1 + L_{sel} + L_{test}$ forward with a shift of $L_{shift} = L_{sel} = 50$ days. In the experiments with data from **DSA** we used 1000 days for the evaluation. Thus, diverse methods were compared according to the average values of $N_{SE} = 19$ sliding experiments. A set of suitable values for λ (0.5), F (51), H_1 (100), L_1 (512), L_{shift} (50), L_{sel} (50), L_{rank} (100), L_{test} (100), m (9), and n (12) were used in subsequent experiments with data from **DSB** and **DSC**.

The employment of a small number of days ($L_{sel} = 50$) to evaluate the Sharpe ratio results in a notable variation of the Sharpe ratio values in $(m \times n)$-D arrays, $\mathbf{Sh}$. For that reason, in subsequent experiments we smoothed each single $\mathbf{Sh}$ array by using a 3×3 window with 1/9 weight values. As a result, we obtained a smaller $(m\text{-}2) \times (n\text{-}2)$-D array, $\mathbf{Sh^*}$, that leads to a much more accurate estimation of the MAS's parameters and enhances portfolio excellence.

In Fig. 10.8 we present a performance evaluation of five portfolio design rules: 19 Sharpe ratio values calculated for $N_{SE} = 19$ overlapping 100-day trading periods during the last 2½ years of the world financial crisis (August 30, 2009 to March 21, 2014; a total of 1000 trading days). In the present analysis we considered the data of $p = 7480$ trading robots employed in automatic trading. The novel intelligent adaptive MAS-based system (dotted blue curve 4) markedly outperformed three of the benchmark methods.

The averages of the $N_{SE} = 19$ Sharpe ratio values for the benchmark rules **A, B,** and **C**, respectively are $Sh_A = 0.14$, 0.92, and 1.37, respectively. On average, the novel MAS results in $Sh_A = 2.79$. The large gap between the benchmark and novel

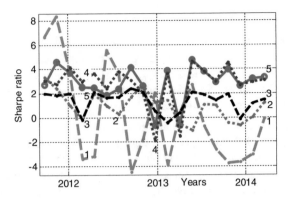

Fig. 10.8 Out-of-sample Sharpe ratios $N_{SE} - 1 = 18$ data shifts for data **TB**, $= 19$; curve 1 the 1/N rule (**BA**); 2 the 1/N rule readapted for each 50-day data shift (**BB**); 3 MAS-based on clustering according to the correlations and classifications applied for agent selection (Raudys 2013) (**BC**); 4 the novel MAS with spectra-based robot clustering, timely agent selection, and the $1/N_{effective}$ fusion of final selected agents; 5 the novel MAS with the mean/variance (Markowitz) rule used for final fusion

methods needs to be explained. One possible reason for this is the fact that the initial pool of trading robots contained a large number of unsuccessful ones. Curve 1 for the classical 1/N rule (gray) shows that very large losses are often observed, where the Sharpe ratio becomes very negative ($Sh_A \approx -4$). The results shown in Fig. 10.8 confirm that during the 2½ years examined many important (and unexpected) changes took place. Hence, it was difficult to react swiftly using traditional trading approaches. The novel adaptive portfolio design schema also gives negative Sharpe ratios.

One attempt to reduce the presence of negative Sh_A values has already been made: we smoothed the values of the matrix **Sh**. Another attempt was made by applying the mean/variance portfolio-management rule in the final fusion of the selected $H_{effective}$ agents. With the regularization of the covariance matrix, done with $\lambda = 0.5$, the mean/variance rule resulted in a vaguely higher performance: instead of $Sh_A = 2.79$ for 1/N fusion we obtained a slightly better result: $Sh_A = 2.84$ (curve 5).

Bayesian learning of portfolio MAS's parameters: experiments using DSC

The much greater usefulness of matrix **Sh*** over **Sh** and the small difference between the efficacy of the 1/N and the mean/variance fusion rules hints that sufficiently good values of the MAS's parameters Dat_{rank}^{best} and $H_{effective}^{best}$ were not found. The probable cause is the small time-span of the validation data ($L_{sel} = 50$). In small sample size situations, one of the potential ways to improve accuracy is to use additional information. In financial trading *useful information* would be *the values of the parameters obtained in previous research* with similar data and similar portfolio design methods. In an attempt to improve accuracy we hypothesized that the agents' and the portfolio's returns change much more quickly as the parameters P_i^{best} and $H_{\text{effective}}^{\text{best}}$, and used the Bayesian learning approach to estimate two of the parameters under discussion.

Denote the 2-D vector $\Theta = [\Theta_1, \Theta_2] = [P_i^{\text{best}}, H_{\text{effective}}^{\text{best}}]$. In Bayesian learning we need to define *apriori* the distribution of the parameter vector Θ, $f_{\text{prior}}(\Theta)$, and conditional density, $f_{\text{cond}}(\hat{\Theta}|\Theta)$, where $\hat{\Theta}$ stands for the sample estimate of vector Θ. Let "se" be the number of independent sample estimates, $\hat{\Theta}_1, \hat{\Theta}_2, \ldots, \hat{\Theta}_{\text{se}}$. The standard statistical theory gives the *aposteriori* distribution of vector Θ

$$f_{\text{aposterior}}\left(\Theta|\hat{\Theta}_1, \hat{\Theta}_2, \ldots, \hat{\Theta}_{\text{se}}\right) = \frac{f\left(\hat{\Theta}_1, \hat{\Theta}_2, \ldots, \hat{\Theta}_{\text{se}}|\Theta\right)f_{\text{prior}}(\Theta)}{\int f\left(\hat{\Theta}_1, \hat{\Theta}_2, \ldots, \hat{\Theta}_{\text{se}}|\Theta\right)f_{\text{prior}}(\Theta)d\Theta}. \quad (10.9)$$

In our experiments we used nine independent estimates, $\hat{\Theta}_1, \hat{\Theta}_2, \ldots, \hat{\Theta}_9$. The sample size is small. In this way, we can use only simple models of $f_{\text{cond}}(\Theta|\Theta)$. We will use 2-D Gaussian density

$$f_{\text{cond}}\left(\hat{\Theta}|\Theta\right) = \text{N}(\hat{\Theta}, \Theta, \Sigma),$$

where Θ stands for the 2-D mean vector, and Σ stands for the (2×2)-D covariance matrix.

To simplify the estimation, we assume

$$f_{\text{prior}}(\Theta) = \text{N}\left(\Theta, \Theta_0, \frac{1}{n_0}\Sigma\right),$$

where Θ_0 is our prior guess about the position of the mean of vector Θ, and n_0 indicates our confidence in the Θ_0 value. If n_0 is high, our confidence is high. If it is low, e.g., $n_0 \leq 1$, our confidence is low. Employment of above assumptions concerning the 2-D Gaussian densities after some algebra results in *aposteriori* distribution.

$$f_{\text{aposterior}}\left(\Theta|\hat{\Theta}_1, \hat{\Theta}_2, \ldots, = \hat{\Theta}_{\text{se}}\right) = \text{N}\left(\left(\Theta, \frac{n_0}{n_0 + n_{\text{se}}}\hat{\Theta}_0 + \frac{n_{\text{se}}}{n_0 + n_{\text{se}}}\sum_{j=1}^{n_{\text{se}}}\hat{\Theta}_i\right), \frac{1}{n_0 + n_{\text{se}}}\Sigma\right).$$

$$(10.10)$$

We are interested in the mean vector of the *aposteriori* distribution. Thus, the new estimate

$$\hat{\Theta}_{\text{new estimate}} = \frac{n_0}{n_0 + n_{\text{se}}}\hat{\Theta}_0 + \frac{n_{\text{se}}}{n_0 + n_{\text{se}}}\sum_{j=1}^{n_{\text{se}}}\hat{\Theta}_i. \quad (10.11)$$

The new, way of estimating based on prior information the parameters P_i^{best} and $H_{\text{effective}}^{\text{best}}$ was tested by using the third, 14686-D data set, **DSC**. As with previous

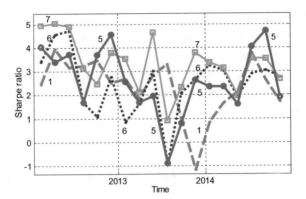

Fig. 10.9 Out-of-sample Sharpe ratios for data **TC**; $N_{SE} - 1 = 17$ data shifts; curve *1* the 1/N rule (A); curve *5* the novel MAS with the mean/variance (Markowitz) rule used for final fusion; curve *6* the newest MAS with the 1/N rule used for final fusion; curve *7* the newest MAS with the mean/variance (Markowitz) rule used for fusion

experiments using **DSB**, in each *out-of-sample sub-experiment* we used $L_1 = 512$ days' history for data clustering as well as forming and ranking the agents2. An additional $L_{sel} = 50$ days were used for the evaluation of the $((m -2) \times (n-2))$-dimensional matrix $\mathbf{Sh}^{**} == ((Sh_{ij}))$, where all Sh_{ij} values were evaluated according to Eq. 10.11. All the remaining parameters of the MAS were the same as in experiments with **DSB**. The exception is $N_{SE} = 18$, since this depends on the data available (1462 days). In Fig. 10.9 we present a performance evaluation of four portfolio design rules: 18 Sharpe ratio values calculated for $N_{SE} = 18$ overlapping 100-day trading periods. The rules **BA** (curve 1) considered novel MAS with $\mathbf{Sh}^*$ in use (curve 5) were used as two benchmark rules. The MAS considered above was tested under two regimes (1/N fusion, curve 6 and the regularized mean/variance fusion, curve 7).

Figure 10.9 shows that the newest MAS mean/variance fusion (mean Sharpe ratio $\overline{Sh} = 3.328$) clearly outperforms both benchmark rules ($\overline{Sh} = 2.34$ and $\overline{Sh} = 2.51$). The newest MAS with 1/N fusion resulted in $\overline{Sh} = 2.47$. We note that in the last series of experiments we introduced very feeble prior information, $n_0 = 1$; the main improvement in the portfolio performance was obtained due to the accumulation of knowledge of the P_i^{best} and $H_{effective}^{best}$ values evaluated in the sequences of previous out-of-sample sub-experiments.

Speaking in general, the sample-based evaluation of older agents2 ranking and selection for final determinations of portfolio weights seems to be a promising research direction for improving ATS in financial markets.

Concluding Remarks

We used the excitable-media wave-propagation model to generate multidimensional data that mimics the economy and financial attributes in rapidly changing environments. Our model is composed of several excitable-media items. Each item is composed of thousands of elements based on cellular automaton similar to single-layer perceptron schema. The elements can be excited by the sum of their incoming signals. After a certain delay, the elements transmit their excitation signals to the nearest and, sometimes, to more distant neighbors. Changes in the model parameters allow periodic, vibrant, and chaotic excitation wave propagation behaviors. The simulation experiments demonstrated that the type of the signal behavior can change overtime. Even without external intervention, the behavior of the time series characters "wanders in space" and from time to time returns to very near some of the previous positions. We showed that the spectral representation of the multidimensional time series allows the wave behavior types to be recognized and the model's parameters to be determined.

Knowledge of synthetic randomly generated time series suggested a way to develop the evolvable portfolio calculation—MAS. The procedure is composed of spectra-based clustering of two-year-old financial time series, the development of second-order, very specific trading agents2 that are capable for good trading for a short period of time, the ranking of agents2, and the weighted summation in the final stage of the MAS.

The main conclusions of our analysis are as follows:

1. Wave-propagation patterns in mutually interacting excitable media models can be characterized by their chaotic, vibrant, oscillating, and near-regular behavior. Diverse behaviors can be recognized by employing the spectral representation of the time series generated by the waves.
2. Portfolio design can be improved by spectral representations of the histories of the financial indicators. It can be argued that portfolio design is associated with the vibrant/chaotic phenomena of which the economic and financial artifact is constituted.
3. With highly specific trading agents, the latest training data interval may be less useful for future predictions than previous well-timed detected data-history segments. The older data, if correctly identified, can be functional for selecting the highly specialized, most profitable trading agents.
4. The multi-agent system of evolving portfolio weights depends on a number of parameters that can and should be learnt from historical data during the creation of the MAS portfolio.

The present study suggests a number of problems to be solved in the future. One of them is the improvement and in-depth analysis of more sophisticated models aimed at generating synthetic time series that mimic automatic trading in the financial markets. Cooperation between oscillating agents as well as between the groups of excitable media models and the development of synthetic time series control methods are just a few directions that can be taken.

In sum, the presented applied research showed that some OAM, PIF and WIM principles can be integrated also in the large scale CA and AI-based MAS modeling. The most important obtained value added comes from the (a) realized example of the simple wavelike communication mechanism between agents and, (b) spectral representation of the multidimensional time series, which allows specific features to be generated that are necessary for recognizing the type of chaos-generating model.

In prospective research, decision-making agents for simplicity represented here by the single-layer perceptron (SLP) model can be replaced with the mind state related OAM approaches. In such a more complex research framework, modeling of agents states dynamics would lead to the observed individual and social decision-making patterns. In this way, the OAM based OSIMAS paradigm can be explored in various fields of research and real life applications.

Chapter 11
Epilogue

The epilogue answers some of the reviewers' comments and wrap ups loose ends, providing a chance to discuss freely about some controversial issues. First of all, I would like to stress that the book draws upon and argues not from methaphysical considerations, but from the firm standpoint of neuroscience research and fundamental physics, leaving just a moderate space for several cutting edge philosophical insights. I do believe that the majority of academic readers will be challenged and intrigued by such a presentation in a positive way.

Clearly, there is a long way to go in proving all the OSIMAS statements and hypotheses. However, until their validity has been proven false, we cannot refute their right to exist. Admittedly, this is how argumentation in science works. Thus, this book is meant to engage its readers with 'out-of-the-box' thinking, which does not have to also go hand in hand with the mainstream view.

One of the main criticisms is the use of some neologisms, i.e., new terms like the 'mind-field', which, following McFadden's CEMI (conscious electromagnetic information) field theory, is understood as a field-like correlate of consciousness (McFadden 2002). I have been accustomed to using it as a convenient abbreviation, which could sound pseudoscientific for somebody outside the bioelectromagnetic brain field context often used in neuroscience (electric fields are measured using an EEG, magnetic fields using an fMRI, and electromagnetic fields using MEG devices). Therefore, using the term 'mind-field' implies bioelectromagnetic brain field throughout the whole book.

Let me explain another new term—'collective consciousness'. Despite how unorthodox it might sound now, at least hypothetically, we could acknowledge the existence of some sort of collective consciousness as an intangible medium, where the intrinsic individual cognitive processes take place, forming a common bioelectromagnetic brain field, which in turn, via feedback, shapes the individual cognitive processes. This is not an excursion into the supernatural, as this type of feedback is common to all living and nonliving levels of self-organization. For instance, field-based coordination has been naturally established as existing a billion years ago as an extremely effective way of self-organization and coordination,

© Springer International Publishing Switzerland 2016   261
D. Plikynas, *Introducing the Oscillations Based Paradigm*,
DOI 10.1007/978-3-319-39040-6_11

e.g., in all multicellular organisms, where the cells mutually communicate via a media of a wide spectrum of fields, starting from the light spectrum and ending with acoustic oscillations. In terms of such capabilities, the brain's neural networks are the most sophisticated of all. Thus, they surely employ field-based mutual coordination and communication mechanisms. In this way, as EEG studies indicate, coherent neural networks produce electromagnetic fields which permeate our craniums and can be detected outside the space of our bodies. Therefore, the assumption of interbrain communication is not science fiction at all.

We are only beginning to understand some brain's functions and capabilities. Therefore, despite how weird it might sound now, we cannot refute the chance that in having a proper understanding and tools, we will sooner or later reveal some sort of quantum or other type of entanglement between the cognitive processes in the brains of individuals and possibly applying to groups of individuals. At this point in time, we can only hypothesize how it might look like in terms of the effects on individuals or social phenomena at large. Hence, this work is one among other pioneering efforts, delving into some aspects of such research.[1]

In this way, we can acknowledge the viewpoint, at least theoretically, that societies can also be understood as global processes emerging from the coherent behavior of the brain and mind processes of the individual members of society. Hypothetically, we can admit that some emergent social behavior could be at least partially induced by the coherent processes taking place between related individual bioelectromagnetic brain fields (some sort a collective consciousness) and therefore, can inherit some degree of wave-like excitation of mental states and their corresponding coherent behavior.

In fact, in this book after some prior argumentation and corresponding literature review, these assumptions were taken for granted or accepted as true without proof. However, as is usually the case while developing new theories, the whole Chap. 4 has been prepared to find a way to prove or falsify these assumptions in an experimental EEG research framework. In fact, new research frontiers are developing in this field of research, like social neuroscience, group neurodynamics, and neuroeconomics.

Social species form organizations that extend beyond the individual. The goal of social neuroscience is to investigate the biological mechanisms that underlie these social structures, processes, and behaviors and the influences between social and neural structures and processes. Such an endeavor is challenging because it necessitates the integration of multiple levels. Mapping across systems and levels (from genome to social groups and cultures) requires interdisciplinary expertise, comparative studies, innovative methods, and integrative conceptual analysis. In

---

[1] For analogy let us recall another well-known example from physics—Einstein's theory of relativity, which started from his hypothesis regarding the laws of time-space continuum, assuming a limited speed of light. At that time, it looked like a weird idea, indeed, but eventually it changed our fundamental understanding about space-time and the matter-energy laws of the Universe. In no way can this book propose similarly radical or ambitious ideas. However, with the help of some "out of the box" thinking, its approach may help to see things from a bit different perspective, which let us hope may open up new prospective multidisciplinary niches of research.

this regard, the book offers some out-of-the-box approaches, extending social neuroscience in terms of some novel conceptual premises, experimental electroencephalography approaches, models, and simulations.

Indeed, to investigate the mutual influence of physical, biological, and social mechanisms, social neuroscientists, ranging from physicists to psychologists, epidemiologists to neurologists, philosophers to neurobiologists, and entomologists to zoologists, have begun to work together in interdisciplinary scientific teams using variety of interdisciplinary and multidisciplinary methods. Developed techniques such as meta-analyses and electroencephalography have seen upgrades that, for instance, permit investigations of the source architecture of the neural substrates of social processes.

Interdisciplinary collaborations are a cornerstone of social neuroscience endeavors. A decade ago, the field was characterized by some as correlating social and cognitive functions with regions of brain activation using functional neuroimaging. The field has always had a broader foundation, however, and social neuroscience is now recognized as drawing on research on animals and humans, with phenomena at multiple levels of organization measured and/or manipulated to go beyond the specification of association across levels to specify the mechanisms operating at and between each of the levels of organization pertinent to an association. The manifold disciplinary expertise for such multilevel investigations is increasingly beyond the reach of individual scholars, so interdisciplinary teams of scientists are increasingly common.

Hence, at a very foundation it can be even attributed to the principles of quantum mechanics. In this sense, the reader might be interested in becoming acquainted with the recently published and cited monograph by Emmanuel Haven and Andrei Khrennikov, "Quantum Social Science" which forms one of the very first contributions in a very novel area of related research, where information processing in social systems is formalized with the mathematical apparatus of quantum mechanics (Haven and Khrennikov 2013).

Perhaps A. Wendt's second monograph, "Quantum Mind and Social Science: Unifying Physical and Social Ontology" (Wendt 2015) would also help to clarify understanding why an underlying assumption in the social sciences, that consciousness and social life are ultimately classical physical/material phenomena, can be challenged by the assumption that consciousness can be, in fact, a macroscopic quantum mechanical phenomenon.

Another recent book, "Quantum Models of Cognition and Decision", written by J.R. Busemeyer and P.D. Bruza (Busemeyer and Bruza 2014), could also shed new light on understanding why probabilistic dynamic systems, using some aspects of quantum theory, can be a way to understand the "contextual" interference effects found with inferences and decisions under conditions of uncertainty or "quantum entanglement", which allows cognitive phenomena to be modeled in nonreductionist ways. Employing these principles drawn from quantum theory allows us to view human cognition and decision in a totally new light.

Moreover, Chap. 5 presents a simulation of human brain EEG signal dynamics using a refined coupled oscillator energy exchange model (COEEM), where the oscillation-based modeling of human brain EEG signal oscillations, using a refined Kuramoto model, not only helps to specify the OAM, but also significantly contributes to EEG prognostication research, which is the major topic of this particular direction of neuroscience research. Besides, Chaps. 6–10 were specifically designed to show how some conceptual OSIMAS ideas and assumptions can find their way into original models and simulation approaches.

To facilitate this process, based on our conceptual and experimental findings, we provided group-wide modeling framework and means for construction of simulation models, which are presented in our online virtual lab simulation platform (there are created 7 original simulation models at http://vlab.vva.lt/; Guest: guest555). This manuscript provides just few examples, however, I am sure, that much more experimental and simulation evidences will appear quite soon as this perspective multidisciplinary field of research is developing very fast.

However, some readers were most probably not sure whether the oscillations are just a tool for simulation or whether the author thinks that the oscillatory nature of agents is also a feature of actual human beings. Let me put it simply, an objective essence of the oscillatory OSIMAS paradigm stems from the fundamental micro (quantum) world. Let us recall that neurons are affected by probabilistic quantum processes of the oscillatory nature. In this way, micro world is linked to the meso world of the individual brain, where aggregated EM (electromagnetic) oscillations of billions of such neurons are experimentally observed, using EEG, fMRI, MEG, PET, or other brain imaging devices, in a form of brain activations or brainwaves. In fact, brainwave patterns are produced by the coherent oscillations of the firing neurons' activities and reflect certain mind-brain states.

However, the missing link, is in the next jump between individual meso and macro world of social systems. The hint is that we do observe some periodic, semiperiodic or chaotic dynamics on the social scale too. For instance, we can objectively observe and measure fluctuations and even cycles of the aggregated social, economic, financial, etc., indicators. However, until now, researchers had no proper means to relate objectively observed characteristics of the mind states of individual people with observed social systems behavior. In essence this manuscript attempts to define and explore conceptual, experimental, and simulation metrics for establishment of such link.

In the book, not accidently for that reason there is employed in mathematics and physics widely adopted oscillations-based modeling approach. It is not very surprising as the best way to describe reality is using its own natural and universal language, so to speak. This language is in a form of the immense spectrum of mutually interacting oscillations.[2] To the author's point of view macro world is not an exception as its underpinning is of the oscillatory nature. A full understanding of this notion alone can render a paradigmatic shift of one's worldview.

---

[2]In fact, all our senses are based on registering and analyzing of oscillations from light to sound frequencies (even smell and taste is based on discerning vibrational modes of molecules).

Nevertheless, we should admit that models, whatever accurate, will always remain just approximate approaches of reality. Hence, in short, the answer to the question 'Oscillations: actual or simulated?'—both: actual and simulated. In this way, I hope manuscript will definitely benefit from a clarification of the simulated role of the OSIMAS paradigm.

Another question—whether we can simulate subjectivity (e.g., commitment, goal-orientations, motivation, job satisfaction, leader–member exchanges, decision-making, consensus, problem-solving activities, perception of fairness and justice, ethical relativism, ethical absolutism, collectivism tendencies, social responsibility, performance, etc.) using OSIMAS approach?

As a matter of fact, not only basic states but also human emotions and even specific thoughts can be discerned using above-mentioned brain imaging devices in real time. In this sense, yes, content of the oscillation becomes a key in the analysis of social dynamics. Let me give an example—there are some applications in the area of team neurodynamics, which empirically investigate in real-time job satisfaction, leader–member exchanges, team decision making, team consensus, problem-solving activities in a group, perception of fairness and justice, etc. Here subjectivity (in terms of people mind states) is hidden in the complex empirical data representations. Admittedly, this is the most objective approach from the experimenters' point of view. Following this line of thought, OSIMAS simulation models attempt to model dynamics of individual and collective consciousness states, which are identified in terms of a chosen set of characteristic (empirically measured) attributes. In this sense, subjectivity is already there (in the empirical data) and there is no need to introduce it superficially according to modeler's view. However, modeler has to make clear distinctions between states in terms of the empirical attributes.

However, simulations of some subjective norms like ethical relativism, ethical absolutism, social responsibility, etc., most probably will be packed with the next layer of subjective abstraction and approximation which have to be dealt with while relating complex empirical data with the complex subjective content.

In sum, empirical data driven simulation of individual and collective (distributed) mind states, using oscillations methodology, is a very useful simulation tool that is capable to explain some emergent and complex social dynamics. In this sense, I do see the OSIMAS paradigm as a powerful way of rendering simulated agents similar to actual individuals. However, qualitative information on the nature of the observed dynamics has to be provided by the modeler in a very beginning of the simulation setup. In this way, subjective elements of the modelers come into play here.

The next question—the meaning of the social system in the book. Admittedly, there are many ways to understand what a social system is and some discussion is required about how this term is understood within the context of this monograph. In general, social systems can be understood as open, self-organizing, living entities that interact with their environment while maintaining an exchange of information, energy, and matter. In this regard, the book follows the science of systems as an interdisciplinary field, where the nature of systems covers many complex systems such as cybernetics, system dynamics, etc. More specifically, the book follows the

German theorist, Niklas Luhmann's approach, where society only works in relation with its environment. It also follows from Humberto Maturana and Francis Varela's idea of autopoiesis, etc.

In the book, social systems are reduced to the economic systems of production and exchange of goods and services as well as allocation of resources in a society, including in this process various institutions, agencies, entities and consumers that maintain information flows and social relations within the system, etc. Hence, I do agree that in the book there is a slight misrepresentation of the social agent and a reduction of the social system to the economic one. Therefore, the book gains in consistency after a clear reference to the economic system (rather than the social system as a whole). Therefore, in some cases instead of the broader "social system" or "social agent" terms I have applied "economic system" and "economic agent" terms.

Hence, the instantiated research framework is pertinent not only to the simulation of the experimentally observed individual cognitive and behavioral patterns, but also to the prospective development of OAM-based multiagent systems, in which the coherence of the individual mind states in social mediums is imperative to maintain social order. This approach leads to distributed cognition in nonlocal multi-agent systems. Simulations under the correct assumptions can help us understand how distributed cognitive processes synchronize and actually work in coherence. However, further empirical research is required so that OSIMAS-based conclusive and experimentally proven social simulation theory is able to evolve.

I have to admit, that because of the immense scope and complexity, there are many unanswered questions left open for discussion and further multidisciplinary research. Thorough additional research needs to be done to examine, in detail, the issues and methods that will help extend our understanding about the fundamental place of consciousness, not only in individual and social life, but also at the scale of entire interconnected Universe. Let us hope that the oscillatory OSIMAS paradigm presented here provides some useful research guidelines, along with explanatory sources, for further far-reaching conceptual, empirical, and simulation studies.

Appendix A: Autopoietic Systems[1]

The interested reader of this monograph may find that there are certain concepts embraced by the oscillation-based multi-agent system (OSIMAS) and oscillation agent model (OAM) that are common to a number of mainstream models of cognition, neurodynamics and sociobiological/cultural systems. As steps to understanding consciousness, such an association of ideas seems quite inevitable, and here one may underscore the important role of (coupled) oscillatory systems as incorporated into processes of massively parallel, and distributed type.

To give a complete acknowledgement of those models concerned by combing through a large number of specific details, is a task that goes way beyond this present contribution. Instead, I will restrict attention to certain theories and concepts with which I have some familiarity, and consider to be germane to the tasking mechanisms of OSIMAS and OAM.

Having said this, the main topics to be covered in this essay are:

1. The Maturana-Varela theory of autopoiesis, leading to Varela's foundations for a theory of neurophenomenology.
2. Global Workspace Theory (GWT)—a principal forerunner in modern theories of cognition.
3. Distributed and Embodied Cognition.
4. Several other approaches inter-related to (1–3) that may also be significant for the OSIMAS and OAM models.

Much of the content of (1)–(4) is increasingly applicable to many mainstream fields of cognition that are currently breaking ground towards a general model of experiential neuroscience. Many of the intrinsic properties are essential contributions to the ongoing task of understanding cognition and consciousness within the framework of structural organization and the evolution of human and social systems, rather much in tune with the drift of ideas of the present monograph. This is thanks to a number of extensive multi-disciplinary connections created over the last

---

[1]The appendix has been cordially contributed by prof. J.F. Glazebrook.

© Springer International Publishing Switzerland 2016
D. Plikynas, *Introducing the Oscillations Based Paradigm*,
DOI 10.1007/978-3-319-39040-6

forty years or so, particularly those that have centered around the core of (applied) phenomenology, as was professed by the school of philosophers such as F. Brentano, E. Husserl, M. Heidegger, and M. Merleau-Ponty.[2]

A.1 Defining Autopoiesis and What It Means

Autopoiesis (from the Greek meaning "self-generating") is a theory conceived for living cellular organisms based upon the circular organization of metabolism and the organism's structural organization. As it was originally proposed by Maturana and Varela (1980):

> ... an autopoietic system is organized as a bounded network of processes and production, transformation and destruction of components which (i) through their interactions and transformations continuously regenerate and realize the network of processes that produced them; (ii) constitute the system as a concrete entity in the space in which the components exist by specifying the topological realization of the system as such a network.

Consequently, the system specifies its own internal states, its domain of variability and its boundary delimited in terms of its sovereign mechanisms. In the sense of a cellular system, this means that the flow of energy and matter onto the membrane boundaries bounds the dynamics of the metabolic network producing the metabolites, the latter constituting that same network together with its boundaries, out of which this flows continues its cycles. This 'circularity' characterizes autopoiesis as a property of autonomy for a living system.

The term *operational closure* was introduced to describe how the product of functioning components in a network would be transformed by other processes into components of the initial process. This concept is ubiquitous to many systems such as AI, ALife, ecosystems, linguistics, economics and social/judicial organization in the form of self-referentiality. In this way, autopoiesis characterizes 'organizationally closed' systems. A basic premise is that the brain and vital systems, as far-from-equilibrium open systems perturbed by an environment, possess this property. Thus, the contextual distinction between use of the terms 'open' and 'closed' is essential here. In fact, the entirety of the system is embodied in its organizational closure; thus the whole is not the sum of its parts; it is instead seen as the organizational closure of its parts, and this is the subtle distinction, following which the system's intrinsic properties are seen to emerge through the interactions of its components via levels of iteration. To a certain extent autonomous systems and control systems are complementary in understanding natural systems, as much as the methods of reductionism and holism can also be seen as complementary

---

[2]The work Thompson (2007a) provides an impressive scholarly account this respect, and since a complete philosophical discussion also flies out of the confines of this Appendix, I shall refer the philosophically-minded reader to this particular reference for a more comprehensive discussion [particularly of case (1)] and how this relates to the mind-body-brain problem.

(in the descriptive framework, such complementarity can be approached by the notion of 'adjointness' in the mathematical theory of categories; see Goguen and Varela (1979) and cf. Baianu et al. (2006).

Despite its growing attraction within the life sciences from the mid 1970s onwards, the theory of autopoietic systems remains to be universally accepted, least of all by experimental biologists who for some time have been greatly preoccupied with DNA, RNA and replication research. Questions have been raised concerning the interpretation of terms such as 'operational closure', 'boundary', and the seemingly implicit sense of 'cognition' that featured in the original theory, matter that might have misconstrued the relationship between 'information' and 'knowledge' (see e.g. Luisi 2003).[3] So it is important to note that the definition of 'autopoiesis', as originally defined in e.g. Maturana and Varela (1980, 1987), has over the years, undergone various adjustments and re-interpretation of meaning since its original conception. Indeed, using his encyclopedic knowledge of cell biology, immunology, and biophysics on the one hand, and neuroscience, cognition and anthropology, on the other hand, Varela spent a large part of his latter career (ending just at the age of 54) towards developing the concept of cognition beyond autopoiesis, to a more widely-based concept with notable epistemic and ontological foundations (see Varela 1996; Varela et al. 1992) as well as other extensive reviews of this work in e.g. Rudrauf et al. (2003), Thompson (2007).

Before outlining the development of ideas, let us consider a sampling of issues that were at stake, and how they were remedied. The importance of the 'boundary' as produced by the network of processes for the sake of separating the system from the non-system, was further highlighted towards a modified definition of the term 'autopoiesis'. As pointed out by (Bourgine and Stewart 2004), in his last work Varela redefined a system as 'autopoietic' if:

(a) it has a semi-permeable boundary;
(b) the boundary is produced from within the system, and
(c) it encompasses reactions that regenerate the components of the system.

By removing the restriction that the components should necessarily be molecular, Bourgine and Stewart (2004) later re-formulated this definition as follows:

> An autopoietic system is a closed network of productions of components that recursively produce the components and the same network that produced them; the network also specifies its own boundary, while remaining open to the flow of matter and energy through it.

Observe that this re-formulation radically separates autopoiesis from cognition, a sense of which was implicit to the original concept as it was pre-supposed with respect to self-organization and structure. The hypothesis of (Bourgine and Stewart 2004) is that a living system is both autopoietic (in their sense) and cognitive. The meaning of 'cognitive' is proposed as follows: Consider two types of interactions between entities called A and B: (a) those that have consequences for the internal

---

[3]How the mechanisms of DNA, RNA, self-replication and protein synthesis can actually be incorporated into cellular autopoietic systems has been studied in e.g. Luisi (1993).

state of the organism (type A), and (b) those that have consequences for the state of the (proximal) environment, or modify the system to its environment (type B). Then the system is cognitive if and only if type A interactions serve to trigger type B interactions in a specific way so as to satisfy a variability constraint.

In a similar way, Atlan and Cohen (1998) view the complex regulatory activities of the immune system as cognitive in the sense that the system's 'cognition' is its innate property of perceiving an incoming signal with an internally structured and hereditary conditioned image of the world, and then selecting a response from a much larger repertoire of possible responses. In other words, the cognitive pattern of recognition-and-response proceeds by an algorithmic combination of an incoming external, and broadly 'sensory' signal, with an internal ongoing activity influenced by this internal image. This is part of the 'immune protocol' that initiates a planned reaction based on a decision that the sensory pattern necessitates a response, so maintaining its emergent complexity as a living system (cf. Cohen and Harel 2007).

Somewhat in consonance with these approaches, Bitbol and Luisi (2004) see matters as follows:

> Cognition is definitely not tantamount to a passive reproduction of some external reality. It is instead mostly governed by the activity of the cognitive system itself. To understand this, one must realize that it is the cognitive structure that selects, and retroactively alters, the stimuli to which it is sensitive. By this combination of choice and feed-back, the organic structure determines (in a way moulds) its own specific environment; and the environment in turn brings the cognitive organization to its full development. The system and the environment make one another: cognition according to Maturana and Varela is a process of co-emergence.

From another perspective, we see that the environment to an extent actualizes the living, and the latter via its intrinsic structure, responds in a way that eventually creates its own world. As pointed out in Luisi (2003), this takes cognition and autopoiesis closer to the realm of phenomenology as proposed in Varela (1996), Varela et al. (1992), in particular, closer to the ideas of philosophers such as Merleau-Ponty (1963), who had proposed:

> ... it is the organism itself–according to the proper nature of its receptors, the thresholds

> of its nerve centers and the movements of the organs–which chooses the stimuli in the physical world to which it will be sensitive. The environment emerges from the world through the actualization or the being of the organism.

A.2   Neurodynamics and Autopoiesis

The circularity of cell systems as described above, has the following homologue in neurophysiology: perturbations from an environmental domain of interactions play into a sensory motor coupling that modulates the dynamics of the nervous system that generates its internal neuronal assemblies as well as those that feed back into the sensory coupling. Likewise, this network of processes determines the system's configuration through processes (of processes) of reciprocal causality, as will be specified below.

This circular causality that is characteristic of autopoietic systems is homologous to the process induced by the combination of neurodynamic with structural effects that create a dynamical system with the following hallmark properties: systemic variation of parameters produces the same in the response mechanism—in other words, simultaneously changing the attractor landscape along with the system's space of states. In this respect, the structural effects influence the topology of spectra and frequency modulation of the prevailing oscillations. An essential difference, however, is that an autopoietic system need not have structural inputs— that is, environmental contingencies do not in themselves define the system's intrinsic dynamics, though the latter can be perturbed by its domain of interaction with the structural dynamics as instrumented by the former. Putting it another way, the environment invokes a reaction from the system whose responses are functions of the latter's internal organization; that is to say, dynamic inputs trigger internal changes and/or the structural effects that define the system's organization. In this way, the property of operational closure, seen as a process of stabilization and preservation of identity, gives rise to an autonomous system that is spatially and functionally distributed; thus, each living organism possesses such a property that is essential for its own survival. In the meantime, this organizational structure remains invariant for as long as its intrinsic regulatory variables and its essential trajectories are not violated, either internally or externally.

As proposed in Maturana and Varela (1987) (cf. Rudrauf et al. 2003), the richness and complexity of the system depends upon its eigenbehaviors: collectively, a measure of its dynamics of self-organization via inter-regulating subsystems leading to enrichment of the system's overall behavior and reproduction while constrained by a variable topology. These factors contribute to: (a) its local dynamics in terms of its immune function, its metabolic and neural networks, and (b) its global dynamics through emergence of its sensory-motor mechanisms towards the environment, while discriminating between self, and nonself. In other words, there is a reciprocal causality between the local interaction rules (such as those for neurophysiological and biochemical interactions), and the global mechanism involving the topological demarcation of shaping processes of diffusion and the local platforms of interactions.

From Varela's point of view, this necessitated a passage beyond seeing the brain as finely-tuned symbolic computer, and beyond second order cybernetics (such as 'observing the observer'), towards the crucial element of incorporating the 'first person experience' on behalf of the prospective subjects (cf. Froese 2011; Rudrauf et al. 2003). Hence, an enactive approach towards neurodynamics was proposed. This arises by blending in the principles of complex systems with those of the immune function. Specifically, the system in managing its sensory motor loops is modulated by continuing patterns of brain activity that induce the specific couplings and synchronous oscillations towards interaction with the environment, while the operational closure of its state dynamics complements computability, and maintains self-organization against the effect of external circumstances. These operations depend upon the brain's interconnectedness, its highly differentiated subnetworks, and its reciprocating neuroanatomical re-entry maps (cf. Edelman and Tononi 2000). Such a blend towards enaction is analogous to how cultural knowledge in anthropology is studied via a background of history, traditions, artifacts, and rituals, etc. The wide-scale distributed form of the vastly numerous interconnections within the brain results then in a range of alterations of neuronal assemblies squaring up to confront the impact of experience.

In this scheme of things the brain is then characterized as an autonomous center for the behavioural organization of the nervous system, the body, and the reaction of these towards the environment, thus comprising dynamically highly structured, multiply-coupled, and mutually embedded subsystems. Consequently, it can be asserted that the mind is not stationed solely within the brain, and cannot be separated from the entirety of the human organism being part-and-parcel with the environment, and neither separated from the space of all possible minds. When we view cognition as an embodied function towards consciousness, the 'mereological fallacy' of attributing the latter property exclusively to the effects of brain-tied neural events, is avoided (Bennett and Hacker 2003) (and in the same way, this fallacy dissipates when cognition is viewed as broadly 'distributed' and 'inferential', in the sense to be made clear below).

On expanding the principles of autopoiesis and cognition, Varela et al. (1992) (see also Froese 2011; Rudrauf et al. 2003) proposed a theory of experiential categories as a descriptive mechanism for the neurophysiological basis for consciousness in terms of highly varied, massively-parallel neuronal assemblies that connect across the landscape of the brain. Among these are included a range of oscillatory systems as generated by (i) recticulo-thalamo-cortical networks, (ii) corticocortical networks, and (iii) cortico-striato-thalamo-cortical networks, as they are observed by EEG and MEG technologies.

It is through these networks, the recorded brain patterns, and spectral distribution of propagating waveforms, that the entirety of neurodynamic processes constitutes a framework towards phenomenology. The combined study of how these networks interact thus leads to the proposed 'enactive cognitivism', a term coined by Varela et al. (1992) towards the neurophysiological foundations for the embodiment of cognition as based upon experiencing the world through the mediation of perceptual sensory loops, together with a mediated internal patterning of brain activity that determines the specific coupling of the system with its environment. This leads to 'cycles of operation': the multi level specific phenomenology of the essential operations that occur within integrated sequences of behavior in which mind, and acts of cognition are inter-engaged. They are cycles that include organismic regulation, ongoing sensory-motor coupling, the cognitive response, and management within the external social world. As multiply-coupled systems of synchronous oscillators, they can be realized as the underlying functions of multi-agent, distributed systems, very similar indeed to those professed by the OSIMAS and OAM models (Plikynas 2010, 2015; Plikynas et al. 2014a, b, 2015; Raudys et al. 2014) (see also Kezys and Plikynas 2014).

A.3 Dynamic Causal Modeling

On closer inspection, suppose we consider two basic types of effects as observed, namely, those that are dynamic and structural. Such effects will clearly depend upon the stability of the system under perturbations, and the subsequent flow nature of the system's trajectories through a continuing alteration of states. This brings us back to a governing principle of autopoietic systems; as pointed out earlier, the environment does not define the system's internal dynamics, instead, the external perturbations trigger internal changes in the system's structural organization. Precisely the same changes are observed in Dynamic Causal Modeling (DCM) (David 2007; Friston et al. 2003) that analyzes fMRI/EEG and MEG data in terms of generative models that are adaptable to self-organization and autonomous dynamics, and as such, are conducive to Bayesian techniques (cf. Friston 2010; Friston et al. 2003).

Consider now a particular case as demonstrated in David (2007). Suppose we start with two basic reciprocating mechanisms. Firstly, we have the generative biophysical model consisting of (1) neuronal variables: synaptic time constraint, synaptic efficacy, inhibition/excitation, and the connectivity of networks. Secondly, we have the prediction of measured data, that is, (2) macroscopic data at the brain level: local field potentials, scalp EEG/MEG and fMRI. 'Forward Problem' is the passage from (1) to (2), that is, from the generative model in predicting the measured data, while the 'Inverse Problem' is that given the measured data, how then can the generative model be estimated by means of a Bayesian analysis of the biophysical parameters?

How one can crack these problems is to some extent dictated by the dynamics of self-organized and dissipative structures as they are studied in the framework of autopoietic systems. One possible strategy is to regard the brain as a nonlinear dynamical systems accommodating inputs, and in return, creating outputs. This principle accounts for the brain's response to perturbations that elicit changes in neuronal activity as simulated in networks, and then measuring the response which is transformed into neuroimaging data using the modality-specific 'forward' model. The systematic change of parameters likewise induces the same in the response mechanism without further input.

How structural effects are proposed for the OAM model of self-organized state dynamics observed by those same technologies, in consonance with the DCM model, can perhaps be seen more specifically in view of (David 2007):

> Structural effects are clearly important in the genesis of induced oscillations because they can produce frequency modulation of ongoing activity that does not entail phase locking to any event. More generally, they play a critical role in short-term plasticity mechanisms observed in neuroimaging, for instance subject's habituation after repetitive stimulation. Activity dependent changes in synaptic activity are an important example of a structural effect that is induced by dynamic effects. This coupling of structural and dynamic mechanisms is closely related to the circular causality that characterizes autopoietic systems … Modulatory effects are expressed as changes in certain aspects of information transfer, by the changing responsiveness of neuronal ensembles in a context sensitive fashion … Here, there

is no distinction between DCMs and autopoietic systems: both receive dynamic inputs, which act as transient perturbations... DCM does not specify operational closure. Instead, structural changes are specified as explicit and direct consequences of particular dynamic inputs ...

The synthesis of DCM methods and autopoiesis thus motivates progressive improvements in designing cognitive experimentation, and at the same time to fathom out and understand the dynamics of the underlying neural (sub)systems (David 2007; Friston et al. 2003).

A.4 Towards Neurophenomenology

The enactive approach to cognition, as we described it above, anticipates Varela's theory of neurophenomenology, (Thompson 2007; Thompson and Varela 2001; Varela et al. 1992), essentially derivable from three basic sources (cf. Froese 2011; Rudrauf et al. 2003):

(i)   the biology of cognition;
(ii) existential phenomenology[4] and
(iii) the Buddhist nature of the contemplative stability of mental states.

These short, but nevertheless profound axioms lead to regarding knowledge as a product of the interface between body and mind, on the one hand, and society and culture on the other, thus postulating the individual's cognitive embodiment in an open-ended, mindful approach to experience. This type of vision is realizable in the practice of meditation, as well as the mindfulness attribute of Buddhist psychology, and is a clear alternative to the solely computational theory of mind in cognition. It furthermore goes against the grain of centuries of Western philosophy which had traditionally emphasized the rational understanding of mind above the nature of human experience within the world; the latter a question that was only fully addressed by the twentieth century school of existentialists.

Again, on emphasizing that mind is not restricted to within the head, Thompson and Varela (2001):

... conjecture that consciousness depends crucially on the manner in which brain dynamics are embedded in the somatic and environmental context of the animal's life, and therefore that there may be no such thing as a minimal internal neuronal correlate whose intrinsic properties are sufficient to produce conscious experience ..

---

[4]Husserl's philosophical axioms of phenomenology did have a significant bearing on the idea, though elements of his theory had been questioned and misjudged on grounds of reductionism and representation, and were somewhat tainted by the influence of the work of H. Dreyfus. After Varela et al. (1992), Thompson (2007a) reviewed these strictures, and then positively re-evaluated Husserl's work on further examination of previously unfathomed resources. He concluded by acknowledging Husserl's contribution in respect of cognition, as well as the broader interpretation of his theory of phenomenology in the light of the traditions of Eastern religious philosophy.

To see this in the context of advances from about 1960 to 1997, cognitive science progressed from "other as problem" to "other as interaction" in terms of *computationalism* → *connectionism* → *enactive embodiment*.

Then from 1997 to 2007, a further progression from "other as interaction" to "other as inter-being" in terms of *neurophenomenology* → *generative neurophenomenology*.

These were the essential building blocks towards Varela's envisioned science of Interbeing (Varela 1999), a task whose complete fulfillment outlived him, unfortunately, but nevertheless provided an inspiration for a later generation to follow (cf. (Froese 2011; Rudrauf et al. 2003), and references therein).

A big question is, how then does this all work in practice? Towards offering any kind of answer, a significant observation is the following (Rudrauf et al. 2003):

> In Neurophenomenology, the subject clearly has a double status: he/she is and acts as a subject in a particular task, but the subject also needs to know about his/her own experience in order to report structural features about his/her own experience. Francisco (Varela) was aware that such a situation raised a fundamental issue: Neurophenomenology requires some degree of self-awareness, even implicit, to provide phenomenological descriptions and structural insights. But this very process of becoming self-aware required by the task could be said paradoxically to introduce unwanted complexity in the data and at the same time to be the very process that science tries to account for.

If then cognition is seen as influenced by states of the body, then one should be aware that there can be unforeseen drawbacks to the experimental methods of EEG and MEG (and other neuroimaging technologies) when they are applied to task analysis. For instance, if a prolonged exposure to a certain stimulus elicits a high variation in response, culminating in higher than expected levels of noise, then it is possible that 'first-person' experiences had not been adequately taken into account. More specifically, the observer at that time, may have neglected the subject's attentional and emotional state, besides an overall preparedness to carry out the experimental task in the first place.

In order to cure matters, several steps may be taken. In particular, by underscoring the importance of progressively and systematically reducing the distance between the subjective and the objective, so that the subjective experience is not altogether a singularly introspective experience, but one that can be outwardly shared under favorable circumstances. Towards achieving this goal, there are several experiments of striking relevance, two of which provide a testing ground for Varela's hypothesis for fusing both the subjective and the objective experience, as well as narrowing the gap between the physical and mental. So, it will be worthwhile to say something about the content of such experiments.

In Varela et al. (1999) (cf. Rudrauf et al. 2003; Thompson and Varela 2001) EEG recordings were taken of subjects who were observing highly contrasted and distortable facial images that are usually recognized as such when the images are placed upright, but can otherwise appear meaningless (they are the so-called 'Mooney faces'). Subjects registered their response by pressing buttons in degrees of perception (that of the image, or otherwise). Two categories were then drawn up:

A 'perception group' that registered large-scale phase-locking 250 ms after appearance of the actual facial image, with large-scale synchrony occurring mainly on the gamma band (at 30–80 Hz), then significant phase scattering, following which a new synchronous assembly emerged. Otherwise, there was a category of 'non-perception', in which there was no visible assembly after an arbitrary image was presented.

Perceptual motor-response patterns from subjects revealed phase synchrony to be significantly varied throughout regions, so suggesting that synchrony need not be restricted to localized sensory characteristics, but ought to be realized in a broader context of cognition that embraces associative memory and emotional disposition, for instance. The outcome again underscores the importance of observing large-scale dynamical patterns across multiple frequency bands, taking heed of first person reports, in contrast to simply analyzing specific neuronal assemblies, again avoiding the trap of the 'mereological fallacy' (Bennett and Hacker 2003).

A related experiment (Lutz et al. 2004) to that of (Varela et al. 1999) studied the empirical qualities of embodied cognition on the basis of 'phenomenological clusters' based upon the subjects' ability of categorizing their experiences through first-person accounts. The subjects themselves were selected on the basis of having had prior instruction in the qualitative awareness of phenomenology, and then variations were studied in a subjective experience by an experiment based on visual perception. This involved a range of abilities and preparedness in mindful attention and emotional self-regulation on behalf of the subjects, in concert with (Lutz et al. 2004; Thompson 2007):

(a) First person data from close analysis of experience with specific first-person methods.
(b) Formal and analytic methods drawn from the theory of dynamical systems.
(c) Neurophysiological data from the measurement of large-scale integrative processes within the brain.

The experiment was based upon how an illusory 3-dimensional figure is seen to emerge from the perceptual fusion of 2-dimensional, randomly distributed dotted images with binocular disparities. On recording the brain patterns observed from EEG signals, subjects were asked to report upon their experience in terms of, (i) phenomenal awareness (experiential categories), (ii) their own private observations (phenomenal invariants), and (iii), their degree of preparation, readiness, and sensitivity of perception. The proposed hypothesis claimed that distinct phenomenological clusters would be characterized by distinct dynamical neuronal signatures prior to stimulation. The outcome of the experiment described in Lutz et al. (2004), is summarized in Thompson (2007) pp. 342–346 as follows:

1. First-person data concerning the subjective context of perception can be related to stable phase-synchrony patterns measured in EEG recordings prior to stimulus.

2. States of preparedness and perception on the basis of how subjects reported them, modulated both behavioral responses and the dynamic neural response post-stimulation.
3. Although variations in phase-synchrony patterns were observed across subjects, in some cases they were stable for certain subjects, thus representing their preparedness and capability in performing the perceptual task.

A.5 Autopoiesis and Socio-cultural Systems

An autopoietic system is a system we have described as organizationally closed and structurally determined. It has been proposed that the system's autopoiesis is preserved within the living state, adaptable only to structural fluctuations for as long as the living entity survives within, and can be structurally coupled to its environment; otherwise there is termination. The autopoiesis of the nervous system, though the latter itself is not strictly autopoietic, assimilates signals from an environment and so creates interactive relations that occur on a sensory surface to be embodied in a pattern of neural activity. For instance, a pin-prick on one's thumb triggers an impulse into the nervous system which registers a structurally-coupled, cognitive sensation of a sharp point having been applied to a recognizable area on the surface of one's skin. Thus, in keeping with the property of cognition, the nervous system creates such domains of interactions permitting its internal components to function by independent means. It is then structurally dependent by affording an innate plasticity, out of which cognitive events constitute a dynamical system, which manifestly creates a neurophysiological basis for phenomenological interactions. Together with making implicit adjustments, the nervous system is then susceptible to learning strategies, and accordingly adapts itself towards broader interactions within the sphere of human self-consciousness (Maturana and Varela 1980; Rudrauf et al. 2003; Varela 1996; Varela et al. 1992).

The autopoietic hypothesis is not necessarily exclusive to the type of paradigm for (biological) living systems as it has been described. N. Luhmann (followed later by others such as Mingers 1994 and Seidl and Cooper 2006) proposed extending the original concept of autopoiesis as one that can be interpreted for general systems (living or otherwise), thus extending multifarious mechanisms of second-order cybernetics which had previously gained some popularity in applied psychology and the social sciences.[5] Such extended interpretations of autopoiesis are professed to apply to a number of disciplines such as linguistics, social institutions, corporations, families and kinship, educational, and legislative systems.

---

[5]We refer to Seidl and Cooper (2006), and references therein, for an exposition and development of Luhmann's work in extending autopoiesis to social and psychic systems, as well as their respective ramifications and consequences. To a degree there is an overlap with earlier theories as proposed by L. Von Bertalanffy and H. Von Foerster (see Seidl and Cooper 2006 loc. cit., and cf. Laszlo 1996).

It can be claimed that such systems are in fact, self-reproducing, self-organized, and selfreferential, each of which is regulated by their respective intrinsic mechanisms of operational closure, and causal reciprocity. I should perhaps remark now, that this association appears closely tied to the order of local-to-global negentropy, self-organized oscillation of processes, and self organizing information, out of which a dynamic complexity emerges as predicted by the OAM and OSIMAS models (see e.g. models Plikynas 2010, 2015; Plikynas et al. 2014a, b, 2015; Raudys et al. 2014) (see also Kezys and Plikynas 2014). At the same time, these latter models can be tailored to describing open systems as determined by the neurocognitive dynamics of their human constituents that, owing to nonlinear synchronization and social binding effects, determine the levels of organization underlying the evolution of social systems.

Thinking of these systems as evolving from the neurophenomenological field, as akin to the behavior of their human constituents, then management of the environment, economic development, the historical and legal culture, all of which contribute to the evolution of any society, follow an analogous patterning of structure as the models here describe. It is close, in essence, to the transition between the mesoscopic to the macroscopic scales, viewing consciousness as a 'mind field' which emerges through a global process to distributed social and ecological systems which condense the collective phenomenology of its interacting human components in environments that are ever increasing in complexity.

The apparent uniqueness of humans in achieving and perpetuating this process, can in part be explained by the plasticity of phenotypes and developmental methods adapted to varying environmental changes, and the representations thereof. It is, according to Sterelny (2003) how the adoption of the 'cognitive niche', from the anthropological viewpoint, allows for inherited epistemic and ecological engineering skills to readily apply to such variations, and how they can be passed on to future generations. What is postulated here is not openly detached from the conceptual sense of a 'Pervasive Information Field' (Plikynas 2010, 2015; Plikynas et al. 2014a, b), and neither from the sense of a self-regenerating autopoietic system; clearly, a question that warrants some further research.

However, when applied to social systems, the concept of autopoiesis inevitably gets detached from the sense of a biological living entity, for which was originally conceived. It is worth making a few points here. As self-reproducing, organizationally closed entities, social systems reproduce themselves on the basis of communication: each communication selects whatever is being communicated out of a stream of possibilities that have already been communicated, be it in the form of utterance, information and understanding, so contributing towards streams of broadcasted contextual information. This is quite in consonance with C. Shannon's original concept of 'information', as grounded on statistical and selective measures. It is also an emergent process in which a sense of understanding is achieved at such a level (of communication), namely, that which is understood in relationship to other possibilities. This enables social systems to reproduce and self-organize, thus creating a society which is sustainable, and in turn, can be reproduced by means of these components. The latter thus communicate something about themselves, and about their environment in relationship to their sovereign categories of awareness and of choice.

Take for instance, how the notion of 'contract' in the corporate world, is effectively a legal communication between two parties (that is, between two systems) that are independently self-interested and motivated accordingly. The operational closure then consists of the respective terms, policies, and decisions that respectively support each party's self-reproduction and structural organization out of which their mutual gain and long-term interests evolve, as well as the means by which these can be preserved. In this way, the 'organization' can be viewed as an autopoietic system that reproduces itself on the basis of decisions, as compactly produced communications, reproducing each other in recurrent phases as distributed across its sphere of influence. Such decisions can be iterated (decisions based on prior decisions, etc.) as procedures internal to the system, and are thus part of the organization's operational closure. By interactively framing the various decision processes, an organization creates its own culture that is exposed to, and interacts with an external social medium which, a priori, is generally a function of some separate 'culture' beyond that of its own corporate type. Just as in the OSIMAS model, this culminates in a process that emerges from the collective conscious and unconscious mind-fields of the organization's closely interacting human membership.

Towards creating its sense of identity, the organization thus condenses its self-descriptions as founded on its collective self-observations, out of which a sense of 'unity' is created; that is, as a function of its organization and its self-referential closure (Seidl and Cooper 2006). Through multiple self-descriptions, social communications, and external observations, the organization's 'reputation' comes about (for an individual, the analogous properties would likely equate with the formation of a character and a personality). The organization emerges from a media of communication through which it reproduces itself through continuing series of decision-making, as was suggested earlier. In keeping with the autopoietic model, its evolution has three factors in common with biological evolution; namely, in terms of (i) the interplay between the evolutionary functions of variation, (ii) selection, and (iii) retention. This evolution is sustainable via an interactional memory that recalls the constitutive uncertainty of previous decisions and extenuating circumstances, and so on. Together with its corporate interaction and collective 'institutionalized cognition', the organization thus shares with society an extended mind and memory, both within itself, and within its environment. To an extent this overlaps significantly with how cognition can be regarded as both distributed embodied; a topic we will focus upon later.[6]

---

[6]At this stage I should point out that it is not my intention in this Appendix to make a complete overhaul of the current brain-mind-body theories that profess evolutionary-type connections between neural organization, development of brain–functional activity in relationship to the environment. But seeing how phenomenology and structuralism are interwoven in cognitive neuroscience cannot be overlooked, I will give a few pointers to the interested reader towards further connections between autopoiesis and oscillatory systems. Of these (Laughlin and D'aquili 1974) adopt a structuralist approach towards acquired models of reality relative to genetically determined attributes of sense, perception and cognition, while viewing the nervous system and associated brain functional mechanisms as self-organizing, much in the same way that was discussed earlier. This 'neurognostic' theory is one interpretation of bioevolutionary/socio-cultural structuralism that affords close ties with autopoietic systems, while occasionally evoking several 'holographic' metaphors. A closer inspection of this theory reveals interpretations within Husserl's transcendental phenomenology that are also compatible with Varela's proposed concept (of neurophenomenology). On the other hand, the holographic approach to studying the function of the brain is more systematically emphasized in K. Pribram's application of the spectral theory of Fourier and Gabor integral transforms in studying episodes of perception and information (Pribram 1991). We can see some relevance to the OSIMAS/OAM models purely on the basis of the implicit oscillatory characteristics in both time and frequency domains as they correspond under these transforms. Pribram's proposal is to view the brain as a self-organizing hologram, while at the same time incorporating the essential task of minimizing elements of surprise for the purpose of maximizing effective information. Possible connections with the OSIMAS/OAM models may be well worth researching into.

Appendix B: Global Workspace Theory[7]

The dynamics of multi-agent systems and sensory mechanisms as they are coupled to massively parallel, distributed processing features of GWT—a framework for studying cognitive processing within a hierarchically modular structure out of which emerges an access to consciousness. Understanding the functional mechanisms of GWT is accessible through its characteristic theatrical and tournament-like metaphors used in describing how a cognitive response is mandated in relationship to information flow and coordination between modules. Accordingly, GWT can capture a broader basis for cognition as realized, for example, in the immune function, epidemiology, neuropathology, machine intelligence, and social/institutional systems. Thus, by regarding cognition as both distributed and embodied, a common meeting ground between GWT and the phenomenological approach to cognition can be attained (cf. Shanahan 2010).

Let us now be more specific regarding the hallmark properties. GWT evolved from a body of selective evidence concerning conscious processing and extends the methodology of some previously known 'blackboard architectures'. The global workspace generalizes the notion of 'blackboard', and accommodates a framework of specialist processors operating competitively towards mandating a cognitive response. The principal axioms of GWT are based upon the following observations (following Baars 1993; Baars and Franklin 2003, 2007; Shanahan 2010; Wallace 2014); see also (Dehaene 2011; Dehaene and Changeux 2005; Dehaene and Naccache 2001) whose slant emphasizes more the neuronal context of the workspace):

1. The brain can be viewed as a collection of distributed specialized networks (processors).
2. Consciousness is associated with a global workspace in the brain—a fleeting memory capacity whose focal contents are widely distributed (via 'broadcasting') to many unconscious specialized networks.
3. Conversely, a global workspace can also serve to integrate many competing and cooperating input networks, thus accounting for global self-amplifying properties of brain activation towards access to consciousness.

---

[7]The appendix has been cordially contributed by prof. J.F. Glazebrook.

© Springer International Publishing Switzerland 2016
D. Plikynas, *Introducing the Oscillations Based Paradigm*,
DOI 10.1007/978-3-319-39040-6

4. Some unconscious networks, called 'contexts', shape conscious contents. For example, unconscious parietal maps modulate visual feature cells that underlie the perception of color in the ventral stream.
5. Such contexts work together jointly to constrain conscious events.
6. Motives and emotions can be viewed as goal contexts.
7. Executive functions work as hierarchies of goal contexts.

Validity of the workspace model is supported by the presence of a significant number of simulations of thalamo-cortical networks which postulate the access to consciousness in terms of activation within a network of connector hubs, and to an extent this brings the model closer to those theories that study the 'neural correlates of consciousness' (Baars et al. 2013). These networks are interconnected in a reciprocal fashion by assemblies of distinct cortical neurons with sprays of transnetwork axons, as seen for instance in Dehaene (2011), Dehaene and Changeux (2005), Dehaene et al. (1998), Dehaene and Naccache (2001) (cf. Shanahan 2010).

This leads to an integrative, distributed architecture in modular form, interconnecting multiply-segregated and specialized regions of the brain in a coordinated fashion, so allowing communication within the infrastructure by means of distributed, massively parallel processors projecting upon neuronal assemblies spanning cortical domains. These processors are capable of assimilating a high load of information for both encapsulated, and unencapsulated processors, through which attention acts to select conscious elements, while working a memory accommodates details for the latter elements. More specifically, Dehaene and Changeux (2005):

> ... the global workspace is the seat of a particular kind of "brain-scale" activity states characterized by the spontaneous activation, in a sudden, coherent and exclusive manner, of a subset of working neurons being inhibited. The entire workspace is globally interconnected in such a way that only one such "workspace representation" can be active at any given time. This all-or-none invasive property distinguishes it from peripheral processors in which, due to local patterns of connections, several representations with different formats may coexist. .... the state of activation of workspace neurons is assumed to be under the control of global vigilance signals, for instance from mesencephalic reticular neurons. Some of these signals are powerful enough to control major transitions between the awake state (workspace active) and slow-wave sleep (workspace inactive). Others provide graded inputs that modulate the amplitude of workspace activation which is enhanced whenever novel, unpredicted, or emotionally relevant signals occur, and conversely, drops when the organism is involved in a routine activity...

B.1 The Global Workspace Small World Network Structure

In recent years there has been wide acceptance of viewing the brain's functional mechanism as modeled on the architecture of a small world network (SWN). The special properties that such a network exhibits, are (following Watts and Strogatz 1998; cf. Barrat et al. 2008; Latora and Marchiori 2001):

1. A significant number of densely interconnected local clusters with high clustering coefficients at the majority of the network edges where the remainder of edges link to these structures.
2. Minimal average path lengths spanning a globally, sparsely connected network between segregated modules.
3. SWNs lie somewhere between ordered and random networks.
4. SWN architecture is enhanced by the addition of widely spanning trans-network linkages between densely clustered subnetworks that sharply decrease the network's average path length while keeping the clustering intact.
5. Relative efficiency and cost minimization in local-to-global processing.
6. SWNs are often seen to be resilient and assortative (that is, when vertices of a given degree have an affinity to connect up with those of the same topological type).

In recent years, the ever-advancing methods of neuroimaging have provided considerable evidence in revealing the brain's functional network architecture as a SWN (see e.g. Bassett et al. 2006; Bassett and Bullmore 2006; Sporns 2010; Sporns et al. 2000). Further significance for GWT is the proposal that, given some degree of optimality, the hierarchically-modular structure of the workspace is one that is indeed compatible with the this type architecture for the brain (Glazebrook and Wallace 2009; Shanahan 2010, 2012). In particular, the workable efficiency in neural coding is conducive to the type of massively parallel computing that the brain undertakes for the purpose of relaying information globally via local and global corticocortical connections.

B.2 The Connective Core

The SWN structure is one special feature. Another concerns how the workspace modular structure can be broken down in terms of networks of distinguished connector hubs which are mapped to a communications infrastructure, which through a process of 'broadcasting', attains to a dynamical system in which local clusters operate in creating a widespread influence. Within the framework of a modular SWN, the workspace is further characterized by its connective core, namely, a functional bottleneck with limited bandwidth that consists of the connector hubs together with a vast network of dense interconnections. Interacting sensory-motor and memory processes are then partitioned into competing coalitions which engage in a vigorous tournament between themselves as they jockey towards dominating the workspace through systematic iterations of local-to-global procedures. They may at the same time form alliances with certain competing coalitions, as founded on 'history' and common goals which are manifestly the evolutionary driven iterations of responding to external signals or threats, for instance, and this is very much in tune with the cognitive property exhibited by the immune function as we had mentioned earlier.

By overcoming certain global constraints, the domineering coalition is ushered through the connective core, and then gains access to center stage (metaphorically, under the spotlight) where it airs a punctuated broadcast towards mandating an instantaneous cognitive reaction.[8] This coalition will then rapidly dissipate in making way for its successor, and so on. During the course of this process, certain sub-coalitions may remain active for the purpose of passing on 'memory' to some other coalition which they may eventually team up with. This is how a cognitive cycle is generated by an ordered sequence of sensory inputs leading to perception and a sense of understanding. In optimal circumstances, it is filtered out for the purpose of sustaining relevant broadcasting with subsequent action and learning then culminating in an effector output (Baars and Franklin 2003, 2007; Franklin and Patterson 2006; Shanahan 2010).

These procedures can be re-interpreted within an information-theoretic context in the following sense. It starts by recalling a provocative duality, as proposed by C. Shannon, linking the properties of an information source with distortion measure, and those of a communication channel (see e.g. Cover and Thomas 2012). This duality is further enhanced if channels can be assigned 'message' costs, and then the task becomes one of finding a source that is suited to the channel at a tolerable level; essentially, this is the essence of the 'tuning' version of the Shannon Coding Theorem (see Wallace 2014; Wallace 2012).

In a formal sense, the channel can be regarded as 'transmitted' by the signal. A dual channel capacity can be defined in terms of a channel probability distribution that maximizes information transmission, once there is assumed a fixed message probability distribution. For a regulatory, interactive system this necessitates a coordination between cooperating low-level cognitive modules of the workspace. Very close to what has just been described, such required patterning may be based upon the network of connections between information sources dual to basic neurophysiological, and conditioned unconscious cognitive modules (UCM). This latter network is of an oscillatory type that functions at a level above a neural learning network, and can be viewed as a type of representation space for the re-mapped network of lower dual cognitive modules (cf. Wallace 2012, 2014).

---

[8]Such punctuated broadcasts are analogous to the intermittent bursts of vocal reactions that can be heard in the course of e.g. an opera, a heated political debate, or by the spectators during a major football game.

B.3 No Free Lunch

Given a set of cognitive regulatory modules that become linked to solve a problem, the famous "no free lunch" theorem of Wolpert and Macready (1997) basically asserts that there exists no generally superior computational function optimizer. There is "no free lunch" in the sense that an optimizer pays for superior performance on some functions with inferior performance on others while under degrees of selective pressure. In particular, neuronal mechanisms (such as action potentials and synapses) run up considerable expenditures of metabolic energy, most of which is invested into processing of information (see e.g. (Sengupta et al. 2013) for a survey of some of the basic neurobiological/physiological details). At the same time, there is the trade-off between wiring costs and topological efficiency, such as when the costs of sustaining useful information at select sites exceeds that for continued listening in to noisy crosstalk. Gains and losses balance precisely, and all optimizers have identical average performance (cf. English 1996).

In other words, an optimizer has to pay for its superior performance on one subset of functions while merely delivering an inferior performance on the complementary subset. This is a particular caveat that should be kept in mind in view of free energy metabolism and inference, since energy provided to neuronal mechanisms is limited; typically, a unit of brain tissue consumes an order of magnitude of metabolic free energy by a factor of ten when compared with any other unit of tissue.

Another slant on this conundrum is to say that a computed (or regulated) solution is the product of the information processing of a problem, and, by a very famous argument, information can never be gained simply by processing. Thus a problem, call it X, is transmitted as a message by an information processing channel Y (typically some computing device), and recoded as an answer. By the above 'tuning theorem' version of the Shannon Coding Theorem, there will be a channel coding of Y which, when properly tuned, can be efficiently transmitted, in a purely formal sense, by the problem; that is, by the transmitted message X. In general, the most efficient coding of the transmission channel, in other words, the optimal algorithm leading to a solution, will necessarily be highly problem-specific. Thus there can be no best algorithm for all sets of problems, although there will likely be an optimal algorithm for any given set.

From the "no free lunch argument", one may conclude that the various challenges facing any cognitive regulatory system—or interacting set of submodules constituting one—must be countered by different arrangements of cooperating 'low level' cognitive modules. One can envisage the phenomenon in terms of a mapping of the network of lower level cognitive modules expressed in terms of the information sources dual to the UCM. Thus, given two distinct problems classes, there must be two different 'wirings' of the information sources dual to the available UCM, with the network graph edges measured by the amount of information crosstalk between sets of nodes representing the dual information sources. An alternative treatment of such coupling can be given in terms of network information theory (see e.g. Cover and Thomas 2012), that in particular, incorporates the effects of an embedding context, as implied by some external information source feeding in signals from the environment. Further elaborations on this particular theme, and others closely related, are studied in Wallace (2014).

B.4 Biased Competition

In keeping with the Atlan-Cohen paradigm for cognition, for instance, continuing patterns of recognition and response entail convoluting an incoming external sensory signal with an internal ongoing operation by means of a system of oscillators, underlying which a flow of mutual information induces a learning process induced by sequences of phase transitions (Wallace 2012, 2014). This principle is very close to the (Bayesian-like) 'inference' patterning based on variational free energy to be discussed later (see also Friston 2012), and affords an equally close association with the global workspace when the network of the UCM is viewed as a functional workspace in its own right.

Although the latter is not strictly identifiable with the connective core, it comprises an oscillatory network that schematically can be mappable to the connective core in relationship to the mutual information of crosstalk that will influence sensorimotor blending, while at the same time being prone to environmental influences. In the sense of the workspace, we recall that rival coalitions operate in a parallel fashion, and may, in a sense, negotiate between each other, while at the same time being disposed to recruiting members from other coalitions. Eventually, a dominant coalition is formed, and its passage into the connective core can be interpreted in a graph-theoretic sense as the emergence of a giant component (Erdös and Rényi 1960); that is, one which captures up the lesser components of the network, while de-activating its competitors by relegating them to the sidelines.

Within the network of the UCM, the giant component, seen as function of a variable topology, is induced by continued symmetry breaking (Glazebrook and Wallace 2009; Wallace 2012), and can be enumerated in terms of the characteristics of the network's architecture (for much of the numeric details, see e.g. Albert and Barabási 2002; Erdös and Rényi 1960).

As one might expect, competition in the workspace is patently biased. To try to connect with what was discussed earlier, we may suggest several reasons for this:

1. We expect that such a bias originates in part from evolutionary driven iterations of instinctive, cognitive, and genetic mechanisms over extensive periods of time. This can be reflected in the evolutionary and cultural bias in human cognitive processes from the anthropological and socio-psychological perspective. This can be mined out from works such as Clark (1998), Sterelny (2003), Thompson (2007), Varela et al. (1992).

2. The process of stochastically driven, intense synchronized firing of neurons, and dynamic reverberation is not sufficient for consciousness, but may be significant

for determining which coalition succeeds in the global access (Maia and Cleeremans 2005; Shanahan 2010). A connectionist interpretation in Maia and Cleeremans (2005) accounts for bias in terms of how feedforward and feedback projections implement selective attention onto the most favored sites while minimizing metabolic free energy costs under evolutionary demands.

3. In cases of cross-frequency coupling and selective communication, biased competition may equate with degrees of selective communication with resonating oscillations between preferred agents (Axmacher et al. 2010; Fries 2005, 2009).

B.5   Meta-Stability and Attractor States

The main faculties of brain function that can be observed (including, in parentheses, the regions served in question) involve the following (as summarized in Shanahan 2010; cf. Dehaene 2011; Dehaene and Changeux 2005; Dehaene and Naccache 2001):

- sensor ($\Rightarrow$ occipito-temporal)
- motor ($\Rightarrow$ frontoparietal)
- affect ($\Rightarrow$ sub cortical structures, including the amygdala and basal ganglia)
- episodic ($\Rightarrow$ medial regions, including the entorhinal cortex and hippocampus)
- working memory ($\Rightarrow$ prefrontal cortex)

Collectively, the amalgamation of these processes enters into the competitive arena when more conscious attention prevails. They contribute towards metastability of the brain's state, namely, as it is displaceable from one attractor niche of its state space to another, while remaining well-poised for releasing swift and fluid responses to incoming stimuli.

Although structured in a serial-like form, the transitions between states are functions of massively parallel computations which are crucial for a rapid turn towards conscious attention. How this compares with the autopoietic property can be seen in way that the states do not simply arise out of their own internal inter-coupling, but are further induced by the brain's own coupling with the body and with the environment.

The methodology of workspace experimentation naturally follows the idea of embodiment by fully considering the phenomenological aspect dependent on introspective subjective reporting, thus going against the grain of the reductionist emphasis on purely objective criteria which was once a mainstay for much of traditional experimental psychology. As argued in Dehaene and Naccache (2001), somewhat in consonance with Varela's viewpoint:

> The first crucial step is to take seriously introspective phenomenological reports. Subjective reports are the key phenomena that a cognitive neuroscience of consciousness purports to study. As such, they constitute primary data that need to be measured and recorded along with other psychophysiological observations.

B.6   Phase Locking and the Connection with OSIMAS

In order to see how aspects of the global workspace model compare with the OSIMAS and OAM models, we can suppose that communicating neuronal groups do so in rhythmic phases of competition and in patterns of coherence as they are modulated by firing rates. The biologically created oscillators underlying cognitive modeling such as provided by the models of Hodgkin-Huxley and Fitzhugh-Nagumo type, as well as a range of low-dimensional spiking-bursting models, provide a spectrum of examples for this purpose (see e.g. Izhikevich 2007). Optimal synchronization here facilitates communication, where a selected group's oscillatory synchronization is all the more stronger when it is distributed coherently— rather like a well-orchestrated choir singing in perfect harmony. It is effectively a two-way function of biased competition from which cortical computations unfold when inputs converge to selectivity in a patterned formation which is supportive of the system's selforganization. However, coherence in oscillations is not in itself sufficient for communication if the resonant frequencies of the transmitting group are out of synch with those of its target.

Typically, phase locking between oscillations from a transmitting group is sent to a receptor group which issues feedback. From the workspace perspective, this may be seen by how a competing pair of neuronal groups (A, B) oscillate at varying frequencies with the intent to influence a third group C. Although both A and B may be themselves coherently oscillating, with mutual information sustaining the communication between the two, they may be jilted out of phase with each other. Consequently, the third party C may eventually entrain to one of the pair, but not to both; or, communication closes down entirely. This will be the case when the channel between A and B closes once the prevailing phase relation between them expires. However, increasing mutual information from A may leap-frog over B to C, thereby capturing C's excitatory phase. This may effect the module in the connector hub that is a function of C's own oscillatory behavior, implying less flexible patterns of coherence in the connective core, so disrupting the necessary cognitive flexibility. Selfsustaining, reverberating patterns that settle in some attractor landscape will stay there until the initial stimulus producing them fades away. Eventually, a competing rival that recognizes the patterned structure evolving within the core, will try to shift this pattern out of its attractor basin into a neighboring one. Such a sequence of events continues to generate episodes of punctuated broadcasting, while initiating excursions between different attractors. References Fries (2005, 2009), Shanahan (2010), Shin and Cho (2013) provide the specific details.

Needless to say, there is much more to be said here at the neurophysiological level. Many, if not all, processes of signal transduction are stochastically driven, and in this way they eclipse their purely deterministic counterparts. For instance, in the case of finite assemblies of ion channels situated within a noisy environment, the cumulative effect of stochastically driven firing and spiking can, at some optimal noise level, enhance signal transduction: a phenomenon we call stochastic resonance (SR) (Hänggi 2002) as depending upon such factors as the type of background noise, weakly coherent inputs, and the capacity for overcoming a threshold towards completing an effective task. A survey of these topics would inevitably go well outside of the scope of the present essay, and would include my own work (with Georgiev, et al.) that studies the possibility of soliton dynamics in microtubular mechanisms, and vibrationally assisted quantum tunneling in synaptogenesis, both of which are susceptible to SR effects (see e.g. (Georgiev and Glazebrook 2006, 2014) and references therein).

B.7   The HPA-Axis

One may propose that enriched environments of a certain kind reciprocate to a greater advantage than others, and the structures governing the co-alignment of the self to existing perturbations evolved in accordance with degrees of mastery over the latter. Take for instance, the hypothalamic–pituitary–adrenal (HPA)-axis. As a principal regulator of the neurophysiology determining the 'flight or fight' mechanism, the HPA-axis can be realized as a cognitive mechanism in the sense of Atlan and Cohen (1998). If there is an arousal of the individual's close environment, then mind and memory engage, evaluate, and then select the appropriate responses—a process interpretable by the mechanisms of the global workspace (cf. Wallace 2012, 2014). The HPA-axis accelerates this process, and possible malfunctioning may induce hyper-reactivity, as, for instance, observed in cases of post-traumatic stress disorder and depression. The latter example may be seen in part as the evolution of a structure conducive to a negative alignment of the self with an external reality, a hypothesis that has been researched in the context of various evolutionary theories (of this disorder) that are founded upon the occurrence of attachment-defeat-loss, diminished opportunities, down-regulation of foraging capability, social/ professional rank, etc., incurred with varying risk factors within a culturally stamped environment (see e.g. Gilbert 2006; Moffitt et al. 2006; Wallace and Wallace 2013). Another partially related example, is the body's blood pressure control system consisting of a network of cognitive systems which compare a set of incoming signals with an internal reference structure in order to select a suitable level of blood pressure from possible levels, implying that an elaborate tumor control strategy, for instance, must be at least as cognitive as the immune system itself (Wallace 2012, 2014).

B.8   The Free Energy Principle and Inference

Much of what has been said so far regarding cognitive functioning depends significantly upon the role of free energy metabolism. Thinking back to the common characteristics of the cognitive processes, we can reasonably assume that a self-organizing system, striving to maintain equilibrium with its environment, does so by minimizing its free energy in order to minimize the long-term, average element of 'surprise'—in other words entropy. This can be seen within the context of autopoiesis, since biological (living) systems innately resist the surprise element in order to preserve and maintain their self-organization. In particular, we can consider the brain as a far from equilibrium open system that is so geared to resist tendencies to disorder in the presence of continuing environmental perturbations, by configuring its sensory mechanisms in order to minimize predictive error, and to then implement error-correction as efficiently and as accurately as possible.

Validation of these principles can be traced back to the striking duality (or homology) as proposed by Feynman (2000) (following to some extent the work of Bennett and Hacker 2003) relating information to free energy. Basically, the idea amounts to saying that computing in any form involves work—or putting it another way, the informational content of a message can be viewed as the work saved by not needing to recompute whatever has been transmitted. Hence, the more complex a cognitive process, as measured by an information source uncertainty, the greater will be the energy consumption, and the lesser the entropy, the more precise the gathered information will be. Thus regarding (thermodynamic) free energy as a measure of available energy for the purpose of implementing effective work, then this homology implies in essence that information emerges as a difference between the world as it is perceived, and what the world actually is.

However, minimizing entropy may be an intractable problem in general. One way of dealing with this difficulty is to replace entropy by a variational free energy bound (Friston 2010; Sengupta et al. 2013). Free energy can then be formulated in terms of a function of the statistical dependence between sensory states, together with a recognition density, so leading to a variational free energy principle that regulates a probabilistic representation of the cause of a particular sensation. It is a means of measuring the information about hidden states that are representable, and not to internal states that represent them. In effect, the hidden states are fictive variables that are essential for constructing such a variational free energy bound in the first place. In minimizing free energy, the probabilities at one level can be optimized given those at the lower level (the principle of empirical priors), and the subsequent dividend is an efficiency of internal representations that will in turn lead to optimizing memory, perception and inference (Clark 2013; Friston 2010).

According to Friston (2010), Sengupta et al. (2013), (variational) free energy, in a more specific sense, can be viewed as a function accessible to an agent in terms of: (i) its sensory states, upon which sensory input is altered probabilistically via action upon the world, and (ii), the recognition density encoded by its internal states, which changes as latter changes. In other words, re-iterating what was said before, minimizing variational free energy leads to minimizing the surprise element, while at the same time maximizing the sensory evidence for the system's capability to update prior beliefs towards predicting future beliefs in relationship to mutual information. In turn, maximizing mutual information leads to maximizing accuracy. The predictive process itself is a hierarchically structured, top-down/bottom up process, where top-down signals will predict and explain away bottom-up signals as are predicted by some 'over-riding policy' or 'hypothesis' by pairing these with a range of predictions across spatial and temporal scales. Interpreted in the global workspace setting, such a 'policy' is effectively the consequence of biased competition on behalf of a winning coalition (as graphically represented by its corresponding giant component) that is capable of dispensing with competing coalitions irrelevant for its policy, and proceeds accordingly. This it does in a highly distributive fashion.

It is through this chain of activity that as intentional species, we can assimilate information from the environment and then make predictions accordingly. These principles are conducive to a Bayesian-like formulation for cortical processing, as professed in Knill and Pouget (2004):

> To use sensory information efficiently to make judgements and guide action in the world, the brain must represent and use information about uncertainty in its computations for perception and action. Bayesian methods have proven successful in binding computational theories for perception and sensorimotor control, and psychophysics is providing a a growing body of evidence that human perceptual computations are 'Bayes optimal'. This leads to the 'Bayesian coding hypothesis': that the brain represents sensory information probabilistically, in the form of probability distributions ..

More generally, and in partly summarizing (Clark 2013) envisages:

> Action-oriented (hierarchial) predictive processing models promise to bring cognition, perception, action, and attention together within a common framework. This framework suggests probability-density distributions induced by hierarchial generative models as our basic means of representing the world, and prediction-error minimization as the driving force behind learning, action-selection, recognition and inference.

Further, this chain of activity determines an essential link between inference and free energy that in some ways characterizes the brain as a Helmholtz-like 'statistical inference machine'; namely, an organism that works through schemata of predictive coding in optimizing the probabilistic representations triggering its sensory input in creating a model of the world to which it belongs (Dayan et al. 1995).[9]

As regards 'cost', this can be interpreted in the framework of (variational) free energy, as a rate of change of 'value'—the latter itself regarded as a function of sensory states. The computational analysis in terms of Bayesian inference patterns (Friston 2010) can be applied to brain functioning in order to optimize these probabilistic representations to the extent of matching sensory input, or altering samples so as to align with predictions (Friston 2010). Within this framework, intermediary agents can alter sensory input by external action, or alter recognition density via adjusting internal states. Thus, the various agents can make predictions with a certain degree of accuracy, and the free energy is simply then reduced by altering the recognition density, and in turn adjusting the proposed conditional expectations (Friston 2010) (cf. Sengupta et al. 2013; Wallace 2012; Wallace and

---

[9]Note that variational free energy does not equate with Helmholtz free energy (Helmholtz and Southall 2005). We refer to e.g. Friston (2010), Sengupta et al. (2013) in which the specific differences are pointed out.

Wallace 2013). Accordingly, variational free energy can be used to convert intricate probability density integration problems to the more familiar instrumentations as, say, representative of machine learning.

B.9 Remarks on the Frame Problem

The central feature of the once elusive Frame Problem is that informationally unencapsulated processes are computationally infeasible, to the extent of deciding upon the actual completion of a given cognitive task. It can be compared with Hamlet's problem of "thinking too much", while short of the prospect of there ever being an effective logical conclusion within near reach (Shanahan and Baars 2005). This problem, for its best part, is a matter encompassed by the ontology of cognition and philosophy, but when tackled in this style, it often overlooks how the brain functions neurophysiologically. Hence the problem can be relegated from metaphysics to one that is a concrete challenge to neuroscientists to try to sort out.[10]

In supporting cognitive fluidity and selected broadcasting, the modus operandi of the workspace provides a solution to the problem. The main observation being that the brain functions in a massively parallel way, and the broadcasting mechanism of the global workspace, the task of filtering out relevant processes from irrelevant ones, can be aptly distributed across specialized processors, so creating a balance within the attractor landscape of segregated and integrated modules through orders of dynamical complexity (Shanahan 2010, 2012). The resulting 'scaffolding' of the architecture conveniently reduces a vast labyrinth of possible connections to more manageable networks capable of assimilating effective, and relevant information. Again assuming a degree of optimal functioning, this effect is achieved in terms of the structure of the workspace that combines serial and parallel information processing, together with analogical reasoning, in order to provide the relevant information flow across the entire system.

---

[10]The Frame Problem has been a topic of much debate with the major protagonists being Fodor and Dennett, and others, but for now I will refer the reader to Shanahan and Baars (2005) that includes some background and history to the problem, as well as the various strictures coming from several camps of the computational-symbolic debate presenting a variety of ontological difficulties of arriving at a solution in the first place (see also Shanahan 2010).

Appendix C: Embodied and Distributed Cognition[11]

C.1 Embodied Cognition

Besides studies such as Shanahan (2010), Thompson (2007), Varela (1996), there are other approaches to abstracting cognition as an 'embodied' function of one's mind-body interaction within an environment as coupled to one's sense of perception, and how this generates a statistically-inferred reaction/response that is both reciprocated, and distributed across task specific resources (cf. Shapiro 2010). When cognition is viewed this way, there are the anticipated nuances of interpretation. For instance, by adopting the viewpoint that processes of cognition are generally deeply-rooted in the individual's reaction towards the world, Wilson (2002) breaks down the concept of embodied cognition into six sub-categories (quote):

1. *Cognition is situated.* Cognitive activity takes place in the context of a real-world environment, and it inherently involves perception and action.
2. *Cognition is time pressured.* We are "mind on the hoof" (Clark 1998), and cognition must be understood in terms of how it functions under the pressure of real-time interaction with the environment.
3. *We off-load cognitive work onto the environment.* Because of limits on our information processing abilities (e.g. limits on attention and working memory), we exploit the environment to reduce the cognitive workload. We make the environment hold or even manipulate information for us, and we harvest that information only on a need-to-know basis.
4. *The environment is part of the cognitive system.* The information flow between mind and world is so dense and continuous that, for scientists studying the nature of cognitive activity, the mind alone is not a meaningful unit of analysis.
5. *Cognition is for action.* The function of the mind is to guide action, and cognitive mechanisms such as perception and memory must be understood in terms of their ultimate contribution to situation-appropriate behavior.

---

[11]The appendix has been cordially contributed by prof. J.F. Glazebrook.

© Springer International Publishing Switzerland 2016
D. Plikynas, *Introducing the Oscillations Based Paradigm*,
DOI 10.1007/978-3-319-39040-6

6. *Off-line cognition is body based.* Even when de-coupled from the environment, the activity of the mind is grounded in mechanisms that evolved for interaction with the environment—that is, mechanisms of sensory processing and motor control.

Each of these categories deserve a little more specification (Wilson 2002): (1) Situated cognition involves interactions with the things that the cognitive activity is about. (2) This distinguishes between cognitive tasks subject to time-pressure (e.g. dealing with potential flooding, fighting forest fires, etc.) and those that are performed at the leisure of the individual (e.g. withdrawing cash from an ATM, playing chess, etc.); different representations are involved. (3) For instance, in playing the game Tetris, how one might apply a purely geometrical strategy in order to simplify a problem, as opposed to performing a series of mental calculations. In this respect, applying a rotation would then be part of the computation. (4) In order to understand cognition, it is important to observe both the situation and the situated (cognitive) aptitude of the individual together as a unified system. (5) How an individual's cognitive framework subserves action in terms of say, perception and memory which invoke sensorimotor activities. (6) Sensorimotor simulations of external activities: mental imagery, working memory, episodic memory, input memory, reasoning and problem solving. The approach then is to regard the mind as a self-organized, embodied dynamical system, as embedded within a social environment, and not exclusively a lone neural mechanism accommodated inside one's head (Thompson and Varela 2001) (cf. Bennett and Hacker 2003). Since these axioms amount to essentially a process of learning through mind-body environment interactions, this process is compatible with a number of theories of developmental psychology/physiology which are based upon the principles of dynamical systems, such as, for instance, those that focus upon infant mobility, developing skills, and paths to learning (see e.g. Port and Gelder 1998; Thelen and Smith 1996). Ultimately, the richer and the more varied is the information provided from the environment, the more complex is the behavioral response.

Towards inter-coordination of mind and body (Wilson and Golonka 2013), view part of the above criteria to be further broken down into four categories, very much in accord with the phenomenological approach to cognition as we discussed earlier: (i) a task analysis from a 'first person' perspective that identifies available task-relevant resources, (ii) identifying these resources as accessible to the individual, (iii) how such resources can be applied to complete the task, and (iv) evaluating how effective the strategy of (iii) is to this extent. In terms of linguistic dynamics (Wilson and Golonka 2013) claim:

> ...The most important similarity is that *from the first person perspective of perceiving, acting language user, learning the meaning of linguistic information, and learning the meaning of perceptual information in the same process* ...

One expects that all of these axioms for embodied cognition fit well with Varela's once proposed idea of 'Interbeing', and likewise for many of the concepts embraced the discussion to follow (distributed cognition, ecological psychology and radical embodied cognition).

How then does one design and implement a neuroscience for a shared world of social interaction beyond that of the individual paradigm? Cognitive neuroscience considers this matter from many perspectives, often in terms of instinctive behaviorial patterns, not simply restricted to humans, but across the entire range of species of non-human animals. The prefrontal regions of the human brain and their connections to other cortical regions are the vital zones of activity in this respect (Frith and Frith 2012). The main task is to extend the more familiar experimental neuroimaging techniques of subjects in isolation, to the more ecological, socially-interactive situations involving shared minds that are embedded within an environment. Insight can be gained from hyperscanning fMRI studies of inter-brain neural sychronization, as observed in a variety of settings, such as the interpersonal transmission of emotions, games of cooperation and trust, and musical ensembles, as surveyed in e.g. Chatel-Goldman et al. (2013) (cf. Kezys and Plikynas 2014; Plikynas 2015; Plikynas et al. 2014). A particular goal here is to realize meta-cognitive processes that crystallize into 'mentalizing'—a means of enhancing how we learn about the world by means of our innate, and very special property of self awareness. This can frequently occur automatically and instinctively, but still affording us the ability to reflect upon our own mental states. It also includes having the choice to relate our own experiences to others, and in turn to learn from their experiences. At this juncture, there is a further significant observation: that common to these dynamic processes of interactions between mind-body and the environment, is the essential property of inference. To be more specific, this brings us back to the free energy principle—such processes are necessarily incumbent upon the brain's capacity for inference, and it is hard to imagine that embodied cognition can dispense with this property, or indeed any other form of cognition if it comes to that (recall that minimizing variational free energy minimizes the element of surprise, and thus maximizes sensory evidence towards realizing a model of the world, according to which this element of surprise can be assessed). Whereas the capacity for inference seems clearly relevant to the fundamental processes underlying the OAM and OSIMAS models, then likewise, the same applies to cognition in its embodied framework (and indeed in its 'radical' form, see below).

C.2 Distributed Cognition

Whereas embodied cognition is a theory of a cognitive system that seeks to understand the deeper levels of cognition in regard to one's sense of self in relationship to and interaction with a social environment, there is a further natural level, and one closely attached, that reaches beyond the individual. This involves not simply the cognitive organization towards the environment, as the previous models have suggested, but includes the nature itself of the interaction between persons, their daily circumstances, the exchange of information between groups, and the means by which available resources aided by a spectrum of cognitive artifacts are employed for implementing tasks of management, exploitation and development (of the environment) beyond that of basic survival needs. Putting it another way, just as in the embodied framework, our cognitive skills are not solely dependent on neuronal architecture, but on how the plasticity of our brains can be modified and influenced by language and by the socio-cultural facets of the prevailing environment (Clark 1998). This extended form of cognition, referred to as *distributed cognition*, can be summarized as follows (Hollan et al. 2000; cf. Hutchins 1996):

- Cognitive processes may be distributed across the members of a social group.
- Cognitive processes may involve coordination between internal and external (material or environmental) structure.
- Processes may be distributed through time in such a way that the products of earlier events can transform the nature of later events.

Keep in mind that these postulates tacitly include, and underscore, the role of operating systems that dynamically regulate their subsystems, and are simultaneously influenced by procedural traditions and social culture. This goes along with the availability and capability in employing a range of artifacts and applicable technologies, as well as exploiting the contextual information internally distributed between groups. Examples that come to mind are navigational, human-cockpit systems, team-coordinated exploration, utilization of environmental resources, general corporative management, interaction via the WWW, traffic control, etc. The diffuse nature of these systems along with their embedding within an ever-changing environments, show that both the distributed and embodied forms of cognition (as surveyed in Clark 1998; Heylighen 1999; Hutchins 1996; Shanahan 2010; Varela et al. 1992), in a similar way to how they are professed for human cognition, are relevant to both the OSIMAS and OAM models as operating systems possessing both shared-mind and memory processes. Quote Hollan et al. (2000):

> ... It is important from the outset to understand that distributed cognition refers to a perspective on all of cognition, rather than a particular kind of cognition... Distributed cognition looks for cognitive processes, wherever they may occur on the basis of the functional relationships of elements that participate together in the process. A process is not simply cognitive because it happens in the brain, nor is a process noncognitive simply because it happens in the interactions between many brains... In distributed cognition one expects to find a system that can dynamically configure itself to bring subsystems into coordination to accomplish various functions. A cognitive process is delimited by the functional relationships among elements that participate in it, rather than by the spatial colocation of the elements... Whereas traditional views look for cognitive events in the manipulation of symbols inside individual actors, distributed cognition looks for a broader class of cognitive events and does not expect all such events to be encompassed by the skin or cranium of an individual...

C.3 Gibsonian Ecological Psychology and the HKB Model

It could be argued that some of the principal ideas that share common ground with embodied and distributed cognition, can be traced back to an earlier theory, namely, one of affordances as proposed in J. Gibson's ecological approach to perception (Gibson 2014) (see also Chemero 2011) that brings to the forefront how information embedded within the environment both determines and modulates cognitive decisions on the part of a living being (for instance, how wind direction can be used for the purpose of sailing a boat, or how fallen trees can be used for crossing a river). The basic principles encompassing affordances can seen to be: (i) that perception is direct, (ii) the environment is perceived for the purpose of functioning, and (iii), which follows from (i) and (ii), is that information necessary for guiding adaptive behavior must be available in the perceived environment. In this way, affordances become directly perceivable environmental opportunities for action.

It goes without saying that how effective the ensuing actions will be, is a function of the individual's cognitive skill and adaptability in regards to the situation in question. This gives rise to the sense of a (ecological) 'niche' as a set of affordances, and the latter then determine the former, while an exploration of the environment simultaneously incurs phenomenal variations.

Since the essence of Gibsonian ecological psychology involves motor-behavior coordination, to an extent influenced by an environment, it encompasses the Haken-Kelso-Bunz (HKB) model (Haken et al. 1985) (see also Chemero 2011) which studies coordination dynamics as typical of many complex biological systems, and so bases cognition within the framework of dynamical systems. The HKB model is geared to predicting patterns of motor behavior by means of coupled nonlinear oscillators, into which noise and symmetry-breaking terms can be incorporated for the purpose of describing spontaneous switching between relative phases, the latter frequently being caused by dynamic instability. This leads to recognizing emergent forms of self-organization in those observed systems relative to the predicted patterns of behavior. Though originally based on seemingly mundane 'finger-wagging' experiments, the HKB model has far reaching consequences for cognition through extension to a range of human (animal) motor-skills, speech, and neural circuitry.

As proposed in Shanahan (2010) the massively-parallel, distributed dynamics of the GW and its integrative faculties, not only contributes to the range of possible affordances, on the one hand, and the co-activation of the underlying neural networks towards cognitive fluidity, on the other, but also provides a mechanism that combines and extends these processes, manifestly through the formation of new and more complex coalitions.

C.4 Radical Embodied Cognition

How radical then is embodied cognition? About 50 years ago, most probably it would have been regarded as such. But in recent years interest has shifted to a subsystem of radical embodied cognitive science (RECS) whose thesis is that structured, symbolic, representational, and computational views, by themselves, are flawed (Chemero 2011). Though not entirely disjoint from computationalism and representationalism, embodied cognition is more effectively studied via noncomputational and nonrepresentational methods that lean more on those of dynamical systems theory. In fact, Thompson and Varela (2001) had previously coined the term "radical embodiment" to describe the neuroscience of conscious experience, but not without certain twists and turns involved (see Chemero 2011; Clark 1998) for a survey of the nit-picking). In the framework of RECS, cognition is explained more from an interactive, dynamic point of view (in keeping with the theme of this Appendix), and incorporates some elements of computationalism. The aim is to find a unified system between agents and environments modeled on nonlinearly coupled dynamical systems beyond the task of threading together the separate components of the system. In highlighting the role of affordances, RECS is derived in part from Gibsonian ecological psychology, and is embraced by the more general structures of embodied cognition. However, certain elements of computationalism and representationalism as indispensable, can be admitted into RECS thus enhancing its viewpoint (see below).

Nevertheless, the essential ingredients of embodied cognition, as we discussed it, still play a vital role. There are two basic observations to be made here. Following (Chemero 2011), we first all recognize affordances as perturbing and influencing sensorimotor coupling (SMC) with the environment. Conversely, given the situation in question, this coupling selects and induces some sort of affordance (via process of perception-action). Secondly, SMC modulates the dynamics of the nervous system within which neural assemblies are co-activated, and these in turn feed back into the SMC (similar to that of an autopoietic process) to eventually register a conscious experience. One is reminded of a token catch-phrase of ecological psychology (Mace 1977): "Ask not what's inside your head, but what your head's inside of".

Because embodied cognition entails dynamical systems (and thus neural and connectionist networks are readily welcomed in), this property is seen to be inherited by RECS, and is realized more fully through, for instance, the dynamics of physiological action exercised in completing a cognitive task as well as the resources available in order to do this. As proposed in Wilson and Golonka (2013): (1) to identify the task to be accomplished, (2) what resources are possessed in order to complete the task, (3) to generate hypotheses concerning the assembly of such resources, and (4) how these resources can be assembled and implemented. These provide axioms for a replacement form of (radical) embodied cognition since they assume intrinsic dynamical processes.

Consider, for instance, how a baseball outfielder prepares for catching a high-flying ball by observing its parabolic trajectory, running within the direction of descent while keeping his/her eye fixed on the ball, so becoming one with it, and then stationing himself/herself accordingly in order to make a successful catch. This affords a relationship between perceptual information (concerning the flight of the ball), and the organism (the outfielder) that replaces the need for an internal simulation of the physics of projectile motion (Chemero 2011; Wilson and Golonka 2013). Besides providing a true-to-life example of RECS, the outfielder scenario is a study in the two-way process between inference and kinetic information. The underlying perceptual mechanisms of this example can be compared with the example of applying dexterous geometrical transformations in the Tetris game so as to skip much of the computational cost.

The HKB model as I mentioned earlier, is one of the principal tools of the theory. Though by incorporating mathematical methods of dynamical systems theory, it more or less extends to all of embodied cognition, radical or otherwise. In short, RECS subsumes embodied cognitive science as well as orders of computationalism, thus creating some appeal, because of: (i) a seemingly unification of disparate theories (Eliminativism and Representationalism), (ii) the tractability of ecological psychology, and (iii) the current trend in cognitive science towards observable dynamical systems in human (animal) interactions.

In this scheme of things, Varela's 'enactive' cognitivism can be subsumed, and hence also the idea of mindfulness of states as we had discussed earlier. But as appealing as many of these modern theories of cognition are (Embodied, Distributed, RECS, and others) one should be aware that they are not without their opponents, some of which belong to the strictly philosophical camps, along with those hard-core reductionist groups within neuroscience (see e.g. Chemero (2011) for a survey of the antagonism, and the role of the major players involved). Of particular importance for the fledgling OSIMAS and OAM models from the neurophysiological viewpoint, is that RECS, in keeping with some foundation on dynamical systems, embraces the role of coupled oscillators [such as the oscillatory models of Fitzhugh-Nagumo type (Izhikevich 2007)] which are capable of implementing representations. Along with classes of relaxation oscillators, these are basically representations of an admissible type; namely, those that are constantly in connection with their targets (Chemero 2011). This also applies to the basic dynamical oscillatory systems used in studying the physiology of infant leg and arm movements as studied in Thelen and Smith (1996), and to an extent how oscillatory mechanisms as observed in rhythmic speech patterns can be tracked by a cortical oscillator (see e.g. Gibson 2014). Here it is interesting to reflect upon the role of the free energy principle from the point of view of inference. Coupled dynamical systems inherently admit orders of inference via covariant interaction between states through input and output. But if we turn to the selective affordances of RECS when considered in the context of neuroscience, there are indeed differing shades of interpretation. For instance, Bruineberg and Rietveld (2014) in spotlighting the free energy principle, point out that the type of inference as displayed by such dynamical systems, is to be regarded as 'minimal' when embedded into the econiche of the human organism. Why this should be so, is basically to reconcile the clash of terminology with the philosophical interpretation (perhaps the more generally understood version of the term in cognitive science) which aims to reach a propositional statement founded on a sequence of premises, a semantic property that is not always realizable in dynamical systems. Though to what order of 'minimality', I should think is open to debate, if the former type of inference does indeed dictate specific choices and action therewith.

Appendix D: Cognition, Categorization, and Adaptive Resonance[12]

D.1 Networks Through Categories and AI

Managing categories of different types and levels, provides a perspective not just restricted to unpredictable attributes, but one that addresses structured information, while viewing the formation of categories as a cultural product of a prevailing language over the course of time. It can be expected that the maximum amount of information assimilated with the least cognitive effort is attainable when perceived world structures are mapped as closely as possible by a categorical framework (the question, as always, is choosing the right framework). The perceived world contains a high correlation structure, and there are categories placed within the taxonomies of concrete objects structured in a way such that generally, there is a level of abstraction at which one can implement the most basic reductions/refinements, prototypes, etc., in terms of a network architecture (symbolic, connectionist or some hybrid version of both of these).

So much so, information-rich modules of perceptual and functional attributes may occur within natural discontinuities, and it is at these discontinuities that basic divisions in categorization are created; these discontinuities can be seen in terms of the by now standard levels: superordinate, basic, and subordinate (e.g. " four-legged animals", "cat", and "Siamese cat", respectively). Thus the 'basic' objects (cf prototypes) comprise the most inclusive categories appropriate to mental imagery that could said to be isomorphic to the representation or appearance of the members of a class as a whole. In this respect, the basic-level lexical acquisition has been hypothesized in Rosch (1975) to be an outcome of word frequencies. A typical means description may combine set-theoretic representations with probabilistic cue-validity on the basis of frequency: namely, a sum of cue validity as taken across all subcategories, implies a cue validity for the total category through word emphasis, sounds, motion, and identifying average shapes or trends. Within the

---

[12]The appendix has been cordially contributed by prof. J.F. Glazebrook.

© Springer International Publishing Switzerland 2016
D. Plikynas, *Introducing the Oscillations Based Paradigm*,
DOI 10.1007/978-3-319-39040-6

context of embodiment, and the enactive viewpoint, a central theme may be summarized as follows (Varela et al. 1992):

> Meaningful conceptual structures arise from two sources: (1) from the structured nature of bodily and social experience and (2) from our innate capacity to imaginatively project from certain well-structured aspects of bodily and interactional experience to abstract conceptual structures. Rational thought is the application of very general cognitive processes – focusing, scanning, superimposition, figure-ground reversal, etc. – to such structures.

This is fine, but doesn't on the surface offer a blueprint for incorporating degrees of admissible representations. The more usual testing ground for embodied cognition is plainly robotics and forms of AI. Robotics with categorical perception based upon visual motion as connected to continuous time, real-valued neural networks (CTRNN) as studied by Beer (2003) (see also Chemero 2011), provide a means of assessing how the nervous system induces agent-object dynamics as modeled by agent-environment systems. Developmental psychology, perception, motor-skills and phenomenology require a certain degree of computation and representations, but no do not categorize these disciplines as being strictly computational or representational in themselves. Likewise, thinking about the past, the future, as well as the environment, are major tasks to be undertaken by a cognitive system, and here is where interaction with the real-world environment may be as much a hallmark of intelligence compared with the habitual manipulation of mental skills. This observation comes to the forefront in the design of completely autonomous robots as proposed in Brooks (1991). The basic principles in this engineered form of AI are : (i) avoiding objects, (ii) wandering, (iii) building maps, (iv) monitoring changes, (v) identifying objects, (vi) planning changes to the world, and (vii) reasoning about the behavior of objects. As pointed out in Varela et al. (1992), this type of approach evoking a structural coupling with the environment, albeit in the context of manufactured intelligence, is not totally out of line with Varela's earlier concept of enactive cognitivism. Here it seems fitting to quote Chemero (2011), p. 186:

> By rejecting representationalism, which is to say, by being radical embodied cognitive scientists, Varela et al. claim to stake out a position that is neither realist nor idealist. On their view, animals and worlds are not separate, so there is no need for animals to represent the world. Without representations, there is nothing besides the world for the animal to interact with, but the worlds that animals–including humans, including philosophers and physicists–interact with are strictly limited, and are determined by sensorimotor capabilities

The questions of realism and idealism are philosophical, and will be set aside for now. On the other hand, one might argue that some representationalism cannot be avoided in enactive (embodied) cognition, and in RECS, in so far that the actual dynamics generate (coupled) oscillatory systems, that by their very nature afford degrees of representations (see Chemero 2011 for an brief excursion into the various arguments for and against).

D.2   Conceptual Blending

Types of mental spaces in the sense of Fauconnier and Turner (2008) provide a logical-linguistic descriptive mechanism for describing concepts embedded within a cluster of related concepts via combining mental spaces within a modular framework that corresponds with co-activated neuronal ensembles. The result, conceptual blending, is a basic cognitive function which blends and progressively compresses different conceptual spaces into a unified concept/situation located within the space of combinations of processes. On the face of it, this may amount to something fairly commonplace, but the process of blending generally leads to creating novel concepts, connotations, as well as representing idiomatic expressions that are peculiar to a culture.

For instance, the mental spaces pertaining to the concepts "user" and "friendly", can be blended to obtain "friendly user" or "user-friendly". Likewise, "boat", "house", and "owner", can be blended in several ways to produce generally distinct meanings, such as "house-boat owner", or "boat-house owner" (who does not necessarily own any of the boats moored within). Just as in the case of, 'affordances', 'conceptual blending' events can realized as a result of integrating parallel distributed networks that project inputs and emerge dynamically, quite in accord with cognitive fluidity as attained by the global workspace. We can expect that radically original blends may very often be those underlying the mechanisms of creative thinking and invention.

D.3   Network Architecture:
The Role of Connectionist Models

The underlying functional mechanisms of connectionist models are close to those of many types of biological systems and neural networks. They are usually recognized as representing mental and cognitive processes via an architecture of learning networks structured on a multiply context dependent system interacting through mechanisms of propagation/activation/learning along with environmental rules, while simultaneously overseen by local-to-global executive regulation. Such models have their foundations in the key concepts and principles of weighted network mechanisms that not only provide structures for learning, but also provide the necessary properties conducive to a semantic memory (Quillian 1968). In this case, subsymbolic distributed representations, as based on ancestral type back-propagation networks having so-called 'hidden layers', implement computations which can fathom out a degree of implicit knowledge from an input stage, and thereafter to assimilate information towards a higher level of ontology. The model also stresses the necessity of feedforward networks to supply the appropriate semantic operations to programs that among other things, are capable of implementing a mode of propositional logic (see e.g. Rogers and McClelland 2004 for specific elaborations).

On the other hand, explicit knowledge may be captured in computational cognitive modelling thanks to symbolic, or localized representations through which a sense of meaning becomes more transparent. Neural networks generally may be devoid of meaning, or convergence to any meaning at all, unless they are supported by an appropriately programmed logical base. It is in this latter case that a given propositional logic program affords the possibility of creating a connectionist network capable of computing the necessary degree of convergence. Such networks may, in addition employ, types of connections which are weighted in such a capacity in in order to represent semantic memory.

Solutions to this kind of problem may be found in connectionist-type theories such as Parallel Distributed Programming (PDP) and Multilayer Perceptron (MLP) (Rogers and McClelland 2004) which, in a Jamesian sense, evolved out of these principles as context-specific, emergent "processes of processes". Accordingly, they can be viewed as massively parallel, combinatorial dynamical systems, which by dint of their functional mechanisms are close to replicating distributed associative memory, and to a certain extent approximate the unconscious specialized processor of the global workspace (Baars 1993), Chap. 1. Keep in mind that this also subsumes the need for systematized, reflexive reasoning which can be achieved using fast phase or rhythmic propagation of the underlying algorithmic patterns of the network architecture. Further, most connectionist models, while frequently templating a neural network, are able to replicate and approximate states of optimality of Bayesian inference and predictive coding as was discussed previously [and this includes the notable 'Helmholtz inference machine' (Dayan et al. 1995)], thus granting them considerable scope in applications (Rogers and McClelland 2004) (cf. Clark 2013; Friston 2010).

D.4 Network Architecture: Psychological and Cognitive Models

As pointed out in Shanahan and Baars (2005), the structural mechanism provided by the global workspace can be matched by psychological models such as Learning and Inference with Schemes and Analogies (LISA) (Hummel and Holyoak 1997) whose distributed representations of concepts and objects of long-term memory, and analogical access, correspond to the parallel unconscious processes in the global workspace. Through these distributed representations, LISA incorporates processes of 'retrieval' (parallel) and 'mapping' (serial), and crucially, the analogical access as based upon memory/familiarity/comparison, much in keeping with activities of the workspace, is inherently competitive. Note that this detaches analogical reasoning from the status of being an exclusive, isolated process. For as far as the workspace mechanism are concerned, such reasoning is amenable to describing a range of cognitive tasks such as perception, selection, and linguistics, and hence it is approximately consistent with modular-type interpretations of the human mind (cf. Franklin and Patterson 2006; Franklin et al. 2012).

How then does one tackle the engineering? Here we can think of four basic steps (Shanahan and Baars 2005) (cf. Baars and Franklin 2003, 2007; Baars et al. 2013): (i) the global workspace model is by its nature, adaptable to parallel hardware, serial processing as well as analogical reasoning which are prima facie requirements for the neurophysiology of cognition, (ii) as indicative of the brain's intrinsic robustness and plasticity, when one particular process collapses, by distributivity, the overall span of the structure weakens only gradually, (iii) consistent with evolutionary constraints, a new process can be welcomed in without having to modify existing ones, and (iv) heterogeneous information processing can be incorporated and integrated. This last property is crucial: the brain possesses the ability to contend with a high bandwidth of information generated by independent modules of specialized processing, thus permitting the workspace to manage informationally unencapsulated processes through distribution of relevance, while at the same time not violating the possible constraints as imposed by degrees of over-riding complexity. This was basically the way of handling the Frame Problem as was briefly surveyed.

At the cognitive level, GWT can be effectively engineered via the domain specific, multi-parallel distributed program of the Learning Intelligent Distribution Agent (LIDA) that has some theoretical common ground with LISA (Baars and Franklin 2007; Franklin and Patterson 2006; Franklin et al. 2012). LIDA, as a principal forerunner in applying GWT, operates via large-scale collections of autonomous agents engaging in iterations of a cognitive cycle of sensory perception and action in responding to an environment, while continually updating how the latter entity can be perceived. It is through such a sensory perceptual learning mechanism with the property of a built-in episodic memory, that the selected representations are realized as the competitors in the workspace, from which a winning coalition will overcome its competitors, and then proclaim a cognitive response.

D.5   Computational-Symbolic Versus Connectionist

It has been argued that a system saturated with symbolic manipulation is not the essence of cognition, and that connectionist methods do not reveal the correct syntactical structures necessary for meaning-to-observation beyond whatever is possible through direct computational procedures as those that can be combined with symbolic/linguistic representations. Thus has been coined the phrase 'a language of thought'. The debate seemingly is one between the broad scope of distributionally represented dynamical systems, on the one hand, and the utility of computationalism, on the other. To some extent matters reduce to a question of 'systematicity': a question of alteration in a representational meaning that can create a corresponding alteration in the internal structure of the semantic content. It might be claimed that mental representations do supposedly exhibit this property, and the appropriate way to model cognition, as was argued by the computationalist groups, is to posit symbolic models within a combinatorially structured space that are quite apart from neuronal structures, and then template the former types into viable computational models. These can sometimes be based on formal, syntactical structures dealing with discrete states, whereas the more dynamic approach of connectionist models consider states varying continuously over time in a distinctively geometric setting. Computationalism is a different strategy compared to representations, employing neural/connectionist networks which are advocated on the basis of experience and learning from environmental stimuli, pattern recognition, and storing information in connections (between neurons), thus leading to an evolving autonomous learning systems.

Connectionist models cannot support compositional semantics. But a careful logical analysis in Chalmers (1997) refutes such a claim: there are areas of psychology that can dispense with symbolic processing for adequate explanations (for sensation and perception, for instance). At the same time it is proposed that distributed representations (of connectionism) can support compositional operations, while not necessarily constrained to classical architectures. Despite this refutation and the sophisticated performance of its architecture as an autonomous representation system, it is argued in Dawson and Shamanski (1994) that PDP may be difficult to interpret, is lacking in biological feasibility, as well as the ability to abstract cognitively, even with the mechanism's patent resemblance to neural circuitry. In view of such claims it seems almost paradoxical that PDPs can exhibit quite interesting behavior, but sometimes in neglect of understanding how the constituent components actually function (a traditional problem of AI). It points to the need for increased performance levels and further abstraction capabilities in PDP (and other connectionist models) towards which future research in this direction is navigated (see e.g. Rogers and McClelland 2004).

In contrast, some supporters of computationalism emphasize syntactical rules and explicit symbolic (mental) representations, because in their opinion, internal mental activity consists of the manipulation of such symbolism which can process information and deliberative reasoning within the framework of the associated networks. These features are apparent in theoretical models such as ACT-R [Adaptive Character of Thought-Rational (Anderson and Lebiere 2014)], and SOAR [State, Operator and Result (Newell 1994)] that combine symbolic and subsymbolic mechanisms as part of a continuum capable of evolving in biological patterns, in the sense that lower level memory structures and subsymbolic processes can induce effects in the corresponding cognitive layers (cf. autopoietic systems). Both procedures exhibit perceptual and goal-oriented chunking mechanisms of information (Gobet et al. 2001). In relationship to the prefrontal parietal network (PPN), chunking has been shown to activate PPN significantly in comparison with a range of cognitively demanding tasks, and in Bor and Seth (2012) there is some suggestion that this type of information gathering which seeks to exploit certain patterns towards optimizing task performances, can be achieved via the multi-parallel, distributed processing of the global workspace. It would be interesting to further explore this connection, and much the same can be said for comparing chunking with the inferentially oriented free energy principle (Friston 2010) (and likewise, e.g. Adaptive Resonance, see below).

Common to the methodology of these studies is the application domain specific subsystems as they are designed for the basis of learning in aspects of cognition, as well as including language and intentionality. This raises the question concerning the use of atomic relationships tied to basic associations as a means of explaining complex relationships. But even with these properties, there is often the problem of adequately programming representations of the real world where certain information patterns may be insufficient. Indeed, there are alternative mean, such as employing stochastic automata to establish semantic meaning to alterations in levels of representation by comparing machine free energy with thermodynamic entropy which measures randomness in a given data flow (Shalizi and Crutchfield 2001); thus 'meaning' can evolve from the information messaging that triggers operations on automata states.

Though connectionists often favor limited sets of very general, non-linear learning mechanisms that are adaptable to a flow of semantic information and causal sensitivity leading to variable network topologies which may also be regulated in part by principles of statistical physics. These include computable functions in the guise of logistic equations, iterative processes and multi-variable calculus which at least partially satisfy the Church-Turing criteria for how the human mind is constrained to computability.

However, these are further observations that may have been sorely overlooked by the symbolists in the continued debate with their connectionist counterparts. A main difference is that connectionist models study patterns of perceptual recognition in the framework of dynamic interaction between the brain, the body, and the environment, thus reflecting once more upon the embodied dynamics of self organizing systems, and how cognition arises in an emergent fashion from the circularity and nonlinearity of sensorimotor and feedback control processes. The use of certain action-oriented representations, however, are unavoidable, and thus connectionist models cannot be divorced from the overall category of the representational theory of mind. Embodied cognition, by embracing certain elements of connectivism, exhibits some form of dynamic computationalism, and so reconciles some old differences between the camps of representational versus non-representational theories of mind (for further perspectives on such distinctions and the how the debate has unfolded, see Chemero (2011), Thompson (2007), Thompson and Varela (2001).

At the end of the day, one is reminded that 'representations' are just what the word says and should be regarded as such. In this respect, it can argued that the contrasting differences between models may dissipate once cognitive processes are considered as: (i) multiply-distributed across processors, (ii) embodied within an environment, (iii) dynamic with respect to mind-body interactions, and (iv) selective, and inferential (to the extent the immune system can be). Living up to these expectations is to some degree achievable by programs such as LISA (relevance) and LIDA (cognitive cycle) that were brought to the table previously.

D.6   Adaptive Resonance Theory

Having delivered a brief sampling of topics concerning categorization, I should add that, although potentially relevant to the dynamical and distributed systems approach to cognition, the descriptive nature of these topics is relatively abstract, and falls somewhat short of a distinct emphasis on oscillating systems and how category recognition can be explicitly realized in terms of neural circuitry.

A somewhat different approach, but one that also appears relevant to the oscillatory functional mechanisms of OSIMAS/OAM, is Adaptive Resonance Theory (ART) (Grossberg 2013). A basic premise of this mainstream theory, is that biological learning involves combinations of spatial/motor processes together with those of cognition/perception, both of which are incumbent on the interplay of predictive mechanisms within the brain in order to achieve and regulate category learning. From this point of view the ART model possesses a number of similarities with Bayesian-type inference procedures, as was mentioned previously.

Towards accomplishing its goals, ART focuses on flows of information through the laminar micro-circuits of the cerebral cortex in order to provide a theoretical and empirical framework for predicting neurophysiological, behavioral, biological and biophysical events in terms of:

visual pathways $\Rightarrow$ recognition coding $\Rightarrow$ self-organization.

This is achievable by learning processes operative in a massively parallel modal architecture of potentially self-organizing resonating states emanating from neural network interactions, that include modulating short and long-term memory patterns, while keeping some degree of vigilant control. Several examples of such resonating states include (Grossberg 2013):

1. Surface-shrouded resonances: predicted to subserve conscious percepts of visual qualia.
2. Feature-category resonances: predicted to subserve recognition of familiar objects and scenes.
3. Item-list resonances: predicted to subserve conscious percepts of speech and language.
4. Spectral-pitch resonances: predicted to subserve conscious percepts of auditory streams.
5. Cognitive-emotional resonances: predicted to subserve conscious percepts of feelings and core consciousness.

The underlying principle is to seek a kind of categorical inference between vigilantly searching through top-down-expectations which are to be paired with bottom-up inputs. Once both sides are optimally matched, then the focus of attention shifts to selecting clusters in the bottom-up input sensory input that are candidates for prediction.

Should the expectations be in tune with the selected inputs, then a resonating state is attained, and the latter in course activates some recognition category. In a less than optimal matching between the top and bottom relays, it is possible that a memory search can be initiated. This is a feature of the oscillatory behavior that synchronizes the firing of resonant neuronal assembles leading to coherent states that will activate categories, and the in-built mechanisms of the model apply predictive error coding to inform the system if the categorization is correct or not. In such circumstances, the vigilance parameters control the quality of the pairing prior to initiating a new category, and the collective process is re-iterated. Eventually, an optimal category is reached, with the bonus that predictive errors are deemed to be minimal, so much so that once all of the interacting neuronal assemblies can be tuned to states that are mutually favorable, the completed cycle of events represents an emergence of a global cooperation.

D.7 Concluding Remarks

Thanks to their close association with dynamical systems, many of the models and concepts surveyed are seen to accommodate certain types of oscillatory (and resonating) mechanisms which emerge between microscopic and macroscopic levels, and so affords seeing the common features as shared by the OSIMAS and OAM models. We have also seen how many of such self-organizing models, such as DCM is formally structured as an autopoietic system, shares the predictive mechanism of Bayesian-type inference which was tied to the 'free energy principle'. Thus it would be interesting, and indeed a challenge for our OSIMAS/OAM colleagues to explore action oriented predictive processing along the lines of free energy principle in terms of oscillating phases between levels (again, Dynamic Causal modeling and references therein perhaps afford some clues towards doing this).

Some of the concepts we have addressed, such as GWT in particular, can be classified as 'modular' theories of mind, and others are perhaps not quite recognized as that in a conventional sense. Also, to the best of my knowledge, none of these models have sailed under the flag of 'mind-field theories', although perhaps the neurophenomenology concept as envisaged by Varela et. al. May have anticipated such a categorization. One might argue for, or against any of the claims; but such arguments will not be taken up here, instead I will simply leave it to the interested reader to make his/her own interpretations and make an assessment accordingly.

References

Abarbanel, H. (2012). *Analysis of observed chaotic data*. Berlin: Springer Science & Business Media.

Acemoglu, D., Ozdaglar, A., & Yildiz, E. (2011). Diffusion of innovations in social networks. In *2011 50th IEEE Conference on Decision and Control and European Control Conference (CDC-ECC)* (pp. 2329–2334).

Acharya, U. R., Chua, E. C.-P., Chua, K. C., Min, L. C., & Tamura, T. (2010). Analysis and automatic identification of sleep stages using higher order spectra. *International Journal of Neural Systems, 20*(6), 509–521.

Achterberg, J., Cooke, K., Richards, T., Standish, L. J., Kozak, L., & Lake, J. (2005). Evidence for correlations between distant intentionality and brain function in recipients: A functional magnetic resonance imaging analysis. *The Journal of Alternative and Complementary Medicine, 11*(6), 965–971.

Adamski, A. G. (2013). Quantum nature of consciousness and the unconscious collective of Carl G. Jung. *NeuroQuantology, 11*(3).

Albert, R., & Barabási, A.-L. (2002). Statistical mechanics of complex networks. *Reviews of Modern Physics, 74*(1), 47–97.

Alivisatos, A. P., Chun, M., Church, G. M., Greenspan, R. J., Roukes, M. L., & Yuste, R. (2012). The brain activity map project and the challenge of functional connectomics. *Neuron, 74*(6), 970–974.

Alkemade, F., & Castaldi, C. (2005). Strategies for the diffusion of innovations on social networks. *Computational Economics, 25*(1–2), 3–23.

Anderson, J. R., & Lebiere, C. J. (2014). *The atomic components of thought*. Abingdon: Psychology Press.

Anokhin, A. P., Müller, V., Lindenberger, U., Heath, A. C., & Myers, E. (2006). Genetic influences on dynamic complexity of brain oscillations. *Neuroscience Letters, 397*(1–2), 93–98.

Araújo, C. H. D., & Castro, P. A. L. de. (2010). Towards automated trading based on fundamentalist and technical data. In A. C. da R. Costa, R. M. Vicari, & F. Tonidandel (Eds.), *Advances in artificial intelligence—SBIA 2010* (pp. 112–121). Berlin, Heidelberg: Springer.

Arndt, C. (2012). *Information measures: Information and its description in science and engineering*. Berlin: Springer Science & Business Media.

Ashwin, P., Coombes, S., & Nicks, R. (2016). Mathematical frameworks for oscillatory network dynamics in neuroscience. *The Journal of Mathematical Neuroscience, 6*(1).

Atlan, H., & Cohen, I. R. (1998). Immune information, self-organization and meaning. *International Immunology, 10*(6), 711–717.

Atmanspacher, H. (2011). Quantum approaches to consciousness. In E. N. Zalta (Ed.), *The Stanford Encyclopedia of Philosophy Archive* (Summer 2011). Retrieved from http://plato.stanford.edu/archives/sum2015/entries/qt-consciousness/

D. Plikynas, *Introducing the Oscillations Based Paradigm*,
DOI 10.1007/978-3-319-39040-6

Axmacher, N., Henseler, M. M., Jensen, O., Weinreich, I., Elger, C. E., & Fell, J. (2010). Cross-frequency coupling supports multi-item working memory in the human hippocampus. *Proceedings of the National Academy of Sciences,107*(7), 3228–3233.

Baars, B. J. (1993). *A Cognitive theory of consciousness* (Reprint ed.). Cambridge: Cambridge University Press.

Baars, B. J., & Franklin, S. (2003). How conscious experience and working memory interact. *Trends in Cognitive Sciences,7*(4), 166–172.

Baars, B. J., & Franklin, S. (2007). An architectural model of conscious and unconscious brain functions: Global workspace theory and IDA. *Neural Networks,20*(9), 955–961.

Baars, B. J., Franklin, S., & Ramsoy, T. Z. (2013). Global workspace dynamics: Cortical "binding and propagation" enables conscious contents. *Frontiers in Psychology, 4*.

Baianu, I. C., Brown, R., Georgescu, G., & Glazebrook, J. F. (2006). Complex non-linear biodynamics in categories, higher dimensional algebra and Łukasiewicz-Moisil topos: Transformations of neuronal, genetic and neoplastic networks. *Axiomathes,16*(1–2), 65–122.

Balanov, A., Janson, N., Postnov, D., & Sosnovtseva, O. (2008). *Synchronization: From simple to complex*. Berlin: Springer Science & Business Media.

Bandini, S., Federici, M. L., & Vizzari, G. (2007). Situated cellular agents approach to crowd modeling and simulation. *Cybernetics and Systems,38*(7), 729–753.

Bandini, S., Manzoni, S., & Vizzari, G. (2004). Multi-agent approach to localization problems: The case of multilayered multi-agent situated system. *Web Intelligence and Agent Systems,2*(3), 155–166.

Bandini, S., Manzoni, S., & Vizzari, G. (2006). Toward a platform for multi-layered multi-agent situated system (mmass)-based simulations: Focusing on field diffusion. *Applied Artificial Intelligence,20*(2–4), 327–351.

Barrat, A., Barthélemy, M., & Vespignani, A. (2008). *Dynamical processes on complex networks* (1st ed.). Cambridge: Cambridge University Press.

Bassett, D. S., & Bullmore, E. (2006). Small-world brain networks. *The Neuroscientist,12*(6), 512–523.

Bassett, D. S., Meyer-Lindenberg, A., Achard, S., Duke, T., & Bullmore, E. (2006). Adaptive reconfiguration of fractal small-world human brain functional networks. *Proceedings of the National Academy of Sciences,103*(51), 19518–19523.

Beer, R. D. (2003). The dynamics of active categorical perception in an evolved model agent. *Adaptive Behavior,11*(4), 209–243.

Benca, R. M., Obermeyer, W. H., Larson, C. L., Yun, B., Dolski, I., Kleist, K. D., et al. (1999). EEG alpha power and alpha power asymmetry in sleep and wakefulness. *Psychophysiology, 36*(4), 430–436.

Benenson, W., Harris, J. W., Stöcker, H., & Lutz, H. (Eds.). (2006). *Handbook of physics* (1st ed. 2002. Corr. 2nd printing 2006 edition). New York: Springer.

Ben-Jacob, E., Cohen, I., & Levine, H. (2000). Cooperative self-organization of microorganisms. *Advances in Physics,49*(4), 395–554.

Bennett, M., & Hacker, P. (2003). *Philosophical foundations of neuroscience* (1st ed.). Malden, MA: Wiley-Blackwell.

Bensoussan, A., Frehse, J., & Yam, P. (2013). *Mean field games and mean field type control theory* (2013th ed.). New York: Springer.

Berestycki, H., Rodríguez, N., & Ryzhik, L. (2013). Traveling wave solutions in a reaction-diffusion model for criminal activity. *Multiscale Modeling & Simulation,11*(4), 1097–1126.

Berger, P. D. H. (1929). Über das Elektrenkephalogramm des Menschen. *Archiv für Psychiatrie und Nervenkrankheiten,87*(1), 527–570.

Billari, F. C. (2006). *Agent-Based computational modelling: Applications in demography, social, economic and environmental sciences*. Abingdon: Taylor & Francis.

Bitbol, M., & Luisi, P. L. (2004). Autopoiesis with or without cognition: Defining life at its edge. *Journal of the Royal Society Interface,1*(1), 99–107.

Blanco, S., Quiroga, R. Q., Rosso, O. A., & Kochen, S. (1995). Time-frequency analysis of electroencephalogram series. *Physical Review E,51*(3), 2624–2631.

Blinowska, K. J., & Malinowski, M. (1991). Non-linear and linear forecasting of the EEG time series. *Biological Cybernetics,66*(2), 159–165.

Bohm, D. (2002). *Wholeness and the implicate order.* Abingdon: Psychology Press.

Bonita, J. D., Ii, L. C. C. A., Rosenberg, B. M., Cellucci, C. J., Watanabe, T. A. A., Rapp, P. E., et al. (2013). Time domain measures of inter-channel EEG correlations: A comparison of linear, nonparametric and nonlinear measures. *Cognitive Neurodynamics, 8*(1), 1–15.

Booth, V., & Diniz Behn, C. G. (2014). Physiologically-based modeling of sleep–wake regulatory networks. *Mathematical Biosciences,250*, 54–68.

Borb, A. A., & Achermann, P. (1999). Sleep homeostasis and models of sleep regulation. *Journal of Biological Rhythms,14*(6), 559–570.

Borbély, A. A., & Achermann, P. (1999). Sleep homeostasis and models of sleep regulation. *Journal of Biological Rhythms,14*(6), 557–568.

Bor, D., & Seth, A. K. (2012). Consciousness and the prefrontal parietal network: Insights from attention, working memory, and chunking. *Frontiers in Psychology, 3.* http://doi.org/10.3389/fpsyg.2012.00063

Bourgine, P., & Stewart, J. (2004). Autopoiesis and cognition. *Artificial Life,10*(3), 327–345.

Brillouin, L. (1953). The negentropy principle of information. *Journal of Applied Physics,24*(9), 1152–1163.

Bronshtein, I. N., Semendyayev, K. A., Musiol, G., & Mühlig, H. (2013). *Handbook of mathematics.* Berlin: Springer Science & Business Media.

Brooks, R. A. (1991). Intelligence without representation. *Artificial Intelligence,47*(1–3), 139–159.

Bruineberg, J., & Rietveld, E. (2014). Self-organization, free energy minimization, and optimal grip on a field of affordances. *Frontiers in Human Neuroscience, 8.*

Bullmore, E., & Sporns, O. (2009). Complex brain networks: Graph theoretical analysis of structural and functional systems. *Nature Reviews Neuroscience,10*(3), 186–198.

Bunge, S., & Kahn, I. (2009). Cognition: An overview of neuroimaging techniques. *Encyclopedia of Neuroscience, 2.*

Busemeyer, J. R., & Bruza, P. D. (2014). *Quantum models of cognition and decision* (Reissue ed.). Cambridge: Cambridge University Press.

Buzsaki, G. (2011). *Rhythms of the brain* (1st ed.). Oxford; New York: Oxford University Press.

Cacioppo, J. T., & Decety, J. (2011). Social neuroscience: Challenges and opportunities in the study of complex behavior. *Annals of the New York Academy of Sciences, 1224*(1), 162–173. http://doi.org/10.1111/j.1749-6632.2010.05858.x

Camurri, M., Mamei, M., & Zambonelli, F. (2007). Urban traffic control with co-fields. In D. Weyns, H. V. D. Parunak, & F. Michel (Eds.), *Environments for multi-agent systems III* (pp. 239–253). Berlin, Heidelberg: Springer. Retrieved from http://link.springer.com/chapter/10.1007/978-3-540-71103-2_14

Cantero, J. L., Atienza, M., Madsen, J. R., & Stickgold, R. (2004). Gamma EEG dynamics in neocortex and hippocampus during human wakefulness and sleep. *NeuroImage,22*(3), 1271–1280.

Centola, D. (2010). The spread of behavior in an online social network experiment. *Science,329* (5996), 1194–1197.

Chalmers, D. (1997). *The conscious mind: In search of a fundamental theory* (1 ed.). Oxford: Oxford Paperbacks.

Chatel-Goldman, J., Schwartz, J.-L., Jutten, C., & Congedo, M. (2013). Non-local mind from the perspective of social cognition. *Frontiers in Human Neuroscience, 7.*

Chemero, A. (2011). *Radical embodied cognitive science* (Reprint ed.). UK: A Bradford Book.

Chua, L. O., Hasler, M., Moschytz, G. S., & Neirynck, J. (1995). Autonomous cellular neural networks: A unified paradigm for pattern formation and active wave propagation. *IEEE Transactions on Circuits and Systems I: Fundamental Theory and Applications,42*(10), 559–577.

Clark, A. (1998). *Being there: Putting brain, body, and world together again.* Cambridge: MIT Press.

Clark, A. (2013). Whatever next? Predictive brains, situated agents, and the future of cognitive science. *Behavioral and Brain Sciences,36*(03), 181–204.

Cohen, I. R., & Harel, D. (2007). Explaining a complex living system: Dynamics, multi-scaling and emergence. *Journal of The Royal Society Interface, 4*(13), 175–182. http://doi.org/10.1098/rsif.2006.0173

Conte, E., Khrennikov, A., Todarello, O., Federici, A., & Zbilut, J. P. (2008). A preliminary experimental verification on the possibility of Bell inequality violation in mental states. *NeuroQuantology,6*(3), 214–221.

Conte, E., Khrennikov, A. Y., Todarello, O., Federici, A., Mendolicchio, L., & Zbilut, J. P. (2009). Mental states follow quantum mechanics during perception and cognition of ambiguous figures. *Open Systems & Information Dynamics,16*(01), 85–100.

Corning, P. A. (1983). *The synergism hypothesis: A theory of progressive evolution.* New York: Mcgraw-Hill.

Cover, T. M., & Thomas, J. A. (2012). *Elements of information theory.* New York: Wiley.

Cruz, C., González, J. R., & Pelta, D. A. (2010). Optimization in dynamic environments: A survey on problems, methods and measures. *Soft Computing,15*(7), 1427–1448.

Daan, S., Beersma, D. G., & Borbély, A. A. (1984). Timing of human sleep: Recovery process gated by a circadian pacemaker. *The American Journal of Physiology,246*(2 Pt 2), R161–R183.

Darley, V., & Outkin, A. V. (2007). *Nasdaq market simulation: Insights on a major market from the science of complex adaptive systems.* River Edge, NJ, USA: World Scientific Publishing Co., Inc.

David, O. (2007). Dynamic causal models and autopoietic systems. *Biological Research, 40*(4), 487–502. http://doi.org/10.4067/S0716-97602007000500010

Dawson, M. R. W., & Shamanski, K. S. (1994). Connectionism, confusion and cognitive science. *Journal of Intelligent Systems,4*, 215–262.

Dayan, P., Hinton, G. E., Neal, R. M., & Zemel, R. S. (1995). The Helmholtz machine. *Neural Computation,7*(5), 889–904.

de Haan, M., & Gunnar, M. R. (2011). *Handbook of developmental social neuroscience.* New York: Guilford Press.

Dehaene, S. (2011). Conscious and nonconscious processes: Distinct forms of evidence accumulation? In V. Rivasseau (Ed.), *Biological physics* (pp. 141–168). Berlin: Springer Basel.

Dehaene, S., & Changeux, J.-P. (2005). Ongoing spontaneous activity controls access to consciousness: A neuronal model for inattentional blindness. *PLoS Biology, 3*(5), e141.

Dehaene, S., Kerszberg, M., & Changeux, J.-P. (1998). A neuronal model of a global workspace in effortful cognitive tasks. *Proceedings of the National Academy of Sciences,95*(24), 14529–14534.

Dehaene, S., & Naccache, L. (2001). Towards a cognitive neuroscience of consciousness: Basic evidence and a workspace framework. *Cognition,79*(1–2), 1–37.

de Hoon, M. J. L., van der Hagen, T. H. J. J., Schoonewelle, H., & van Dam, H. (1996). Why Yule-Walker should not be used for autoregressive modelling. *Annals of Nuclear Energy,23*(15), 1219–1228.

De Paoli, F., & Vizzari, G. (2003). Context dependent management of field diffusion: An experimental framework. *In WOA*, 78–84.

DeMiguel, V., Garlappi, L., & Uppal, R. (2009). Optimal versus naive diversification: How inefficient is the 1/N portfolio strategy? *Review of Financial Studies,22*(5), 1915–1953.

Dijk, D. J. (1999). Circadian variation of EEG power spectra in NREM and REM sleep in humans: Dissociation from body temperature. *Journal of Sleep Research,8*(3), 189–195.

Dijk, D.-J., & von Schantz, M. (2005). Timing and consolidation of human sleep, wakefulness, and performance by a symphony of oscillators. *Journal of Biological Rhythms,20*(4), 279–290.

Di Paolo, E., & De Jaegher, H. (2012). The interactive brain hypothesis. *Frontiers in Human Neuroscience, 6.*

Durante, F., & Foscolo, E. (2013). An analysis of the dependence among financial markets by spatial contagion. *International Journal of Intelligent Systems,28*(4), 319–331.

Düzel, E., Penny, W. D., & Burgess, N. (2010). Brain oscillations and memory. *Current Opinion in Neurobiology,20*(2), 143–149.

Edelman, G. M., & Tononi, G. (2000). *A universe of consciousness: How matter becomes imagination.* New York: Basic Books.

Engel, G. S., Calhoun, T. R., Read, E. L., Ahn, T.-K., Mančal, T., Cheng, Y.-C., et al. (2007). Evidence for wavelike energy transfer through quantum coherence in photosynthetic systems. *Nature, 446*(7137), 782–786.

English, T. M. (1996). Evaluation of evolutionary and genetic optimizers: No free lunch. In *Evolutionary programming* (pp. 163–169). Cambridge, MA: MIT Press.

Epstein, J. M. (2006). *Generative social science: Studies in agent-based computational modeling.* Princeton: Princeton University Press.

Erdös, P., & Rényi, A. (1960). On the evolution of random graphs. *Publications of the Mathematical Institute of the Hungarian Academy of Sciences, Series A,5*, 17–61.

Fauconnier, G., & Turner, M. (2008). *The way we think: Conceptual blending and the mind's hidden complexities.* New York: Basic Books.

Feynman, R. (2000). *Feynman lectures on computation* (1 ed.). Boulder: Westview Press.

Fingelkurts, A. A., & Fingelkurts, A. A. (2001). Operational architectonics of the human brain biopotential field: Towards solving the mind-brain problem. *Brain and Mind,2*(3), 261–296.

Fingelkurts, A. A., Fingelkurts, A. A., Ermolaev, V. A., & Kaplan, A. Y. (2006). Stability, reliability and consistency of the compositions of brain oscillations. *International Journal of Psychophysiology: Official Journal of the International Organization of Psychophysiology,59* (2), 116–126.

Franklin, S., & Patterson Jr, F. G. (2006). The LIDA architecture: Adding new modes of learning to an intelligent, autonomous, software agent. *Pat, 703*, 764–1004.

Franklin, S., Strain, S., Snaider, J., McCall, R., & Faghihi, U. (2012). Global workspace theory, its LIDA model and the underlying neuroscience. *Biologically Inspired Cognitive Architectures, 1*, 32–43.

Freeman, W. J., & Vitiello, G. (2006). Nonlinear brain dynamics as macroscopic manifestation of underlying many-body field dynamics. *Physics of Life Reviews,3*(2), 93–118.

Fries, P. (2005). A mechanism for cognitive dynamics: Neuronal communication through neuronal coherence. *Trends in Cognitive Sciences,9*(10), 474–480.

Fries, P. (2009). Neuronal gamma-band synchronization as a fundamental process in cortical computation. *Annual Review of Neuroscience,32*(1), 209–224.

Friston, K. (2010). The free-energy principle: A unified brain theory? *Nature Reviews Neuroscience,11*(2), 127–138.

Friston, K. (2012). Self-organisation, inference and cognition: Comment on "Consciousness, crosstalk, and the mereological fallacy: An evolutionary perspective" by Rodrick Wallace. *Physics of Life Reviews,9*(4), 456–457.

Friston, K. J., Harrison, L., & Penny, W. (2003). Dynamic causal modelling. *NeuroImage,19*(4), 1273–1302.

Friston, K. J., & Price, C. J. (2001). Dynamic representations and generative models of brain function. *Brain Research Bulletin,54*(3), 275–285.

Frith, C. D., & Frith, U. (2012). Mechanisms of social cognition. *Annual Review of Psychology,63* (1), 287–313.

Froese, T. (2011). From second-order cybernetics to enactive cognitive science: Varela's turn from epistemology to phenomenology. *Systems Research and Behavioral Science,28*(6), 631–645.

Fröhlich, H. (1968). Long-range coherence and energy storage in biological systems. *International Journal of Quantum Chemistry,2*(5), 641–649.

Fukunaga, K. (2013). *Introduction to statistical pattern recognition.* Cambridge: Academic Press.

Fulcher, B. D., & Jones, N. S. (2014). Highly comparative feature-based time-series classification. *IEEE Transactions on Knowledge and Data Engineering,26*(12), 3026–3037.

Fulcher, B. D., Phillips, A. J. K., & Robinson, P. A. (2008). Modeling the impact of impulsive stimuli on sleep-wake dynamics. *Physical Review E, 78*(5), 051920.

Gallagher, S. (2006). *How the body shapes the mind* (New ed.). Oxford, New York: Oxford University Press.

Georgiev, D. D., & Glazebrook, J. F. (2006). Dissipationless waves for information transfer in neurobiology-some implications. *Informatica,30*, 221–232.

Georgiev, D., & Glazebrook, J. F. (2014). Quantum interactive dualism: From Beck and Eccles tunneling model of exocytosis to molecular biology of SNARE zipping. *Biomedical Reviews, 25*(0), 15. http://doi.org/10.14748/bmr.v25.1038

Gibson, J. (2014). *The ecological approach to visual perception: Classic edition* (1 ed.). New York; Hove, England: Psychology Press.

Gilbert, P. (2006). Evolution and depression: Issues and implications. *Psychological Medicine,03*, 287–297.

Glansdorff, P., & Prigogine, I. (1971). *Thermodynamic theory of structure, stability and fluctuations.* New York: Wiley.

Glass, A. M., & Cardillo, M. J. (2012). *Holographic data storage.* In H. J. Coufal, D. Psaltis, & G. T. Sincerbox (Eds.). Berlin; New York: Springer (Softcover reprint of the original 1st ed. 2000 edition).

Glazebrook, J. F., & Wallace, R. (2009). Small worlds and Red Queens in the global workspace: An information-theoretic approach. *Cognitive Systems Research,10*(4), 333–365.

Gobbini, M., Koralek, A., Bryan, R., Montgomery, K., & Haxby, J. (2007). Two takes on the social brain: A comparison of theory of mind tasks. *Journal of Cognitive Neuroscience,19*(11), 1803–1814.

Gobet, F., Lane, P. C. R., Croker, S., Cheng, P. C.-H., Jones, G., Oliver, I., et al. (2001). Chunking mechanisms in human learning. *Trends in Cognitive Sciences, 5*(6), 236–243.

Goguen, J. A., & Varela, F. J. (1979). Systems and distinctions; duality and complement arity†. *International Journal of General Systems,5*(1), 31–43.

Gong, P., Nikolaev, A. R., & van Leeuwen, C. (2007). Intermittent dynamics underlying the intrinsic fluctuations of the collective synchronization patterns in electrocortical activity. *Physical Review E, 76*(1), 011904.

Grinberg-Zylberbaum, J., & Ramos, J. (1987). Patterns of interhemispheric correlation during human communication. *International Journal of Neuroscience,36*(1–2), 41–53.

Grossberg, S. (2013). Adaptive resonance theory: How a brain learns to consciously attend, learn, and recognize a changing world. *Neural Networks: The Official Journal of the International Neural Network Society, 37*, 1–47.

Haas, A. S., & Langer, E. J. (2014). Mindful attraction and synchronization: Mindfulness and regulation of interpersonal synchronicity. *NeuroQuantology, 12*(1).

Haggard, P. (2005). Conscious intention and motor cognition. *Trends in Cognitive Sciences,9*(6), 290–295.

Haken, H., Kelso, J. A. S., & Bunz, H. (1985). A theoretical model of phase transitions in human hand movements. *Biological Cybernetics, 51*(5), 347–356.

Hameroff, S., & Penrose, R. (2014). Consciousness in the universe: A review of the "Orch OR" theory. *Physics of Life Reviews,11*(1), 39–78.

Hänggi, P. (2002). Stochastic resonance in biology how noise can enhance detection of weak signals and help improve biological information processing. *ChemPhysChem,3*(3), 285–290.

Hansmann, U., Merk, L., Nicklous, M. S., & Stober, T. (2003). *Pervasive computing: The mobile world.* Berlin: Springer Science & Business Media.

Haven, E., & Khrennikov, A. (2013). *Quantum social science.* Cambridge, New York: Cambridge University Press.

Haykin, S. (1998). *Neural networks: A comprehensive foundation* (2nd ed.). Upper Saddle River, NJ: Prentice Hall.

Hazarika, N., Chen, J. Z., Tsoi, A. C., & Sergejew, A. (1997). Classification of EEG signals using the wavelet transform. In *1997 13th International Conference on Digital Signal Processing Proceedings, 1997. DSP 97* (Vol. 1, pp. 89–92).

Helbing, D. (2013). Globally networked risks and how to respond. *Nature,497*(7447), 51–59.

Helbing, D., & Balietti, S. (2013). *How to do agent-based simulations in the future: From modeling social mechanisms to emergent phenomena and interactive systems design* (SSRN Scholarly Paper No. ID 2339770). Rochester, NY: Social Science Research Network. Retrieved from the Santa Fe Institute Archive: http://www.santafe.edu/research/working-papers/abstract/51b331dfecab44d50dc35fed2c6bbd7b/

Heylighen, F. (1999). Collective intelligence and its implementation on the web: Algorithms to develop a collective mental map. *Computational & Mathematical Organization Theory,5*(3), 253–280.

Hoffmann, A. O. I., Jager, W., & Von Eije, J. H. (2007). Social simulation of stock markets: Taking it to the next level. *Journal of Artificial Societies and Social Simulation, 10*(2).

Hollan, J., Hutchins, E., & Kirsh, D. (2000). Distributed cognition: Toward a new foundation for human-computer interaction research. *ACM Transactions on Computer-Human Interaction,7* (2), 174–196.

Ho, M.-H. R., Ombao, H., & Shumway, R. (2005). A state-space approach to modelling brain dynamics. *Statistica Sinica,15*, 407–425.

Horne, J. A., & Ostberg, O. (1976). A self-assessment questionnaire to determine morningness-eveningness in human circadian rhythms. *International Journal of Chronobiology,4*(2), 97–110.

Hou, S. P., Haddad, W. M., Meskin, N., & Bailey, J. M. (2015). A mechanistic neural field theory of how anesthesia suppresses consciousness: Synaptic drive dynamics, bifurcations, attractors, and partial state equipartitioning. *The Journal of Mathematical Neuroscience (JMN), 5*(1).

Hu, H., & Wu, M. (2010). Current landscape and future direction of theoretical & experimental quantum brain/mind/consciousness research. *Journal of Consciousness Exploration & Research, 1*(8).

Hummel, J. E., & Holyoak, K. J. (1997). Distributed representations of structure: A theory of analogical access and mapping. *Psychological Review,104*(3), 427–466.

Hutchins, E. (1996). *Cognition in the Wild* (New ed.). UK: A Bradford Book.

Izhikevich, E. M. (2007). *Dynamical systems in neuroscience*. Cambridge: MIT Press.

Jacobs, J., Kahana, M. J., Ekstrom, A. D., & Fried, I. (2007). Brain oscillations control timing of single-neuron activity in humans. *The Journal of Neuroscience,27*(14), 3839–3844.

John, E. R. (2001). A field theory of consciousness. *Consciousness and Cognition,10*(2), 184–213.

John, E. R. (2002). The neurophysics of consciousness. *Brain Research Reviews,39*(1), 1–28.

Kadanoff, L. P. (2009). More is the same; phase transitions and mean field theories. *Journal of Statistical Physics,137*(5–6), 777–797.

Kahana, M. J. (2006). The cognitive correlates of human brain oscillations. *The Journal of Neuroscience,26*(6), 1669–1672.

Kauffman, S. A. (1993). *The origins of order: Self organization and selection in evolution*. Oxford: Oxford University Press.

Keller-Schmidt, S., & Klemm, K. (2012). A model of macroevolution as a branching process based on innovations. *Advances in Complex Systems,15*(07), 1250043.

Kezys, D., & Plikynas, D. (2014). Prognostication of human brain EEG signal dynamics using a refined coupled oscillator energy exchange model. *Neuroquantology,12*(4), 337–349.

Kim, S.-H., Faloutsos, C., & Yang, H.-J. (2013). Coercively adjusted auto regression model for forecasting in Epilepsy EEG. *Computational and Mathematical Methods in Medicine,2013*, e545613.

Knill, D. C., & Pouget, A. (2004). The Bayesian brain: The role of uncertainty in neural coding and computation. *Trends in Neurosciences,27*(12), 712–719.

Krishnanand, K. N., & Ghose, D. (2008). Glowworm swarm optimization for simultaneous capture of multiple local optima of multimodal functions. *Swarm Intelligence,3*(2), 87–124.

Kuramoto, Y. (2012). *Chemical oscillations, waves, and turbulence*. Berlin: Springer Science & Business Media.

Kurooka, T., Yamashita, Y., & Nishitani, H. (2000). Mind state estimation for operator support. *Computers & Chemical Engineering,24*(2–7), 551–556.

Laszlo, E. (1995). *The interconnected universe: Conceptual foundations of transdisciplinary unified theory*. Singapore: World Scientific.

Laszlo, E. (1996). *The systems view of the world: A holistic vision for our time* (2nd ed.). Cresskill, NJ: Hampton Pr.

Latora, V., & Marchiori, M. (2001). Efficient behavior of small-world networks. *Physical Review Letters,87*(19), 198701.

Laughlin, C. D., & D'aquili, E. G. (1974). *Biogenetic structuralism*. New York: Columbia University Press (Monograph Collection (Matt - Pseudo)).

Libet, B. (1994). A testable field theory of mind-brain interaction. *Journal of Consciousness Studies,1*(1), 119–126.

Libet, B. (2006). Reflections on the interaction of the mind and brain. *Progress in Neurobiology,78*(3–5), 322–326.

Lindenberger, U., Li, S.-C., Gruber, W., & Müller, V. (2009). Brains swinging in concert: Cortical phase synchronization while playing guitar. *BMC Neuroscience,10*(1), 22.

Litvak-Hinenzon, A., & Stone, L. (2007). *Epidemic waves, small worlds and targeted vaccination*. arXiv:0707.1222 [nlin, Q-Bio]. Retrieved from http://arxiv.org/abs/0707.1222

Liu, Y., Chen, X., & Ralescu, D. A. (2015). Uncertain currency model and currency option pricing. *International Journal of Intelligent Systems,30*(1), 40–51.

Llinas, R., & Ribary, U. (1998). Temporal conjunction in thalamocortical transactions. In H. Jasper, L. Descarries, V. Castellucci, & S. Rossignol (Eds.), *Consciousness: At the frontiers of neuroscience*. Hagerstown: Lippincott-Raven.

Loewenstein, G., Rick, S., & Cohen, J. D. (2008). Neuroeconomics. *Annual Review of Psychology,59*(1), 647–672.

Luisi, P. L. (1993). Defining the transition to life: Self-replicating bounded structures and chemical autopoiesis. *Thinking about Biology*, 17–39.

Luisi, P. L. (2003). Autopoiesis: A review and a reappraisal. *Naturwissenschaften,90*(2), 49–59.

Lutz, A., Greischar, L. L., Rawlings, N. B., Ricard, M., & Davidson, R. J. (2004). Long-term meditators self-induce high-amplitude gamma synchrony during mental practice. *Proceedings of the National Academy of Sciences of the United States of America,101*(46), 16369–16373.

Mace, W. M. (1977). James J. Gibson's strategy for perceiving: Ask not what's inside your head, but what your head's inside of. *Perceiving, Acting, and Knowing*, 43–65.

MacLennan, B. J. (1999). Field computation in natural and artificial intelligence. *Information Sciences,119*(1–2), 73–89.

Maia, T. V., & Cleeremans, A. (2005). Consciousness: Converging insights from connectionist modeling and neuroscience. *Trends in Cognitive Sciences,9*(8), 397–404.

Malik, K. (2002). *Man, Beast, and Zombie: What science can and cannot tell us about human nature* ((F First ed.) ed.). New Brunswick, NJ: Rutgers University Press.

Mamei, M., & Zambonelli, F. (2006). *Field-based coordination for pervasive multiagent systems*. Berlin: Springer Science & Business Media.

Marcer, P., & Schempp, W. (1998). The brain as a conscious system∗. *International Journal of General Systems,27*(1–3), 131–248.

Martin, P. M. V., & Martin-Granel, E. (2006). 2,500-year evolution of the term epidemic. *Emerging Infectious Diseases,12*(6), 976–980.

Matsuoka, K. (2011). Analysis of a neural oscillator. *Biological Cybernetics,104*(4–5), 297–304.

Maturana, H. R. (1980). *Autopoiesis and cognition: The realization of the living*. Berlin: Springer Science & Business Media.

Maturana, H. R., & Varela, F. J. (1980). *Autopoiesis and cognition: The realization of the living*. Berlin: Springer Science & Business Media.

Maturana, H. R., & Varela, F. J. (1987). *The tree of knowledge: The biological roots of human understanding*. Boston, MA: New Science Library/Shambhala Publications.

McCauley, J. L. (2004). *Dynamics of markets: Econophysics and finance*. Cambridge: Cambridge University Press.

McFadden, J. (2002). The conscious electromagnetic information (CEMI). *Field Theory,9*(8), 45–60.

Mehta, P., & Gregor, T. (2010). Approaching the molecular origins of collective dynamics in oscillating cell populations. *Current Opinion in Genetics & Development,20*(6), 574–580.

Meijer, D. K. F., & Jakob, K. (2013). Quantum modeling of the mental state: The concept of a cyclic mental workspace. *Syntropy,1*, 1–41.

Merleau-Ponty, M. (1963). *The structure of behaviour*. Beacon, Boston: Beacon Press.

Milovanovic, D. (2014). *Quantum holographic criminology: Paradigm shift in criminology, law, and transformative justice*. Durham, North Carolina: Carolina Academic Press.

Mingers, J. (1994). *Self-Producing systems: Implications and applications of autopoiesis*. Berlin: Springer Science & Business Media.

Moe, G. K., Rheinboldt, W. C., & Abildskov, J. A. (1964). A computer model of atrial fibrillation. *American Heart Journal,67*(2), 200–220.

Moffitt, T. E., Caspi, A., & Rutter, M. (2006). Measured gene-environment interactions in psychopathology concepts, research strategies, and implications for research, intervention, and public understanding of genetics. *Perspectives on Psychological Science,1*(1), 5–27.

Moioli, R. C., Vargas, P. A., & Husbands, P. (2012). Synchronisation effects on the behavioural performance and information dynamics of a simulated minimally cognitive robotic agent. *Biological Cybernetics,106*(6–7), 407–427.

Möller-Levet, C. S., Archer, S. N., Bucca, G., Laing, E. E., Slak, A., Kabiljo, R., et al. (2013). Effects of insufficient sleep on circadian rhythmicity and expression amplitude of the human blood transcriptome. *Proceedings of the National Academy of Sciences, 110*(12), E1132–E1141.

Montagnier, L., Aissa, J., Giudice, E. D., Lavallee, C., Tedeschi, A., & Vitiello, G. (2011). DNA waves and water. *Journal of Physics: Conference Series,306*, 012007.

Müller, K.-R., Tangermann, M., Dornhege, G., Krauledat, M., Curio, G., & Blankertz, B. (2008). Machine learning for real-time single-trial EEG-analysis: From brain-computer interfacing to mental state monitoring. *Journal of Neuroscience Methods,167*(1), 82–90.

Müller-Schloer, C., & Sick, B. (2006). Emergence in organic computing systems: Discussion of a controversial concept. In L. T. Yang, H. Jin, J. Ma, & T. Ungerer (Eds.), *Autonomic and trusted computing* (pp. 1–16). Berlin, Heidelberg: Springer.

Murugappan, M., Rizon, M., Nagarajan, R., & Yaacob, S. (2008). EEG feature extraction for classifying emotions using FCM and FKM. In *Proceedings of the 7th WSEAS International Conference on Applied Computer and Applied Computational Science* (pp. 299–304). Stevens Point, Wisconsin, USA: World Scientific and Engineering Academy and Society (WSEAS).

Nagpal, R., & Mamei, M. (2004). Engineering amorphous computing systems. In *Methodologies and software engineering for agent systems* (pp. 303–320).

Neumann, J. V. (1955). *Mathematical foundations of quantum mechanics*. Princeton: Princeton University Press.

Newandee, D. A., & Reisman, S. S. (1996). Measurement of the electroencephalogram (EEG) coherence in group meditation. In *Bioengineering Conference, 1996, Proceedings of the 1996 IEEE Twenty-Second Annual Northeast* (pp. 95–96).

Newell, A. (1994). *Unified theories of cognition*. Massachusetts: Harvard University Press.

Nielsen, L. S., Danielsen, K. V., & Sørensen, T. I. A. (2011). Short sleep duration as a possible cause of obesity: Critical analysis of the epidemiological evidence. *Obesity Reviews: An Official Journal of the International Association for the Study of Obesity,12*(2), 78–92.

Nummenmaa, L., Glerean, E., Viinikainen, M., Jääskeläinen, I. P., Hari, R., & Sams, M. (2012). Emotions promote social interaction by synchronizing brain activity across individuals. *Proceedings of the National Academy of Sciences,109*(24), 9599–9604.

Nunez, P. L., & Srinivasan, R. (2005). *Electric fields of the brain: The neurophysics of EEG,* (2nd ed.). Oxford; New York: Oxford University Press.

Nunez, P. L., Wingeier, B. M., & Silberstein, R. B. (2001). Spatial-temporal structures of human alpha rhythms: Theory, microcurrent sources, multiscale measurements, and global binding of local networks. *Human Brain Mapping,13*(3), 125–164.

Onton, J., & Makeig, S. (2006). Information-based modeling of event-related brain dynamics. In C. Neuper & W. Klimesch (Eds.), *Progress in brain research* (Vol. 159, pp. 99–120). Amsterdam: Elsevier.

Oppenheim, J., & Wehner, S. (2010). The uncertainty principle determines the nonlocality of quantum mechanics. *Science,330*(6007), 1072–1074.

O'Reilly, E. J., & Olaya-Castro, A. (2014). Non-classicality of the molecular vibrations assisting exciton energy transfer at room temperature. *Nature Communications, 5.*

Orme-Johnson, D., Dillbeck, M. C., Wallace, R. K., & Landrith, G. S. (1982). Intersubject EEG coherence: Is consciousness a field? *International Journal of Neuroscience,16*(3–4), 203–209.

Orme-Johnson, D. W., & Oates, R. M. (2009). A field-theoretic view of consciousness: Reply to critics. *Journal of Scientific Exploration,23*(2), 139.

Ormerod, P. (1997). *The death of economics.* New York: Wiley.

Osipov, G. V., Kurths, J., & Zhou, C. (2007). *Synchronization in oscillatory networks.* Berlin: Springer Science & Business Media.

Panko, R., & Panko, J. (2010). *Business data networks and telecommunications* (8th ed.). Upper Saddle River, NJ, USA: Prentice Hall Press.

Penrose, R. (2007). *The road to reality: A complete guide to the laws of the universe* (Reprint ed.). USA: Vintage.

Perc, M., Gómez-Gardeñes, J., Szolnoki, A., Floría, L. M., & Moreno, Y. (2013). Evolutionary dynamics of group interactions on structured populations: A review. *Journal of The Royal Society Interface,10*(80), 20120997.

Perry, R. B. (1996). *The thought and character of William James.* Nashville: Vanderbilt University Press.

Pessa, E., & Vitiello, G. (2004). Quantum noise induced entanglement and chaos in the dissipative quantum model of brain. *International Journal of Modern Physics B,18*(06), 841–858.

Peters, E. E. (1996). *Chaos and order in the capital markets: A new view of cycles, prices, and market volatility.* New York: Wiley.

Phillips, A. J. K., Czeisler, C. A., & Klerman, E. B. (2011). Revisiting spontaneous internal desynchrony using a quantitative model of sleep physiology. *Journal of Biological Rhythms,26* (5), 441–453.

Phillips, A. J. K., & Robinson, P. A. (2007). A quantitative model of sleep-wake dynamics based on the physiology of the brainstem ascending arousal system. *Journal of Biological Rhythms,* 22(2), 167–179. http://doi.org/10.1177/0748730406297512

Phillips, A. J. K., & Robinson, P. A. (2008). Sleep deprivation in a quantitative physiologically based model of the ascending arousal system. *Journal of Theoretical Biology,* 255(4), 413–423. http://doi.org/10.1016/j.jtbi.2008.08.022

Phillips, A. J. K., Robinson, P. A., Kedziora, D. J., & Abeysuriya, R. G. (2010). Mammalian sleep dynamics: How diverse features arise from a common physiological framework. *PLOS Computational Biology,* 6(6), e1000826. http://doi.org/10.1371/journal.pcbi.1000826

Pikovsky, A., Rosenblum, M., & Kurths, J. (2003). *Synchronization: A universal concept in nonlinear sciences.* Cambridge: Cambridge University Press.

Pizzi, R., Fantasia, A., Gelain, F., & Rossetti, D. (2004). Non-local correlation between human neural networks on printed circuit board. In *Toward a Science of Consciousness Conference, Tucson, Arizona.*

Plikynas, D. (2010). A virtual field-based conceptual framework for the simulation of complex social systems. *Journal of Systems Science and Complexity,23*(2), 232–248.

Plikynas, D. (2015). Oscillating agent model: Quantum approach. *NeuroQuantology., 13*(1).

Plikynas, D., Basinskas, G., Kumar, P., Masteika, S., Kezys, D., & Laukaitis, A. (2014a). Social systems in terms of coherent individual neurodynamics: Conceptual premises, experimental and simulation scope. *International Journal of General Systems, 43*(5), 434–469.

Plikynas, D., Basinskas, G., & Laukaitis, A. (2014b). Towards oscillations-based simulation of social systems: A neurodynamic approach. *Connection Science, 27*(2), 188–211.

Plikynas, D., Masteika, S., Basinskas, G., Kezys, D., & Pravin, K. (2015). Group neurodynamics: Conceptual and experimental framework. In H. Liljenström (Ed.), *Advances in cognitive neurodynamics* (Vol. IV, pp. 15–20). Dordrecht: Springer.

Plikynas, D., Raudys, A., & Raudys, S. (2015). Agent-based modelling of excitation propagation in social media groups. *Journal of Experimental & Theoretical Artificial Intelligence,27*(4), 373–388.

Plikynas, D., & Raudys, S. (2015). Towards nonlocal field-like social interactions: Oscillating agent based conceptual and simulation framework. In D. Secchi & M. Neumann (Eds.), *Agent-based simulation of organizational behavior: New frontiers of social science research* (pp. 237–263). Berlin: Springer.

Popp, F. A., Chang, J. J., Herzog, A., Yan, Z., & Yan, Y. (2002). Evidence of non-classical (squeezed) light in biological systems. *Physics Letters A,293*(1–2), 98–102.

Port, R. F., & van Gelder, T. (Eds.). (1998). *Mind as motion: Explorations in the dynamics of cognition.* UK: A Bradford Book.

Poslad, S. (2009). *Ubiquitous computing: Smart devices, environments and interactions* (1st ed.). Chichester, UK: Wiley.

Pribram, K. H. (1991). *Brain and perception: Holonomy and structure in figural processing* (1st ed.). Hillsdale, NJ: Lawrence Erlbaum Associates.

Pribram, K. H. (1999). Quantum holography: Is it relevant to brain function? *Information Sciences,115*(1–4), 97–102.

Quillian, M. (1968). *Semantic memory.* In M. Minsky (Ed.), *Semantic information processing* (pp. 227–270). Cambridge, MA: MIT Press.

Quiroga, R. Q., Rosso, O. A., Başar, E., & Schürmann, M. (2001). Wavelet entropy in event-related potentials: A new method shows ordering of EEG oscillations. *Biological Cybernetics,84*(4), 291–299.

Radin, D. I. (2004). Event-related electroencephalographic correlations between isolated human subjects. *The Journal of Alternative and Complementary Medicine,10*(2), 315–323.

Raudys, A., Plikynas, D., & Raudys, S. (2014). Novel automated multi-agent investment system based on simulation of self-excitatory oscillations. *Transformations in Business & Economics,13*(2), 42–59.

Raudys, S. (2001). *Statistical and neural classifiers: An integrated approach to design.* Berlin: Springer Science & Business Media.

Raudys, S. (2004a). Information transmission concept based model of wave propagation in discrete excitable media. *Nonlinear Analysis: Modeling and Control, 9*(3), 271–289.

Raudys, Š. (2004b). Survival of Intelligent Agents in Changing Environments. In L. Rutkowski, J. H. Siekmann, R. Tadeusiewicz, & L. A. Zadeh (Eds.), *Artificial Intelligence and Soft Computing - ICAISC 2004* (pp. 109–117). Springer Berlin Heidelberg.

Raudys, S. (2008). A target value control while training the perceptrons in changing environments. In *2013 International Conference on Computing, Networking and Communications (ICNC)* (Vol. 3, pp. 54–58). Los Alamitos, CA, USA: IEEE Computer Society.

Raudys, S. (2013). Portfolio of automated trading systems: Complexity and learning set size issues. *IEEE Transactions on Neural Networks and Learning Systems,24*(3), 448–459.

Raudys, S., & Raudys, A. (2010). Pairwise costs in multiclass perceptrons. *IEEE Transactions on Pattern Analysis and Machine Intelligence,32*(7), 1324–1328.

Raudys, Š., Raudys, A., & Pabarškaitė, Ž. (2014). Sustainable economy inspired large-scale feed-forward portfolio construction. *Technological and Economic Development of Economy,20*(1), 79–96.

Raudys, S., Raudys, A., & Plikynas, D. (2014). Multi-agent system based on oscillating agents for portfolio design. In *2014 14th International Conference on Intelligent Systems Design and Applications (ISDA)* (pp. 134–139).

Reilly, F., & Brown, K. (2011). *Investment analysis and portfolio management* (10th ed.). Mason, OH: South-Western College Pub.

Reimers, M. (2011). Local or distributed activation? The view from biology. *Connection Science,23*(2), 155–160.

Robinson, R. (2009). Exploring the "global workspace" of consciousness. *PLoS Biology,7*(3), e1000066.

Roenneberg, T., Daan, S., & Merrow, M. (2003). The art of entrainment. *Journal of Biological Rhythms,18*(3), 183–194.

Rogers, E. M. (2010). *Diffusion of innovations* (4th ed.). New York: Simon and Schuster.

Rogers, T. T., & McClelland, J. L. (2004). *Semantic cognition: A parallel distributed processing approach*. Cambridge: MIT Press.

Rosch, E. (1975). Cognitive representations of semantic categories. *Journal of Experimental Psychology: General,104*(3), 192–233.

Rossi, C., Foletti, A., Magnani, A., & Lamponi, S. (2011). New perspectives in cell communication: Bioelectromagnetic interactions. *Seminars in Cancer Biology,21*(3), 207–214.

Rubinov, M., & Sporns, O. (2010). Complex network measures of brain connectivity: Uses and interpretations. *NeuroImage,52*(3), 1059–1069.

Rudrauf, D., Lutz, A., Cosmelli, D., Lachaux, J.-P., & Le Van Quyen, M. (2003). From autopoiesis to neurophenomenology: Francisco Varela's exploration of the biophysics of being. *Biological Research,36*(1), 27–65.

Sawyer, R. K. (2005). *Social emergence: Societies as complex systems*. Cambridge: Cambridge University Press.

Schelter, B., Winterhalder, M., Eichler, M., Peifer, M., Hellwig, B., Guschlbauer, B., et al. (2006). Testing for directed influences among neural signals using partial directed coherence. *Journal of Neuroscience Methods, 152*(1–2), 210–219.

Schrödinger, E. (1955). *WHAT is life? The physical aspect of the living cell* (Fifth Printing ed.). Cambridge: Cambridge University Press.

Schwartz, S. A., & Dossey, L. (2010). Nonlocality, intention, and observer effects in healing Studies: Laying a foundation for the future. *EXPLORE: The Journal of Science and Healing,6* (5), 295–307.

Seidl, D., & Cooper, R. (2006). *Organisational identity and self-transformation: An autopoietic perspective*. Aldershot, Hants, England ; Burlington, VT: Ashgate Pub Co.

Sengupta, B., Stemmler, M. B., & Friston, K. J. (2013). Information and Efficiency in the nervous system—A synthesis. *PLoS Computational Biology,9*(7), e1003157.

Servat, D., & Drogoul, A. (2002). Combining amorphous computing and reactive agent-based systems: A paradigm for pervasive intelligence? *Proceedings of the First International Joint Conference on Autonomous Agents and Multiagent Systems: Part 1* (pp. 441–448). New York, NY, USA: ACM.

Shalizi, C. R., & Crutchfield, J. P. (2001). Computational mechanics: Pattern and prediction, structure and simplicity. *Journal of Statistical Physics,104*(3–4), 817–879.

Shanahan, M. (2010). *Embodiment and the inner life*. Oxford: Oxford University Press.

Shanahan, M. (2012). The brain's connective core and its role in animal cognition. *Philosophical Transactions of the Royal Society of London B: Biological Sciences,367*(1603), 2704–2714.

Shanahan, M., & Baars, B. (2005). Applying global workspace theory to the frame problem. *Cognition,98*(2), 157–176.

Shannon, C. E. (2001). A mathematical theory of communication. *SIGMOBILE Mobile Computing and Communications Review,5*(1), 3–55.

Shapiro, L. (2010). *Embodied cognition*. New York: Routledge.

Sheldrake, R. (2009). *Morphic resonance: The nature of formative causation*. Rochester, Vt.: Park Street Press. (4th Edition, Revised and Expanded Edition of A New Science of Life edition).

Shen, W.-M., Salemi, B., & Will, P. (2002). Hormone-inspired adaptive communication and distributed control for CONRO self-reconfigurable robots. *IEEE Transactions on Robotics and Automation,18*(5), 700–712.

Shin, D., & Cho, K.-H. (2013). Recurrent connections form a phase-locking neuronal tuner for frequency-dependent selective communication. *Scientific Reports, 3.*

Sissors, J. Z., Baron, R. B., & Smith, D. L. (2010). *Advertising media planning* (7th ed.). New York: McGraw-Hill Education.

Skeldon, A. C., Dijk, D.-J., & Derks, G. (2014). Mathematical models for sleep-wake dynamics: Comparison of the two-process model and a mutual inhibition neuronal model. *PLoS ONE, 9* (8).

Spach, M. S. (1997). Discontinuous cardiac conduction: Its origin in cellular connectivity with long-term adaptive changes that cause arrhythmias. In *Discontinuous conduction in the heart* (pp. 5–51). Armonk, NY: Futura Publ. Company, Inc.

Spivey, M. (2008). *The Continuity of mind* (1st ed.). New York: Oxford University Press.

Sporns, O. (2010). *Networks of the brain.* Cambridge, MA: MIT Press.

Sporns, O., Tononi, G., & Edelman, G. M. (2000). Theoretical neuroanatomy: Relating anatomical and functional connectivity in graphs and cortical connection matrices. *Cerebral Cortex,10*(2), 127–141.

Stančák, A, Jr, & Kuna, M. (1994). EEG changes during forced alternate nostril breathing. *International Journal of Psychophysiology,18*(1), 75–79.

Standish, L. J., Kozak, L., Johnson, L. C., & Richards, T. (2004). Electroencephalographic evidence of correlated event-related signals between the brains of spatially and sensory isolated human subjects. *The Journal of Alternative & Complementary Medicine,10*(2), 307–314.

Stapp, H. P. (2009). *Mind, matter and quantum mechanics* (3rd ed., 2009 edition). Berlin: Springer.

Steele, A. J., Tinsley, M., & Showalter, K. (2008). Collective behavior of stabilized reaction-diffusion waves. *Chaos: An Interdisciplinary Journal of Nonlinear Science,18*(2), 026108.

Sterelny, K. (2003). *Thought in a hostile world: The evolution of human cognition.* Hoboken: Wiley-Blackwell.

Sternberg, R. J. (2011). *Cognitive psychology* (6th ed.). Belmont, CA: Cengage Learning.

Stevens, R., Galloway, T., Wang, P., Berka, C., Tan, V., Wohlgemuth, T., et al. (2012). Modeling the neurodynamic complexity of submarine navigation teams. *Computational and Mathematical Organization Theory, 19*(3), 346–369.

Sun, R., & Hélie, S. (2013). Psychologically realistic cognitive agents: Taking human cognition seriously. *Journal of Experimental & Theoretical Artificial Intelligence,25*(1), 65–92.

Šušmáková, K., & Krakovská, A. (2008). Discrimination ability of individual measures used in sleep stages classification. *Artificial Intelligence in Medicine,44*(3), 261–277.

Tarlacı, S. (2010). A historical view of the relation between quantum mechanics and the brain: A neuroquantologic perspective. *NeuroQuantology, 8*(2).

Tesfatsion, L., & Judd, K. L. (2006). *Handbook of computational economics: Agent-based computational economics.* Amsterdam: Elsevier.

Thaheld, F. H. (2005). An interdisciplinary approach to certain fundamental issues in the fields of physics and biology: Towards a unified theory. *Biosystems,80*(1), 41–56.

Thatcher, R. W. (2010). Validity and reliability of quantitative electroencephalography. *Journal of Neurotherapy,14*(2), 122–152.

Thelen, E., & Smith, L. (1996). *A dynamic systems approach to the development of cognition and action* (Reprint edition). UK: A Bradford Book.

Thompson, E. (2007). *Mind in life: Biology, phenomenology, and the sciences of mind.* Massachusetts: Harvard University Press.

Thompson, E., & Varela, F. J. (2001). Radical embodiment: Neural dynamics and consciousness. *Trends in Cognitive Sciences,5*(10), 418–425.

Travis, F., & Arenander, A. (2006). Cross-sectional and longitudinal study of effects of transcendental meditation practice on interhemispheric frontal asymmetry and frontal coherence. *The International Journal of Neuroscience,116*(12), 1519–1538.

Travis, F. T., & Orme-Johnson, D. W. (1989). Field model of consciousness: EEG coherence changes as indicators of field effects. *International Journal of Neuroscience,49*(3–4), 203–211.

Übeyli, E. D. (2009). Combined neural network model employing wavelet coefficients for EEG signals classification. *Digital Signal Processing,19*(2), 297–308.

Valente, T. W. (1996). Network models of the diffusion of innovations. *Computational & Mathematical Organization Theory,2*(2), 163–164.

van Rijn, A. C. M., Peper, A., & Grimbergen, C. A. (1990). High-quality recording of bioelectric events. Part 1. Interference reduction, theory and practice. *Medical & Biological Engineering & Computing, 28*(5), 389–397.

Vannini, A., & Di Corpo, U. (2008). Quantum models of consciousness. *Quantum Biosystems,2*, 165–184.

Varela, F. J. (1996). Neurophenomenology: A methodological remedy for the hard problem. *Journal of Consciousness Studies,3*(4), 330–349.

Varela, F. J. (1999). Steps to a science of Interbeing: Unfolding the Dharma implicit in modern cognitive science. *The Psychology of Awakening*, 71–89.

Varela, F. J., Rodriguez, E., George, N., Lachaux, J. P., Martinerie, J., & Renault, B. (1999). Perception's shadow: Long-distance synchronization in the human brain. *Nature,397*, 340–343.

Varela, F. J., Thompson, E., & Rosch, E. (1992). *The embodied mind: Cognitive science and human experience* (New ed.). Cambridge: The MIT Press.

Vitiello, P. D. G. (2001). *My double unveiled: The dissipative quantum model of brain*. Amsterdam ; Philadelphia, PA: John Benjamins Publishing Company.

Vizzari, G., Manenti, L., & Crociani, L. (2013). Adaptive pedestrian behaviour for the preservation of group cohesion. *Complex Adaptive Systems Modeling,1*(1), 1–29.

von Helmholtz, H., & Southall, J. P. C. (2005). *Treatise on physiological optics*. USA: Courier Corporation.

Wackermann, J., Seiter, C., Keibel, H., & Walach, H. (2003). Correlations between brain electrical activities of two spatially separated human subjects. *NeuroScience Letters,336*(1), 60–64.

Wallace, R. (2012). Consciousness, crosstalk, and the mereological fallacy: An evolutionary perspective. *Physics of Life Reviews,9*(4), 426–453.

Wallace, R. (2014). *Consciousness: A mathematical treatment of the global neuronal workspace model* (2005th ed.). Berlin: Springer.

Wallace, R., & Wallace, D. (2013). *A mathematical approach to multilevel, multiscale health interventions: Pharmaceutical industry decline and policy response*. Singapore: World Scientific Publishing Company.

Wang, X., Tao, H., Xie, Z., & Yi, D. (2012). Mining social networks using wave propagation. *Computational and Mathematical Organization Theory,19*(4), 569–579.

Watts, D. J., & Strogatz, S. H. (1998). Collective dynamics of "small-world" networks. *Nature,393* (6684), 440–442.

Wendt, A. (2015). *Quantum mind and social science: Unifying physical and social ontology*. Cambridge: Cambridge University Press.

Werth, E., Achermann, P., & Borbély, A. (1997). Fronto-occipital EEG power gradients in human sleep. *Journal of Sleep Research,6*(2), 102–112.

Wijermans, N., Jorna, R., Jager, W., van Vliet, T., & Adang, O. (2013). CROSS: Modelling crowd behaviour with social-cognitive agents. *Journal of Artificial Societies and Social Simulation,16* (4), 1.

Wilson, A. D., & Golonka, S. (2013). Embodied cognition is not what you think it is. *Frontiers in Psychology*, 4.

Wilson, M. (2002). Six views of embodied cognition. *Psychonomic Bulletin & Review,9*(4), 625–636.

Wolpert, D. H., & Macready, W. G. (1997). No free lunch theorems for optimization. *IEEE Transactions on Evolutionary Computation,1*(1), 67–82.

Wooldridge, M. (2009). *An introduction to multiagent systems* (2nd ed.). Chichester, U.K: Wiley.

Yokoi, H., Mizuno, T., Takita, M., & Kakazu, Y. (1995). Amoeba searching behavior model using vibrating potential field. In *SICE '95. Proceedings of the 34th SICE Annual Conference. International Session Papers* (pp. 1297–1302).

Young, H. P. (2006). The diffusion of innovations in social networks. In *The economy as an evolving complex system III: Current perspectives and future directions* (pp. 267–282).

Zacks, J. (2008). Neuroimaging studies of mental rotation: A meta-analysis and review. *Journal of Cognitive Neuroscience,20*(1), 1–19.

Zador, A. M., Dubnau, J., Oyibo, H. K., Zhan, H., Cao, G., & Peikon, I. D. (2012). Sequencing the connectome. *PLoS Biology,10*(10), e1001411.

Zhang, Y., & Wu, Y. (2011). How behaviors spread in dynamic social networks. *Computational and Mathematical Organization Theory,18*(4), 419–444.

Zhang, Z., & Zhang, C. (2004). *Agent-based hybrid intelligent systems: An agent-based framework for complex problem solving*. Berlin: Springer Science & Business Media.

Zhou, S.-A., & Uesaka, M. (2006). Bioelectrodynamics in living organisms. *International Journal of Engineering Science,44*(1–2), 67–92.

Printed in the United States
By Bookmasters